Poems in Steel

Monographs in German History

POEMS IN STEEL

National Socialism and the Politics of Inventing
from Weimar to Bonn

Kees Gispen

Volume 6 of Monographs in German History

Berghahn Books
NEW YORK · OXFORD

Published in 2002 by

Berghahn Books

www.berghahnbooks.com

Library of Congress Cataloguing-in-Publication Data

Gispen, Kees, 1943-
    Poems in steel : National socialism and the politics of inventing from Weimar to Bonn /
Kees Gispen.
        p. cm. -- (Monographs in German history ; v. 6)
    Includes bibliographical references and index.
    ISBN 1-57181-242-3 (alk. paper) -- ISBN 1-57181-303-9 (alk. paper: pbk.)
        1. Inventions--Germany--History--20th century. 2. Inventors--Germany--History--20th
century. 3. Patents--Germany--History--20th century. 4. National socialism. 5.
Germany--Politics and government--20th century. I. Title. II. Series.

T26.G3 G57 2001
609.43'09'04--dc21   2001025625

British Library Cataloguing in Publication Data

A catalogue record for this book is available from the British Library.

Printed in the United States on acid-free paper.

TABLE OF CONTENTS

TABLES AND FIGURES

Abbreviations

AC	*(Zeitschrift für) angewandte Chemie*
ACID	Arbeitgeberverband der Chemischen Industrie Deutschlands
AfA	Allgemeine freie Angestelltenschaft
AfT	Amt für Technik
AftW	Amt für Technische Wissenschaften
AHR	*American Historical Review*
ALR	*American Law Reports, Annotated*
AOG	Arbeitsordnungsgesetz (Gesetz zur Ordnung der nationalen Arbeit of 20 Jan. 1934)
Arbeitnordwest	Arbeitgeberverband für den Bezirk der nordwestlichen Gruppe des Vereins Deutscher Eisen- und Stahlindustrieller
AS Duisberg	Akten-Sammlung Duisberg, Bayer Archive
BAK	Bundesarchiv Koblenz
BAM	Bundesarbeitsministerium
BAP	Bundesarchiv, Abteilungen Potsdam
BASF	Historisches Archiv der BASF AG
Bayer	Bayer Archive
BA-Zwi	Bundesarchiv, Zwischenarchiv St. Augustin
BDA	Bund Deutscher Architekten, also Bundesvereinigung der Deutschen Arbeitgeberverbände
BDC	Berlin Document Center
BDCI	Bund Deutscher Civil-Ingenieure
BJM	Bundesjustizministerium
Bl.f.PMZ	*Blatt für Patent-, Muster- und Zeichenwesen*
BNSDJ	Bund nationalsozialistischer deutscher Juristen
BPV	Bayerischer Polytechnischer Verein

BUDACI	Bund angestellter Chemiker und Ingenieure, e.V.
Bundesblätter	*Bundesblätter: Mitteilungen des Bundes angestellter Chemiker und Ingenieure*
BUTAB	Bund der technischen Angestellten und Beamten
CEH	*Central European History*
DAF	Deutsche Arbeitsfront
DAG	Deutsche Angestelltengewerkschaft
DGB	Deutscher Gewerkschaftsbund
DHV	Deutschnationaler Handlungsgehilfen-Verband
DIHK	Deutsche Industrie-und Handelskammer
DINTA	Deutsches Institut für technische Arbeitsschulung (German Institute for Technical Labor Training)
DM	Deutsches Museum
DTV	Deutscher Techniker-Verband
DTZ	*Deutsche Techniker-Zeitung*
DVO	Durchführungsverordnung
DVSGE / DVGRU	Deutscher Verein für den Schutz des gewerblichen Eigentums; after 1945: Deutsche Vereinigung für gewerblichen Rechtsschutz und Urheberrecht
DWV	Deutscher Werkmeisterverband
GDM	Gesamtverband Deutscher Metallindustrieller
GEDAG	Gesamtverband Deutscher Angestellten-Gewerkschaften
GRUR	*Gewerblicher Rechtsschutz und Urheberrecht*
GuG	*Geschichte und Gesellschaft*
HA	Historisches Archiv
HK	Handelskammer
Hoechst	Historisches Archiv der Hoechst AG
IEEE	Institute of Electrical and Electronic Engineers
IHK	Industrie- und Handelskammer
ILO	International Labor Office, Geneva
IPQ	*Industrial Property Quarterly*
JCH	*Journal of Contemporary History*
JEH	*Journal of Economic History*
JMH	*Journal of Modern History*
JPOS	*Journal of the Patent Office Society*
JSH	*Journal of Social History*

MAN	Maschinenfabrik Augsburg-Nürnberg
NGug	Nachlaß Guggenheimer, K65, MAN, Werksarchiv Augsburg
NSDAP	Nationalsozialistische Deutsche Arbeiterpartei
OKW	Oberkommando der Wehrmacht (Supreme Command of the Armed Forces in Nazi Germany)
RAB	*Reichsarbeitsblatt*
RAM	Reichsarbeitsministerium
RDI	Reichsverband der Deutschen Industrie; after 1933: Reichsstand der Deutschen Industrie
RDT	Reichsbund Deutscher Technik
RFM	Reichsfinanzministerium
RGBl	*Reichsgesetzblatt*
RJM	Reichsjustizministerium
RLM	Reichsluftfahrtministerium
RMBM	Reichsministerium für Bewaffnung und Munition
RMI	Reichsministerium des Innern
RMRK	Reichsministerium für Rüstung und Kriegsproduktion
RPA	Reichspatentamt
RTV	Reichstariffvertrag für die Angestellten der Chemischen Industrie, 1920
RWM	Reichswirtschaftsministerium
RWR	Reichswirtschaftsrat
RWWA	Rheinisch-Westfälisches Wirtschaftsarchiv
SAA	Siemens Archives
SPEEA	Society of Professional Engineering Employees in Aerospace (formerly Seattle Professional Engineering Employees Association)
T&C	*Technology and Culture*
TgR	*Technik und gewerblicher Rechtsschutz: Zeitschrift zur Förderung des Erfindungs- und Markenschutzes; Mitteilungen des Verbandes Beratender Patentingenieure*
ULA	Union der leitenden Angestellten
VBM	Verband Bayerischer (and also Berliner) Metallindustrieller
VDA	Verband Deutscher Arbeitgeberverbände

VDC	Verein Deutscher Chemiker
VDDI	Verband Deutscher Diplom-Ingenieure
VDDIZ	*Zeitschrift des Verbandes Deutscher Diplom-Ingenieure*
VDE	Verein Deutscher Elektrotechniker
VDEh	Verein Deutscher Eisenhüttenleute
VDESI	Verein Deutscher Eisen- und Stahlindustrieller
VDI	Verein Deutscher Ingenieure
VDIZ	*Zeitschrift des Vereins Deutscher Ingenieure*
VDMA	Verein Deutscher Maschinenbau-Anstalten
VDPA	Verband Deutscher Patentanwälte
VELA	Vereinigung leitender Angestellten, e.V.
VfW	Verwaltung für Wirtschaft, or Economic Administration in Frankfurt
VO	Verordnung
ZAG	Zentralarbeitsgemeinschaft (Central Working Community)
ZENDEI	Zentralverband der deutschen Elektrotechnischen Industrie
ZfI	*Zeitschrift für Industrierecht*

To

Jeannie, Fiona, and Adrienne

For their patience, love, and understanding
and for making everything worthwhile

Viele meinen, Erfinden sei heute nichts mehr anderes, als auf den Grundpfeilern technischer Erkenntnisse systematisch weiter entwickeln. Erfinden sei daher nichts als fleißiges Suchen auf realer Grundlage, planmäßiges Durcharbeiten eines Fachgebietes und emsiges Forschen auf diesem.

Man kann aber nur lächeln, wenn diese vielen glauben, das Erfinden wäre mit materiellen Permutationen bis ins Unendliche abzutun. Gewiß, auch auf diesem Wege kommen Neuerungen zustande, ja sogar die Mehrzahl von ihnen; Neuerungen von großer Brauchbarkeit, Neuerungen, denen man den Fleiß ihres Bearbeiters ansieht. Solche Neuerungen durch systematische Forschung zu suchen und zu finden, dazu bedarf es freilich technischer Vorbildung und technischer Sonderpraxis. Aber alle diese fleißigen technischen Arbeiter, sie lassen sich niemals unter die wirklichen schöpferischen Erfinder einreihen, unter die Großen des Geistes und des Herzens, die ihre Dichtungen in Stahl und Materie mit der gleichen Inbrust gestalten, wie ihre Kameraden die ihren auf den Gebieten der Kunst. Denn wahrlich, ein technisch schöpferischer Geist, ein richtiger Erfinder von Format, er ist nicht niedriger einzuschätzen, als ein bedeutender Dichter oder Komponist.

–Dipl.-Ing. Robert Hans Walter, "Die Erfinderbetreuung bei der Semperit AG," *Berichte der Arbeitsgemeinschaft "Fragen des gewerblichen Rechtsschutzes" und "Gefolgschaftserfinderrecht" des Amtes für technische Wissenschaften der Deutschen Arbeitsfront, no. 1 (September 1941): 26.*

Acknowledgments

In the course of writing this book I accumulated many debts, and it is a pleasure to acknowledge them here. In the early stages I benefited from discussions with John Connelly, Dill Hunley, Wolfgang König, Peter Lundgreen, Jens Müller-Kent, and Jim Retallack. Later on I had the opportunity to test my ideas in conversations with Ann Allen, Ken Barkin, Geoffrey Giles, Ann Goldberg, Konrad Jarausch, Vernon Lidtke, David Lindenfeld, Don McKale, and Don Schilling. I received stimulating critiques of early articulations for conference papers from Volker Berghahn, Alan Beyerchen, Hans-Liudger Dienel, Jeffrey Herf, Hans Mommsen, Ron Smelser, and Helmuth Trischler. Roger Chickering invited me to present my arguments at his Georgetown seminar, where I profited from insightful questions and comments by his graduate students and by Wolfram Fischer and Rüdiger vom Bruch, both of whom inspired me with their enthusiasm for the subject.

During a semester at the Friedrich-Schiller University in Jena in 1996 I benefited immeasurably from the hospitality of Jürgen Bolten and Susanne Kirchmeyer as well as the friendship, encouragement, and support of Steffen Höhne and Dirk van Laak. I am grateful to Lutz Niethammer for letting me try out my ideas in his weekly Wednesday colloquium and take in his critique and comments by Tanja Bürgel, Rainer Gries, and especially Rüdiger Stutz; they all helped me rethink and improve my arguments. I would like to thank Rudolf Boch, for inviting me to Chemnitz in the fall of 1997 to speak about the German patent system, and Hans-Peter Müller and his colleagues of the Johann Beckmann Collegium, for giving me an opportunity to present my research at a conference on inventing in Kassel in 1998.

I owe a special and great debt of gratitude to Ray Stokes, who has known about this project almost from its inception and invested a great deal of time in challenging me to do more, to think outside the box, and to write better history. If I have succeeded and address a broader audience than I had originally envisaged, it is in large measure because of Ray.

I am also much indebted to Robert Haws, who remained a steadfast believer in this undertaking and listened to the author's intermittent ramblings about it with patience and a great sense of humor. Readers familiar with my earlier work will recognize that the current study grew out of my interest in one particular aspect of the professionalization of German engineers before 1914—their struggle with management over patent rights. To devote a whole book to such a footnote of history might well seem foolhardy, as, indeed, was pointed out to me on several occasions. If nonetheless I persisted, it was in no small measure thanks to the early interest shown by two non-historians, Jean Guyton Gispen and Joel Greenberger, both of whom liked the topic immediately and believed it was worth pursuing.

I received invaluable financial assistance from a variety of sources. The University of Mississippi on several occasions provided summer support, enabling me to travel to the archives. The German Academic Exchange Service (DAAD) awarded me a summer research and travel grant in 1991. The National Science Foundation supported this study with grant no. SBE-9213569, which gave me a chance to spend the academic year 1992/93 reading, researching, and writing drafts of the first six chapters.

I am deeply grateful to the many librarians and archivists without whom this study would never have materialized. I especially thank Frau Fromm at BASF in Ludwigshafen; the archival staff at Bayer Leverkusen; David Marwell in his capacity as Director of the former Berlin Document Center; Herr Dr. Lenz, Herr Scharmann, Frau Schoettler, and Herr Dr. Trumpp of the Bundesarchiv in Koblenz; the archivists at the Bundesarchiv in Potsdam; and Herr Gabriel at the Bundesarchiv-Zwischenarchiv St. Augustin. I also thank Herr Heinrich of the Deutsches Museum in Munich; Frau Wex at Hoechst in Frankfurt a. M.; Herr Wittmann of MAN in Augsburg; Herr Dr. Wessel at Mannesmann in Düsseldorf; Herr Boehner and Herr Dr. Schoen at the Siemens archives in Munich; and especially Herr Dr. Goldrian, at the time Siemens's chief patent director and latter-day successor to some of the principal characters in this study. I received expert and courteous assistance from the staff at the National Archives at College Park, Maryland, and the Rheinisch-Westfälisches Wirtschaftsarchiv in Cologne. Finally, I am enormously appreciative of all the help and professionalism shown by three indefatigable interlibrary loan librarians at the University of Mississippi: Lisa Harrison, Anne Johnson, and Martha Swan.

Considering all the suggestions and assistance I received, and all the work and time that went into this book, one would like to think that the sixteenth-century motto of Christopher Plantin might still apply: "*Un Labeur courageux muni d'humble Constance / Resiste à tous assauts par douce pacience.*" One can hope, but no work of scholarship is beyond criticism, and no amount of help from others should obscure that I alone bear the responsibility for the content of this book.

K. G.

Oxford, Mississippi
October 2000

Introduction

The role of National Socialism—origins, actions, and legacy—in the development of German society continues to be a source of endless fascination. I approach this theme by focusing on the "politics of inventing," a problem at the intersection of politics, law, and technology that for the better part of a century occupied the German community of inventors, industrial scientists, patent experts, business executives and, intermittently, the country's political leaders. The specific problem I address, and with which my protagonists wrestled, concerns the nature of technological creativity and its rewards: the rights of the inventor in the age of corporate capitalism. This is a problem still with us today. It centers on questions of professional recognition and personal dignity for engineers, researchers, and scientists in industry. It is about the chances of the independent inventor—often a small operator—in a world dominated by big business and organized capital. In short, it is about the balance between the rights of the individual and the power of the large organization.

Today, in the United States, the image that comes to mind when these things are mentioned is the instant Silicon Valley fortune or the newly minted Microsoft millionaire. It would be well to remember, though, that for every success story there are a thousand lives of quiet desperation—the countless salaried scientists and engineers, the untold inventors, tinkerers, and small entrepreneurs whose inventions are appropriated by their employers and large corporations as a matter of course, and who toil in obscurity, augmenting the stock of technological capital without special compensation or public acknowledgment. This study is about them, their struggle for professional recognition and a just financial reward, and what their story tells us about the history of twentieth-century Germany.

The politics of inventing first emerged in the 1860s, when German engineers, industrialists and political leaders began to contem-

Notes for this section begin on page 11.

plate the need for a new patent system.[1] Conflict and debate over the Patent Code, which when it was adopted in 1877 did not even bother to mention the inventor, subsided only around 1960. By that time, the cumulative weight of reforms in years past as well as sheer exhaustion finally caused the issue to drop below the threshold of political visibility. The intervening decades witnessed an on-again, off-again struggle over the position of the inventor in the patent system—its basic design, its details, its relation to labor law, its significance for the country's technological culture and economy—and who lost or gained by it.

Organized efforts to make the Patent Code of 1877 more inventor-friendly began in the Wilhelmine era. Interrupted by World War I, those efforts resumed but were botched in Weimar, decided in the Third Reich, and finally concluded in the Federal Republic. I touch on the earlier phases of this story, but concentrate on developments from Weimar to the early Federal Republic. I pay special attention to the languages of inventing used by the contending parties: the scientific-rationalistic discourse employed by big business, on the one hand, and the affinities between the poetic vocabulary of inventor-rights advocates and the ideology of National Socialism, on the other.[2]

*   *   *

Besides the intrinsic interest suggested by its theme and chronology, the struggle over inventor rights is instructive for two additional reasons. It addresses a problem in the history of technology and it sheds light on the role of the Nazi regime in transforming German society from Weimar to Bonn. A common theme in the history of technology is the ambivalence of technological progress. We are the unquestioned beneficiaries of the countless great and small advances that have made life so much more convenient and comfortable in the last century and a half. But we worry deeply about technological disasters such as Chernobyl, Three-Mile Island, the Challenger space shuttle tragedy, and irreparable pollution of the natural environment. Of particular concern is the tendency of technological systems or organizations to become so large and rigid that humans—though they constructed them in the first place—lose control. Huge systems and gigantic projects seem to acquire a life of their own, become immune to further human manipulation, and, enduring on the strength of their own momentum, gradually petrify.[3]

How real are such concerns? To be sure, there is plenty of evidence of undeserved persistence, ossification, and disaster in the realm of technology. But some technological systems and countries seem better at warding off problems than others. Japanese-made automobiles and consumer electronics, for instance, are widely considered to be superior in quality to American ones. Railroads have remained a great deal more viable in Europe than in the United States. Similar observations have been made with respect to German machine tools, automobiles, complex machinery, and capital-equipment goods, relative to their American counterparts.[4] There are many reasons for such differences, but one of them has to do with the ability to preserve or revive the human control, flexibility, and dynamism of large systems—in other words, to manage them better. One crucial ingredient in the management of technology is the human factor, specifically the motivation and the incentive-and-reward mechanism of the technologists who make up such a big part of the system.

The story of how and why Nazi Germany restructured the patent system and adopted pro-inventor measures shows that it is by no means easy to change what Thomas Hughes calls the "conservative momentum of technology."[5] But it also shows that adjustments in the organization of technological production can succeed in keeping large systems competitive, in making them capable of adaptation and evolution. The remarkable staying power of West Germany's mostly older industries in the postwar-war era, which built in part on the changes made by the Nazis, clearly indicates this.[6] It is an important lesson for all those interested in strengthening industrial competitiveness and preserving technological vitality in the modern bureaucratic world.

Contrast between Germany and the United States brings up another, closely related consideration. The question is not merely the extent to which we are the beneficiaries or victims of technology, but also its masters or slaves. Karl Marx's vision of human liberation by modern science and technology never came true in the United States, it has been argued. The dialectical conflict between the forces of production (technological know-how) and the relations of production (the capitalist power hierarchy) failed to materialize, and Marx's two opposing forces combined instead into a single juggernaut. "The history of modern technology in America," David Noble has written, "is of a piece with that of the rise of corporate capitalism." Modern engineering, the source of technological progress, became an integral part of the capitalist system, and the "professional engineers who emerged during the second half of the [nineteenth] century in America as the foremost agents of modern technology became as well the agents of corporate capital."[7]

The historical development Noble describes and analyzes was not limited to the United States. It occurred in all societies where corporate capitalism emerged dominant. Even so, there were important national variations on the common theme, and not everywhere did engineers "resolve the tension between the dictates of the capitalist system and the social potentials implicit in technological development" as perfectly and rapidly as Noble argues happened in America.[8] In Germany, the process whereby technologists became the willing agents of capital was more agonizing, stayed contested longer, and remained less complete than in the United States.

The imperfect integration of German technologists into the capitalist system manifested itself in a variety of ways. One was nineteenth-century German technical education's resistance to becoming the handmaiden of capitalist industry; it defined itself instead as a variation of classical German *Bildung* and *Kultur*.[9] Another was the politics of inventing, a struggle that was fought with great bitterness in Germany—producing notable entitlements for its engineering labor—but one that never took place in the United States. Such differences point to variations in the degree to which engineers—and nations—embraced capitalism. With the politics of inventing as our measure, it becomes possible to see that German technological culture not only had a more difficult time becoming fully capitalist, but also preserved more remnants of anticapitalist sentiment than the United States ever possessed.[10] The precise meaning of that sentiment is discussed more fully in the next chapter, but again it centered on resistance to the subordination of the technological imagination to commercial criteria, financial calculation, and business rationality. This became particularly pronounced in the 1920s, when large bureaucratic corporations moved to routinize technological progress by turning inventors into substitutable labor but simultaneously suffered from stagnation and a lack of entrepreneurial vision. The politics of inventing therefore also involved a struggle against the rationalization of industry—in the sense of resisting Max Weber's bureaucratization and demystification of the world—in order to recapture technological dynamism.[11]

* * *

Historians have long debated an issue known as the *Sonderweg*, or Germany's special path. The basic question is whether, and if yes to what extent, the Nazi regime can be explained as a consequence of Germany's "divergence from the west"—of conditions such as the "unpolitical German," the tradition of "illiberalism," the failed effort to establish parliamentary government in 1848 or the 1860s, the position of the Army and the bureaucracy, and many other German "peculiarities" reaching back far into the nineteenth century. For a long time, the *Sonderweg* interpretation ruled, but prevailing opinion nowadays holds that it was vastly exaggerated and mostly wrong.[12]

Much of the effort to discredit the older interpretation has gone into demonstrating that Imperial Germany was rather more normal as a capitalist-bourgeois society than scholars used to think. This conclusion cuts two ways. It tends to reduce the weight of long-term causes of National Socialism, making it more of a contingency, a product of problems in the 1920s and early 1930s. Another inference, however, is to trace the roots of Nazism not to the strength of preindustrial traditions, insufficient liberalism, and resistance to the advent of modernity, but rather to its opposite—to the power of bourgeois and capitalist institutions in Imperial Germany.

Studying the politics of inventing and the history of the German patent system suggests that German peculiarities, whether of the industrial or preindustrial variety, are not quite as irrelevant as some have maintained. The German Patent Code of 1877 that plays such a large role in the following chapters certainly differed from its Western counterparts, and it possessed a number of regressive and illiberal features absent elsewhere that helped pave the road to 1933. The illiberal features of Germany's patent system were consciously adopted because they served the most modern of industrial capitalist purposes and had little to do with preindustrial traditions. At the same time, the enactment of a bill with those socially reactionary features was also part of the larger rejection of economic and political liberalism in the late 1870s. The political dynamism generated by the German Patent Code sprang from the organizational and technological modernity of an intentional authoritarianism, which in turn cannot be explained without reference to economic and political conditions in the first part of the nineteenth century.

As the intricacies of this amalgam suggest, the debate over Germany's *Sonderweg* can easily degenerate into something like the question of which came first, the chicken or the egg. It has long been known that certain aspects of Germany's preindustrial heritage were exploited for the purposes of a highly modern industrial capitalism. With respect to the Patent Code, this produced an institutional framework that, relative to other countries, was unusually favorable to big business. In that sense Germany was exceptional rather than normal. But it is also true that German exceptionalism in this instance hinged more on the strength than on the weakness of its bourgeois institutions.[13]

It was precisely in its capacity as an instrument of big business and the upper bourgeoisie that the patent system became a bridge of continuity from the Empire to Weimar to Hitler. It set in motion a chain of events that caused deep resentment among middle-class technical professionals, first in Wilhelmine Germany and then in the Weimar Republic, against the excessive powers of capital. Political agitation and failed reform efforts as well as the intransigence and internal divisions of big business over the inventor issue played directly into the hands of the Nazis, helping Hitler climb to power as the would-be savior of an exploited and unprotected middle class.

But the story does not end there, nor did Germany's ultracapitalist patent system stop making history in 1933. On the contrary, the Patent Code became a lightning rod for social-revolutionary energies and disaffection due to the rigidities and perceived technological stagnation of the Weimar era.[14] While many other revolutionary pressures, such as those associated with the Stormtroopers or "Aryan physics," came to nothing, inventor reform only gained momentum.[15] This inspired an intriguing reversal of fronts, in which the dictatorship, National Socialist rhetoric to the contrary notwithstanding, enacted a new Patent Code that borrowed much from older Anglo-American, liberal models favoring the individual inventor. The underlying idea was to revitalize Germany's economy and technology by unleashing the creative genius of industrial scientists and inventors from the shackles of capital: in other words, to carry out social-legal reform as economic and technology policy.[16] Corporate leaders did their best to preserve the reactionary modernity of the old patent law, but as early as 1934 the regime ruled against them. Subsequent measures during the war moved inventor policy closer to the employer-inspired movement of labor rationalization begun in the second half of the 1920s, but officials in the Nazi party and German Labor Front made sure this development did not undermine their central achievement of protecting the inventor against capitalist exploitation. The findings of this study, therefore, do not readily fit interpretations concerning the post-1933 alliance of "entrepreneurs ... with the Nazi leaders" and the "power cartel" of Nazis, big business, and the Army. With respect to inventors and their rights, there was no "free hand accorded to industry," nor was the "reordering of labor relations" in this case the "basis of a positive relationship between the Nazi government and 'big business.'"[17]

The affinities between Nazi social policy and economic policy—more precisely, the 1934 Law for the Regulation of National Labor and the German Labor Front's interest in social policy as a means of fostering industrial efficiency—have received considerable scholarly attention. The origins of Nazi social policy have been traced to capitalist strategies of rationalization (Taylorism, Fordism, the science of work, industrial psychology) aimed at increasing labor productivity in Weimar, which then continued in modified form after 1933.[18] To be sure, the conflicts between industry and Nazi social planners, who in the Third Reich came to represent the material interests of labor, have also been noted.[19] But there is broad agreement that regardless of such tensions an underlying and unspoken consensus existed between capitalism and Nazism on the basis of the productivity-social policy nexus.[20] There is also recognition that this nexus survived the Third Reich. Some innovations, such as wage calculation methods, transferred directly to West Germany.[21] More generally, the regime's social policies contributed to the destruction of separate class milieus and the emergence of a qualitatively different type of "atomized society" after 1945—a society characterized by heightened concern with private life; comprehensive social security legislation; and a population of achievement-oriented, individualistic workers focused on mass consumption and leisure.[22] Postwar German society, which the Nazis did so much to create, is sometimes linked—either implicitly or explicitly—with the unchecked growth of an instrumental rationality cut loose from its ethical moorings, a "dialectic of the Enlightenment," and the consolidation of an American-style regime of capitalist power conjoined with technocratic domination.[23]

This study traces another lineage of the postwar order. Instead of equating engineers with capitalist management, which is the premise of most work on social rationalization in industry, I focus on the politics of inventing to highlight the tension between them. This tension assumed different forms between 1933 and 1945. One was the political opportunism of inventor-engineers such as Wernher von Braun and Ferdinand Porsche, who used the possibilities created by the Nazi dictatorship to realize radical technological ambitions that would never have succeeded under a technologically more conservative regime of rational, profit-orientated capitalism.[24] Another manifestation of the same tension was the regime's enthusiasm for patent reform and inventor policy, which had little to do with capitalist ways of thinking about technological innovation, or with capitalist methods of stimulating the productivity of engineering and inventing labor. On the contrary, that enthusiasm was rooted in a potent critique of industrial capitalism, especially when capitalist rationality served as a vehicle for power over engineers and inventors.

Capitalist power over inventors had two dimensions, which corresponded to the two faces of inventor policy. Technologically, inventor policy had its origins in the frustrated professional ambitions of engineers, inventors, and others who believed that Germany's capitalist management in Weimar was insufficiently dynamic and put caution and conservatism ahead of invention and innovation. Patent reform and pro-inventor measures were designed to solve this problem. Socially, inventor policy carried on the struggle of the "little man" squeezed between big business and big labor. Inventors and their spokesmen in the Third Reich effectively used a language of inventing that centered on poetry, heroism, and genius to create the space for their own survival as a social group, as middle-class professionals with specific economic entitlements.[25] This was a struggle for social emancipation from big capitalism (and for escape from Marxist leveling). If in the end it did not succeed in achieving all its objectives, it was by no means a failure either.

After World War II, industrialists did their best to undo the dictatorship's inventor and patent policies and hoped to return to the legal situation of the Weimar Republic. While the totalitarian aspects of Nazi inventor policy were soon dismantled, its most important elements survived the Third Reich. Stripped of their racial dimension, they govern Germany's patent system to this day. In the end, therefore, it was the Nazi dictatorship that demolished this particular bastion of capitalist power. Demolition, moreover, was coupled with construction, which means that in this case the regime's role in the modernization of German society went beyond the mere destructiveness with which it is typically associated.[26]

To acknowledge Nazism's constructive role is not to deny that Hitler's dictatorship was fundamentally barbaric, nor to argue that the revisionist view of National Socialism as modernization by design is without problems. It is to suggest, however, that an impulse toward societal reform and the pursuit of a more modern, technologically dynamic, equitable, and efficient *Volksgemeinschaft* ("community of the people") of consumers constitute at least one part of the solution to the Nazi conundrum.[27] The existence of these energies, together with the notion of "polycracy," provide an answer to the question of how the regime could leave the positive legacy that it did in the politics of inventing.

The Nazi regime is often described as a polycratic regime, to indicate that it was characterized by chaos, heterogeneity, ad hoc solutions, opportunism, fragmentation, and the simultaneous pursuit of many different, incompatible projects by Hitler's various lieutenants and different Nazi factions. In principle, elements of social revolution and modernization in this configuration could readily coexist with all sorts of other, contrary developments, because the regime as a whole is seen as a bundle of contractions held together only by the charismatic *Führer* principle. In practice, however, the polycratic school of thought inclines to the view that the Third Reich's chaotic system of rule worked against modernization.[28]

Other interpretations of National Socialism stress the overriding importance of biology, racism, and Hitler's socially revolutionary visions. Hitler's domestic policy objectives, in this view, were not merely to get rid of the Jews and other "biological undesirables," but, more importantly, also to destroy the inherited cleavages of status and class and to build a new, more fluid system of equal opportunity and social mobility for the remaining, purified community of true Germans. Whatever modernization took place in the Third Reich was therefore not unintended consequence but the implementation of program.[29]

The answer rests with a synthesis of the polycratic and racial-reformist interpretations. Socioeconomic policy-making always possessed a dynamic and a constituency of its own (typically, the mostly middle-class and lower-middle-class, self-described "losers" of the Wilhelmine and Weimar periods), and was an integral part of the larger Nazi phenomenon. Insofar as Hitler and other true believers had an overarching vision, it amalgamated race and socioeconomic activism. The regime aimed at the creation of a racially homogeneous community of *Volksgenossen* (the Nazi counterpart of citizens), for whom a comprehensive social welfare state as well as substantial social equality, high productivity, and mass consumption would be implemented from above. The ultimate objective, in the words of one Nazi technocrat, was the "establishment of a true people's community of achievement."[30] Needless to say, this community would be one without Jews, "Gypsies," and other "biological undesirables," and it entailed war, territorial conquest, and massive "demographic engineering" in the East.[31] But it would also be a good and socially more equitable society—albeit authoritarian and dictatorial—for the remaining "real" Germans at the center. And in fact, so it was for many ordinary German people, if the findings of *Alltagsgeschichte* and the positive memories of life in the Third Reich teach us anything.[32]

In other words, Nazi ideology both in theory and practice linked the creation of a new social order that prefigured aspects of the post-1945 comprehensive welfare state and consumer society with the elimination of Jews and other "racial inferiors" said to embody the excesses of capitalism and Marxism. There was no contradiction between the two. To be forward-looking and remake society as a "healthy organism" included thinking and acting in racial-biological categories. Creating the racially homogeneous community therefore was the other side of implementing progressive or regenerative social and economic measures, and vice versa.[33]

This was the vision. In reality, the connection between race and social reform was a postulate of the ideological imagination rather than a logical necessity. For that reason, and because of the chaos associated with polycracy, the links between racism and social policy were not inseparable. This was especially true in specialized areas of policy-making away from the center of power. Racial measures often were simply tacked on in more or less artificial ways— just as, say, researchers in the former German Democratic Republic would preface essentially nonideological scholarship with perfunctory professions of Marxism-Leninism.[34] To be sure, in the Third Reich those connections had devastating consequences, but conceptually they could easily be broken, and in fact were easily broken after the war, when both German and occupation authorities cleaned up a mass of Nazi laws and regulations for continued use in the postwar world.

In this way, biology and Social Darwinism were exorcised from post-1945 political discourse, while many of the socially reformist elements of Nazi policy (and their human executors) survived the regime to lay the groundwork for a fundamentally different political order.[35] So it was with patent reform and inventor policy. The dictatorship bequeathed its legacy of pro-inventor measures as a foundation stone that West Germany would use to build more stable management-labor relations and more dynamic technological progress than Weimar ever could. It is the ironies of this development—and the light it sheds on German history from the Empire to the Federal Republic—that make the politics of inventing a story worth retelling.

The narrative is divided into four parts. Part one presents the main lines of the argument in comparative perspective, with an emphasis on theoretical and conceptual considerations. It gives the nineteenth-century background, provides an introduction to the debate about the nature of inventing and the notion of technological culture, makes the connection with National Socialism in Weimar and the Third Reich, and establishes continuities beyond 1945 (chapter 1). Part two examines the various stages and dimensions of the inventor-rights struggle between management and inventors in the Weimar Republic. The basic theme here is the process whereby the language and demands of inventor-rights activists became the Nazi inventor agenda after 1933, discontent with Weimar forging a mindset in which reformist activism grew all but indistinguishable from faith in salvation by National Socialism (chapters 2–6). Part three describes the continuation of the conflict in the Third Reich. It pays special attention to the anticapitalist impulses of newly empowered inventor "revolutionaries" and the conditions that made possible the implementation of large parts of their ideas, even as other parts had to be modified or abandoned. The focus then widens to address the relationship between the regime's inventor measures and Albert Speer's "production miracle" of 1942–44 (chapters 7–12). Part four explores the politics of inventing after 1945 and the adaptation of Nazi inventor measures to conditions in the Federal Republic. It concludes with an assessment of the significance of inventor legislation for West Germany's political and technological cultures (chapter 13).

Notes

1. For background and early patent conflicts, see Alfred Heggen, *Erfindungsschutz und Industrialisierung in Preußen 1793–1877* (Göttingen, 1975).
2. For the inventor-rights struggle before 1914, see the author's *New Profession, Old Order: Engineers and German Society, 1815–1914* (Cambridge, 1989), pt. 3.
3. Thomas P. Hughes, *American Genesis: A Century of Invention and Technological Enthusiasm 1870–1970* (New York, 1989), 459–72, esp. 470–71; Dirk van Laak, *Weiße Elephanten: Anspruch und Scheitern technischer Großprojekte im 20. Jahrhundert* (Stuttgart, 1999); James C. Scott, *Seeing Like a State: How Certain Schemes to Improve the Human Condition Have Failed* (New Haven, 1998).
4. Gary Herrigel, "Industry as a Form of Order: A Comparison of the Historical Development of the Machine Tool Industries in Germany and the United States," in *Governing Capitalist Economies: Performance and Control in Economic Sectors*, ed. J. Rogers Hollingsworth, Philippe Schmitter, and Wolfgang Streeck (New York, 1992), 97–128; Richard J. Overy, "The Economy of the Federal Republic since 1949," in *The Federal Republic of Germany since 1949: Politics, Society and Economy*

before and after Unification, ed. Klaus Larres and Panikos Panayi (London, 1996), 3–34, esp. 15–22. Also, Herrigel, *Industrial Constructions: The Sources of German Industrial Power* (Cambridge, 1996). Herrigel attributes the historical success of German industry to its regional diversity and the parallel development of its small and medium business sectors, on the one hand, and large integrated firms, on the other.

5. Hughes, *American Genesis*, 461.

6. On the relationship between the Third Reich's socioeconomic policies and postwar economic success: Simon Reich, *The Fruits of Fascism: Postwar Prosperity in Historical Perspective* (Ithaca, 1990); Neil Gregor, *Daimler-Benz in the Third Reich* (New Haven, 1998), 1–15, 247–52; Alan Kramer, *The West German Economy, 1945–1955* (New York, 1991), 17–30, 172, 182–96; Wolfram Fischer, "The Role of Science and Technology in the Economic Development of Modern Germany," in *Science, Technology, and Economic Development: A Historical and Comparative Study*, ed. William Beranek, Jr., and Gustav Ranis (New York, 1978), 71–113, esp. 101–4; Raymond G. Stokes, "Technology and the West German Wirstschaftswunder," *T&C* 32, 1 (1991): 1–22. More broadly on German technological culture: Joachim Radkau, *Technik in Deutschland: Vom 18. Jahrhundert bis zur Gegenwart* (Frankfurt a. M., 1989).

7. David F. Noble, *America by Design: Science, Technology, and the Rise of Corporate Capitalism* (New York, 1977), 84–109, xxii–xxiii.

8. Noble, *America by Design*, xxiii. For a more nuanced view of the same development, see Edwin T. Layton, Jr., *The Revolt of the Engineers: Social Responsibility and the American Engineering Profession* (Cleveland, 1971).

9. See Gispen, *New Profession*, pts. 1 and 2; Karl-Heinz Manegold, *Universität, Technische Hochschule und Industrie: Ein Beitrag zur Emanzipation de Technik im 19. Jahrhundert unter besonderer Berücksichtigung der Bestrebungen Felix Kleins* (Berlin, 1970); Wolfgang König, *Künstler und Strichezieher: Konstruktions- und Technikkulturen im deutschen, britischen, amerikanischen und französischen Maschinenbau zwischen 1850 und 1930* (Frankfurt a. M., 1999). Comparable tendencies existed in the United States but were less powerful than in Germany; cf. Noble, *America by Design*, chs. 2–3, and literature cited there.

10. Anticapitalism among engineers is discussed by Karl-Heinz Ludwig, *Technik und Ingenieure im Dritten Reich* (Düsseldorf, 1974), 15–102; premodern elements in German society: Ralf Dahrendorf, *Society and Democracy in Germany* (Garden City, 1969); Norbert Elias, *The Germans: Power Struggles and the Development of Habitus in the Nineteenth and Twentieth Centuries* (New York, 1996); good introductions to German economic thought: Avraham Barkai, *Nazi Economics: Ideology, Theory, and Policy*, trans. Ruth Haddass-Vashitz (New Haven, 1990), esp. 71–105; A. J. Nicholls, *Freedom with Responsibility: The Social Market Economy in Germany 1918–1963* (Oxford, 1994), 1–89; David F. Lindenfeld, *The Practical Imagination: The German Sciences of State in the Nineteenth Century* (Chicago, 1997).

11. For the broader political and philosophical meaning of such resistance: Ernst Nolte, *Three Faces of Fascism: Action Française, Italian Fascism, National Socialism*, trans. Leila Vennewitz (New York, 1965), 429–54; Tilla Siegel, "It's Only Rational: An Essay on the Logic of Social Rationalization," *International Journal of Political Economy* 24, 4 (winter 1994–95): 35–70, 53.

12. David Blackbourn and Geoff Eley, *The Peculiarities of German History: Bourgeois Society and Politics in Nineteenth-Century Germany* (New York, 1984), 1–35, esp. 20–28; Geoff Eley, "What Produces Fascism: Preindustrial Traditions or a Crisis

of a Capitalist State?" *Politics and Society* 12, 1 (1983): 53–82; idem, "Is There a History of the *Kaiserreich?*" in *Society, Culture and the State in Germany, 1870–1930,* ed. Geoff Eley (Ann Arbor, 1996), 1–42; Detlev J. K. Peukert, *Inside Nazi Germany: Conformity, Opposition and Racism in Everyday Life* (New Haven, 1987), 45; James Retallack, *Germany in the Age of Kaiser Wilhelm II* (New York, 1996).

13. Heggen, *Erfindungsschutz,* 135.

14. Exemplary discussion of Nazi efforts to overcome perceived conservatism in the automobile industry: Hans Mommsen with Manfred Grieger, *Das Volkswagenwerk und seine Arbeiter im Dritten Reich* (Düsseldorf, 1996), 51–113.

15. On the Stormtroopers, see Conan Fischer, *Stormtroopers: A Social, Economic, and Ideological Analysis 1929–1935* (London, 1983); Richard Bessel, *Political Violence and the Rise of Nazism: The Storm Troopers in Eastern Germany 1925–1934* (New Haven, 1984); Mathilde Jamin, *Zwischen den Klassen: Zur Sozialstruktur der SA-Führerschaft* (Wuppertal, 1984). On Aryan physics: Alan D. Beyerchen, *Scientists under Hitler: Politics and the Physics Community in the Third Reich* (New Haven, 1977).

16. The idea that social policy and social reform were key to making German industry more productive and efficient is a thread that runs from the rationalization movement in Weimar through the Third Reich into the Federal Republic. Explored here with specific reference to inventor policy, social policy is analyzed in many works, including Mary Nolan, *Visions of Modernity: American Business and the Modernization of Germany* (New York, 1994); Tim Mason, *Social Policy in the Third Reich: The Working Class and the "National Community,"* trans. Joan Broadwin, ed. Jane Caplan, intr. Ursula Vogel (Providence, 1993); Carola Sachse, *Siemens, der Nationalsozialismus und die moderne Familie: Eine Untersuchung zur sozialen Rationalisierung in Deutschland im 20. Jahrhundert* (Hamburg, 1990); Ronald Smelser, *Robert Ley: Hitler's Labor Front Leader* (Oxford, 1988); idem, "Die Sozialplanung der Deutschen Arbeitsfront," in *Nationalsozialismus und Modernisierung,* ed. Michael Prinz and Rainer Zitelmann, 2d ed. (Darmstadt, 1994), 71–92; Matthias Frese, *Betriebspolitik im "Dritten Reich": Deutsche Arbeitsfront, Unternehmer und Staatsbürokratie im Dritten Reich* (Paderborn, 1991); Tilla Siegel, *Leistung und Lohn in der nationalsozialistischen "Ordnung der Arbeit"* (Opladen, 1989); idem, "Wage Policy in Nazi Germany," *Politics & Society* 14, 1 (1985): 1–51; idem, "It's Only Rational" ; Thomas von Freyberg and Tilla Siegel, *Industrielle Rationalisierung unter dem Nationalsozialismus* (Frankfurt a. M., 1991).

17. Making the case for the "power cartel" and labor relations on management's terms: Ian Kershaw, *The Nazi Dictatorship: Problems and Perspectives of Interpretation,* 3d ed. (London: 1993), 49–50; Heinrich August Winkler, "Der entbehrliche Stand: Zur Mittelstandspolitik im 'Dritten Reich,'" *Archiv für Sozialgeschichte* 17 (1977), 1–40; Alan Milward, "Fascism and the Economy," in *Fascism: A Reader's Guide: Analyses, Interpretations, Bibliography,* ed. Walter Laqueur (Berkeley, 1976), 399, 408; Rüdiger Hachtmann, *Industriearbeit im "Dritten Reich": Untersuchungen zu den Lohn- und Arbeitsbedingungen in Deutschland 1933–1945* (Göttingen, 1989), 13–23, 302–9; Siegel, "Wage Policy"; Barkai, *Nazi Economics,* 17–20; Peukert, *Inside Nazi Germany,* 43, 245.

18. E.g., Nolan, *Visions of Modernity,* 179–205, 227–35; Frese, *Betriebspolitik,* 93–113, 449–54.

19. E.g., Mason, *Social Policy,* 208–24; Frese, *Betriebspolitik,* passim.

20. E.g., Gregor, *Daimler-Benz,* 1–15; and most of the literature cited in note 16.

21. Siegel, "Wage Policy"; idem, "It's Only Rational"; idem, *Leistung und Lohn*, esp. 13–18.
22. Peukert, *Inside Nazi Germany*, 236–42.
23. Hans Dieter Schäfer, "Amerikanismus im Dritten Reich," in *Nationalsozialismus und Modernisierung*, 199–215, esp. 214–15; cf. also Siegel, *Leistung und Lohn*, 13–18, 272; general critiques of modernity: Zygmunt Bauman, *Modernity and the Holocaust* (Ithaca, 1989), viii–xiv; Max Horkheimer and Theodor W. Adorno, *Dialectic of Enlightenment* (New York, 1975); also, Alan Beyerchen, "Rational Means and Irrational Ends: Thoughts on the Technology of Racism in the Third Reich," *CEH* 30, 2 (1997): 386–402.
24. Michael Neufeld, *The Rocket and the Reich: Peenemünde and the Coming of the Ballistic Missile Era* (New York, 1995); Mommsen, *Volkswagenwerk*, 25–176.
25. Although the Nazi regime's inventor policy applied to white-collar and blue-collar workers, it inevitably benefited the latter more than the former. Its effect therefore ran counter to the leveling of differences between them in the Third Reich; cf. Michael Prinz, *Vom neuen Mittelstand zum Volksgenossen: die Entwicklung des sozialen Status der Angestellten von der Weimarer Republik bis zum Ende der NS-Zeit* (Munich, 1986), 328–36.
26. Nazism as destruction: Jürgen Kocka, "1945: Neubeginn oder Restauration?" in *Wendepunkte Deutscher Geschichte 1848–1945*, ed. Carola Stern and Heinrich A. Winkler (Frankfurt a. M., 1979), 141–68; Kershaw, *Nazi Dictatorship*, 140–9; Peukert, *Inside Nazi Germany*, 42–46, 174, 175–83, 240–42, 243–49, esp. 247; Gregor, *Daimler-Benz*, 251–52. See also Gerhard A. Ritter's excellent *The Transformation of German Society: Continuity and Change After 1945 and 1989/90* (Berkeley, 1996).
27. On Nazi impulses toward societal reform ("social revolution" and "modernization" debate), see David Schoenbaum, *Hitler's Social Revolution: Class and Status in Nazi Germany, 1933–1939* (New York, 1967); Dahrendorf, *Society and Democracy*; Rainer Zitelmann, *Hitler: Selbstverständnis eines Revolutionärs* (Hamburg, 1987); Prinz and Zitelmann, *Nationalsozialismus und Modernisierung*; Prinz, *Vom neuen Mittelstand*; Smelser, *Robert Ley*, esp. 306–7; idem, "Sozialplanung"; Thomas Saunders, "Nazism and Social Revolution," in *Modern Germany Reconsidered*, ed. Gordon Martel (London, 1992), 159–77; Reich, *Fruits of Fascism*; Barkai, *Nazi Economics*; Milward, "Fascism and the Economy"; for modernization/rationalization in a more sinister light: Tilla Siegel, "Rationalizing Industrial Relations: A Debate on the Control of Labor in German Shipyards in 1941," in *Reevaluating the Third Reich*, ed. Thomas Childers and Jane Caplan (New York, 1993), 139–60, esp. 155; idem, "Wage Policy"; Hachtmann, *Industriearbeit*; a highly critical assessment of the modernization thesis: Norbert Frei, "Wie modern war der Nationalsozialismus?" *GuG* 19 (1993): 367–87. On the shift toward a racial-biological paradigm: Michael Burleigh and Wolfgang Wippermann, *The Racial State: Germany 1933–1945.* (Cambridge, 1991); Charles Maier's "Foreword" and Tim Mason's "Whatever Happened to 'Fascism'?" in *Reevaluating the Third Reich*, xi–xvi, 253–62; Paul Weindling, *Health, Race and German Politics Between National Unification and Nazism, 1870–1945* (Cambridge, 1989); Kershaw, *Nazi Dictatorship*, 131–49, 202–6.
28. Leading protagonists of the polycratic interpretations are Hans Mommsen and Martin Broszat; Kershaw, *Nazi Dictatorship*, passim.
29. Zitelmann, *Hitler*; Smelser, *Robert Ley*, Prinz, *Vom neuen Mittelstand*.
30. Robert Kahlert, *Erfindertaschenbuch* (Berlin, 1939), 6; Smelser, *Robert Ley*.

31. Christopher R. Browning, *The Path to Genocide: Essays on Launching the Final Solution* (Cambridge, 1992), 18, 20, 22, 24.

32. Ulrich Herbert, "Good Times, Bad Times: Memories of the Third Reich," in *Life in the Third Reich*, ed., intr. Richard Bessel (New York, 1987),97–111; Peukert, *Inside Nazi Germany*, passim; Burleigh and Wippermann, *Racial State*.

33. Burleigh and Wippermann, *Racial State*, 4; Kershaw, *Nazi Dictatorship*, 184–89; cf. also Brian Jenkins and Spyros A. Sofos, "Nation and Nationalism in Contemporary Europe: A Theoretical Perspective," in *Nation and Identity in Contemporary Europe*, ed. Brian Jenkins and Spyros A. Sofos (London, 1996), 9–32.

34. Raul Hilberg, *Victims, Perpetrators, Bystanders: The Jewish Catastrophe 1933–1945* (New York, 1994) 2–74; Hans Mommsen, "The Realization of the Unthinkable: The 'Final Solution of the Jewish Question' in the Third Reich," in his *From Weimar to Auschwitz*, trans. Philip O'Connor (Princeton, 1991), 224–53.

35. Reich, *Fruits of Fascism*, 316; Ronald Smelser, "Die Sozialplanung," 72–92, esp. 82; Ritter, *Transformation*.

Part I

THE PROBLEM

Chapter 1

THE INVENTOR IN GERMAN LAW AND HISTORY

A Comparative Perspective

"*To* promote the progress of science and useful arts," nations have long used incentives known as patents.[1] A patent may be defined as the exclusive right, granted by a government, to manufacture, use, or sell an invention for a certain number of years. There are a number of different theories justifying patents, each having different consequences for patent legislation. All modern theories, however, agree that patents benefit society because the advantages of technological progress induced by temporary monopoly, and the dissemination of new ideas resulting from their public disclosure, outweigh the disadvantages of granting exclusive rights. Patents spur technological progress because, in Abraham Lincoln's famous words, they "add ... the fuel of interest to the fire of genius."[2] Without patents there is little incentive to invent. Competitors are free to copy technological advancements, depriving inventors and innovators of any reward for their efforts. Absence of patent protection can also lead to secrecy. This increases the risk that technological advances are lost upon the inventor's death or that progress remains isolated and sporadic, secrecy being antithetical to the free flow of information on which the interdependent process of technological advancement depends.[3]

Patent codes and procedures vary widely from country to country. As far as basic organization is concerned, however, patent systems generally belong to one of two types. The one most widely used

Notes for this section begin on page 49.

is the *first-to-file* system, in which a patent is issued to the person who is the first to apply for it. The invention, rather than the inventor, is the primary focus of attention. The first-to-file system speeds disclosure and reduces the likelihood of protracted and expensive litigation over authorship. Most nations, including Germany, administer patents under some version of the first-to-file principle.[4]

The other, less common type of patent system is based on the *first-to- invent* principle, which entitles the first and true inventor to a patent. It centers on the inventor, as opposed to the invention. No one can file for a patent who is not the original inventor, and the claim of a second inventor who makes the same invention later but applies for a patent earlier than the first inventor (known as "interference") is invalid. The United States and Canada are the most important of the few countries whose patent systems are based on the first-to-invent principle.

The first-to-invent system is older and used to be more popular than the first-to-file system. In addition to governing the United States Code, it served as the foundation of British and French patent legislation from the seventeenth and eighteenth centuries until after World War II. Like the American Code, the original British and French patent systems dated from preindustrial or early industrial times, when inventing was still largely a matter of individual entrepreneurship, private effort, and, by today's standards, small proportions.[5] Geared toward those economic-technological circumstances, the first-to-invent system is a typical manifestation of classical liberalism, with its emphasis on economic deregulation and the rights and responsibilities of the individual. The English Statute of Monopolies of 1624, for instance, prohibits all monopolies and special business privileges as economically harmful, but exempts patents for inventions on the grounds that they increase overall wealth just as does the free-market system. Patents and the market therefore complement each other as related instruments of economic progress in the 1624 statute. In fact, the bond between them is so close that one recent observer concludes: "it is no historical accident that the world's first anti-monopoly law ... is also the first patent law ..."[6] In the liberal tradition, the right to a patent is based on a combination of the economic principle with the natural-law idea of "intellectual property," by which the true and first inventor is entitled to his creation just as a composer or writer would be.[7] The first-to-invent principle treats inventions as the product of individual genius and creativity, analogous to artistic and literary works, and grants the inventor broad license to maximize profits and litigate his claims in court.

Despite the extensive protection that the first-to-invent system offers in principle, most inventors in the United States today are shielded poorly or not at all against the loss of their inventions. Nor does the patent system today act as the incentive it was meant to be, as far as the majority of inventors are concerned. The bureaucratization of industry and the emergence of organized inventing in corporate and government laboratories since the last quarter of the nineteenth century have turned the majority of inventors into salaried employees. The employer now comes between the inventor and the invention and lays claim to the rights of the employee's intellectual product.[8]

In theory, the first-to-invent system presents the employer with formidable obstacles to appropriating the results of the employee's inventive genius. In the United States, "the law considers an invention as the property of the one who conceived, developed, and perfected it, and establishes, protects, and enforces the inventor's rights until he has contracted away those rights."[9] Without such a special contract, only the rights to those inventions the employee was hired to invent pass directly to the employer. Other work-related inventions the employee makes on his own initiative, even if using the employer's time and materials, equipment, or labor, remain his property subject to a "shop right," which gives the employer a nonexclusive license to use the invention. Inventions made at home and unrelated to the job also belong to the employee, even if they fall within the scope of the employer's business.[10] To circumvent the obstacle represented by the Patent Code, employers routinely require scientists and engineers to sign special patent agreements, stipulating that as a condition of employment any inventions or improvements made by the employee during employment shall belong to the employer.[11]

The permissible scope of pre-employment patent agreements has always been extremely broad in the United States. An employee can bargain away the right to ideas, concepts, inventions, know-how, proprietary information, or patents in exchange for employment. It is not uncommon for such contracts to require assignment of all inventions, regardless of time, location, or subject matter, and extending for six months or more beyond termination of employment. Only a few states have statutes restricting the employer's right to an employee's invention that is unrelated to the employer's business. Some companies have voluntary invention-incentive plans, offering employees a nominal sum or special recognition when a patent is applied for or awarded. Sometimes they have spe-

cial invention-award systems or profit-sharing plans, but such programs are relatively rare and more rarely generous. While most universities (and under certain conditions the federal government)
typically offer the employee attractive royalty arrangements, most
companies in the private sector pay their employees little, if anything, for their inventions.[12] In sum, as far as the salaried inventor
in industry is concerned, the intention of the first-to-invent principle at the heart of the United States Patent Code has been turned on
its head.

In contrast to the United States, most European nations and
Japan have legislation to protect the salaried inventor. Sometimes
these laws merely restrict the scope of patent agreements, preserving the rights of employees by making it illegal for employers to
require the wholesale assignment of all inventions. In other cases,
the restrictions go a step further by mandating compensation for
certain types of employee inventions. Some nations, including Germany and Sweden, prohibit patent agreements altogether and have
comprehensive statutes determining all the rights in all possible
types of employee inventions.[13] There have been sporadic attempts—
all unsuccessful—to introduce similar legislation in the United
States since the early 1960s. In 1962 Congressman George Brown of
California introduced bills that would have outlawed pre-employment patent agreements. Arkansas Senator John McClellan in the
1960s held hearings to determine whether legislation was needed to
rekindle the salaried employee's inventive energies and accelerate
technological progress. In the mid-1970s, Senator Gary Hart proposed legislation that would have severely restricted the use of
patent agreements. In 1981, Congressman Robert Kastenmeier
sponsored another bill limiting the permissible scope of pre-inven-
tion patent agreements. During the 1970s and into the early 1980s,
California representative John Moss sponsored various resolutions—
all modeled on the West German employee-invention statute—dis-
tinguishing between free inventions, which would belong to the
employee outright, and service inventions, which would be assigned
to the employer subject to compensation.[14]

Those initiatives failed to become law for at least three reasons.
One centers on the pervasive influence of capitalist ideology in the
United States and the disproportionate power of corporate management vis-à-vis labor. The imbalance is especially pronounced with
regard to the relations between management and salaried professionals such as engineers, who have been more heavily influenced
by the individualist ethos of American-style professionalism than

their European or Japanese counterparts. American engineers have been taught to view themselves as part of management.[15] Their degree of militancy, solidarity, and unionization therefore has always been low, and despite widespread discontent with patent agreements, American engineering associations have never pursued the issue with any real passion.[16] Salaries and living standards for employed engineers have been sufficiently high to blunt most threats of collective action. Meanwhile, employers have consistently and effectively lobbied against reform of the status quo.[17]

A second reason is that the conditions for innovation in the United States, in spite of the invention-stifling potential of corporate patent practices, remain more favorable than in most other countries. The patent always goes first to the actual inventor, who may have to assign his rights to the employer but still receives the name recognition and resumé-building credit important for career mobility and self-esteem. Large bureaucratic companies may waste much of their engineering talent, but it is not uncommon for engineers and industrial scientists in their most creative years to switch jobs or start their own business. There has also long existed an elaborate network of private, semiprivate, and government research institutes, as well as contract and independent research by universities and special agencies. United States patent fees have always been low compared with those of other countries, and venture capital for promising technologies has been readily available. There exists a thriving culture of small, entrepreneurial firms nurturing innovation in the United States, supported by capital markets much more flexible and aggressive than in Germany. Stock options are a common form of incentive and compensation for critical employees not readily available in Germany—certainly not in the decades before World War II.[18] While the percentage of patents issued to independent inventors in the United States has long been declining as a proportion of the total, historically the rate has been far higher than in Germany. In the early 1950s, the number of patents issued to individuals was above 40 percent of the total in the United States and less than 25 percent in Germany. By the early 1970s those figures had dropped to 23.5 percent and to somewhere between 10 and 20 percent respectively.[19]

Finally, the United States has amassed such an enviable record of technological achievements that, paradoxically, its very success gave rise to a powerful anti-inventor ideology. The many technological triumphs in and after World War II made it easy to believe that the systematization and routinization of invention by the large organization—the origins of which reach back to the turn of the century—had fully succeeded in supplanting the older mechanism of generating innovation through individual incentive and the independent inventor. As one particularly bold exponent of this view put it in 1947, "invention is a term that is not essential" for modern technological progress. The "orderly process of basic research, technical research, design, and development, or approximately this process as conducted in organized laboratories has become so powerful that the individual inventor is practically a thing of the past." According to the same author, "the vast majority of new things today come from organized scientific effort, not from a single inventor. Consequently, invention is not considered to have a place in a critical examination of ... research and development."[20] In this view, division of labor and bureaucratization make invention predictable and rationalize innovation. Most engineers and industrial scientists are reduced to obedient automatons, "captives" who are "told by business executives what problems to work on."[21] These specialists, in the words of one concerned insider, "are primarily hired for their competence in certain limited fields, outside of which they are not encouraged to go, or even to satisfy their curiosity. This is partly to prevent them from wasting their time."[22] It was the natural corollary of such a system that special incentives to encourage creativity were unnecessary, even counterproductive.[23]

When critics of corporate inventor policy in the United States began to cast about for a different approach, they invariably pointed to Germany. Of all the countries that have intervened in the relations between management and engineering labor, Germany has wrestled with the problem longest and come up with the most comprehensive solution. By many accounts it is also the best and the fairest solution. The West German Act on Employee Inventions of 1957 reconciles the conflicting interests of management and salaried inventors by treating the employee as the original owner of the invention and then granting the employer the right to acquire it, subject to compensation and name recognition.

The 1957 statute revolves around two key issues: (1) which inventions the employer can claim, and 2) how the employee is remunerated for claimed inventions.

Claimable inventions. The employee must report all inventions made during the time of employment. These are then divided into "service inventions," which the employer may claim subject to compensation, and "free inventions," which belong to the employee. The scope of service inventions is broad, being defined as all inventions made during the period of employment that are (a) either related to the employee's responsibilities, or (b) based in some measure on the company's in-house experience.[24] All other inventions are free, though the employer is entitled to purchase a "shop right" (internal use) in them.[25]

Remuneration. By claiming a service invention, the employer automatically becomes liable for compensation and must file for a patent, which, if granted, lists the employee as the inventor. The amount of compensation is calculated by a complicated formula. The formula, which revolves around the invention's economic value and the size of the employee's share in making it, is said to yield awards ranging from 2 to 7 percent of the profit from the invention.[26]

Despite occasional grumbling about the law's financial and administrative burdens, German employers on balance recognize its usefulness. There is broad agreement that it boosts employee morale and helps foster creativity and innovation. As far as employees are concerned, a common complaint has to do with supervisors being unjustly named as co-inventors and with insufficient information about the statute. Most employees, however, are said to look upon it with favor. It secures an important measure of professional status and, with some luck, generates a significant amount of extra income.[27]

Werner von Siemens and the Peculiarities of the German Patent Code

It is ironic that the advantageous position of German salaried inventors today, like the unsatisfactory one of their American counterparts, is a consequence of historical developments that began at the opposite end of the scale. In the United States, the original intent of the patent system to encourage innovation by privileging the inventor has effectively been reversed, as far as employees are concerned. In Germany, enlightened inventor policies are the legacy of a patent system that was designed in the last quarter of the nineteenth century with the aim of de-emphasizing the inventor as much as possible.

Germany's Patent Code dates from 1877, a time when the industrialization of invention, bureaucratization of industry, and transformation of entrepreneurial into organized capitalism were beginning to take shape. Science-based industry was in its infancy.[28] Classical liberalism with its concern for individual rights had been dethroned as the reigning ideology, and collectivist social theories, conservative as well as radical, were moving center stage. Tariffs and other forms of protection for industry and agriculture were replacing the free-trade principles that had guided economic policy until the mid-1870s.[29] The 1877 Patent Code reflected these intellectual and socioeconomic changes.

The greatest flaw of the pre-1877 patent situation in Germany was fragmentation and lack of uniformity. The nature of patent protection and theory differed from state to state, ending at the border of jurisdictions that no longer bore much relationship to the patterns of commerce in the unified market of the German Customs Union. Patent laws were an odd mixture of French ideas of the natural rights of the inventor, Manchesterian free-trade principles, and the mercantilist practice of granting special privileges to spur economic activity. Some states followed the French practice of registering and approving all patent claims without examination, which resulted in numerous overlapping and conflicting rights. Other states had strict or impenetrable examination procedures and rejected all but a handful of claims. There was no uniform period of protection from state to state or even within states. In Prussia, for example, the maximum duration of a patent was fifteen years, but the typical length was three years, which in most cases was far too short to move an invention from conceptualization to market. Following French practice, the majority of German patent laws did not call for public disclosure, exposing inventors and businessmen to the danger of involuntary patent infringement and perpetuating industrial secrecy. Patents also constituted an obstacle to inter-state commerce, as different monopolies of manufacture and exclusion in the different German states prevented the free flow of goods.[30]

Largely as a consequence of such problems, a strong anti-patent movement had emerged about 1850. There were similar movements in other countries, but with the exception of the Netherlands, which in 1869 discarded patent protection altogether, none was as powerful as the German one. The leading adherents of the German anti-patent movement were economists and high civil servants. Its ideological underpinnings were the Manchesterian free-trade doctrines that controlled German economic policy from mid-century to the early 1870s. Adam Smith, Jeremy Bentham, and John Stuart Mill had all condoned patenting as sound economic policy and a natural complement to other aspects of liberal economics. Their German disciples, however, increasingly came to see patents as harmful monopolies that restrained trade and impeded economic growth.

Patent opponents also argued that Germany was an economic fol-
lower nation that acquired much of its technology by unlicensed
borrowing and copying from abroad and was therefore better off
without patent protection.

Along with this denial of the economic justification for patents,
the natural-law theory of the inventor's inherent right in his invention came under attack. Logical difficulties—such as the contradiction between the theory of an original property right in inventions
and the requirement of imposing time limits on patents, or granting
a patent to the first inventor but depriving all subsequent, including
genuine, inventors—were advanced to discredit the natural-law theory of intellectual property. It was argued that the purpose of government policy was not to enrich individual inventors but to benefit
society as a whole; that patents harmed technological advancement
and increased prices; that inventing was not analogous to artistic or
literary creation but rested on scientific discovery, which explained
why the same invention was often made by different inventors at
approximately the same time in different places.[31]

The anti-patent movement was especially strong in Prussia,
where it succeeded in reducing to a trickle the number of patents
issued. If the depression of 1873-79 had not intervened, they would
have been eliminated altogether. But the deep economic crisis of the
1870s turned the tide and handed victory instead to a pro-patent
lobby that had emerged in the 1860s. Its leading figure was Werner
Siemens, the brilliant inventor-entrepreneur at the head of the
Siemens electrical company and the individual who did more to
shape the 1877 Patent Code than anyone else.[32]

In contrast to the free traders, Siemens was convinced German
industry could not survive without strong patent protection. But he
shared the anti-patent faction's views about the shortcomings of the
existing patent system, especially its critique of the natural-law theory underpinning the inventor's right. He also agreed that by requiring neither disclosure nor implementation, and by interpreting
inventions as natural property rights entitled to secrecy, the old
statutes gave inventors excessive privileges that harmed economic
growth. This perspective did much to narrow the gap between the two
sides and helped prepare the ground for passage of the 1877 statute.[33]

Although he was Germany's greatest living inventor, in the campaign for a new patent system Siemens acted first and foremost as
a leader of business, a visionary industrialist intent on pressing science and invention into the service of capital. Siemens was determined to execute what David Noble has called the "wedding of

science to the useful arts," and his design for a new patent system was an integral part of this larger vision. Siemens used all his powers to secure a patent system that would cement the position of big business—not inventors—and facilitate the acquisition of the rights in inventions by industrialists and investors.[34] As he put it in a speech to the Association of German Engineers in 1883: "The interests of inventors and the interests of industry are not always the same. The interest of the inventor may only be furthered to the extent that it promotes the interest of industry, and when both interests come into conflict, the law must always put the latter first."[35] Siemens justified this position with a new patent philosophy: the purpose of patents was to make it possible for industry to develop, "and not for inventors to make a lot of money."[36] Neither the Anglo-American blend of economic motives and copyright analogies nor French theories of natural law and intellectual property were the proper basis of a modern patent system. The only legitimate focus of patent legislation was the interest of the national economy—specifically, the speediest possible dissemination of new ideas and the willingness of industry to invest in the risky business of developing new technology.

A patent system based on this purely economic principle would have at least four features, according to Siemens. First, it should have a patent examiner—lacking in French and much current German law—to make a technical and historical investigation of the patent claim and determine its merits. (To merit a patent, an invention should be new, useful, and non-obvious.) Examination would prevent the accumulation of worthless, identical, or contradictory patents and spur industrial activity by reducing the fear of litigation. Second, a patent system should have high and steeply rising annual fees, to discourage inventors from holding on to monopolies that were not financially rewarding. Inventors should either realize and market their invention or get out of the way and let someone else try. In either case, high and progressive fees meant that there would be no accumulation of useless patents. (This was known as the self-cleansing principle.) Third, patent rights should be limited not just by a cap on the number of years they remained in effect, but also by the threat of compulsory licensing and compulsory working. If the patent holder did not attempt to bring the invention to market in a meaningful way, someone else should have the right to force the granting of a license to effect commercialization.

Fourth, patenting should be based not on the first-to-invent principle, but on the first-to-file principle or "registrant's right"

(*Anmelderrecht*). Siemens and his supporters argued that this rule would accomplish a number of important objectives. It would get rid of the harmful practice of keeping technological advances secret. It would eliminate the problem of inventor procrastination and force the invention's fastest possible public disclosure and dissemination by rewarding the first person to report the invention. It would free the Patent Office from the difficult and time-consuming task of determining priority and reduce litigation on that point. Finally, it would obviate the need for assignment of an invention by an employed inventor to the employer (who acquired the patent simply by filing) and so eliminate a potential source of conflict over this question.

When the new patent statute went into effect in 1877, it bore Werner Siemens's heavy imprint. Sometimes described as the "Charta Siemens," the German Code pioneered the economic or incentive principle that is the basis of most current patent systems and theories.[37] The new system did have many advantages. It caused German patents to become highly regarded, forced rapid disclosure and dissemination, cut down on secrecy and litigation, and spurred investment in technological innovation. In this sense, adoption of the patent statute was undoubtedly a major step into the future. It created a stable legal platform that promoted predictability and calculability. The new code was therefore ideally suited to the dawning age of corporate capitalism; to the large technological systems based on electricity, chemistry, physics, and engineering that are the mainstay of life in industrial society; and to the huge investments of capital needed to realize those massive projects. The code's role in making Germany an industrial powerhouse has been stressed time and again.[38]

At the same time, the new German patent system went a long way toward expropriating and subjugating the inventor. In this sense, the statute must be judged an unprogressive, illiberal, even reactionary piece of legislation, in which the needs of German industry were a euphemism for capitalist domination, and social power was redistributed to the detriment of the individual.[39] Specifically, the rules concerning annual increases in maintenance fees, compulsory working, and forcible licenses were major obstacles for independent inventors and small businessmen.[40] More often than not, these groups lacked the resources to maintain a patent for the full fifteen years or hold out against large corporations and capitalists with deep pockets. Their disadvantage, in turn, had consequences for the nation's political and technological cultures. It helps explain, for instance, why Wilhelmian and Weimar Germany could generate so much discontent among engineers, inventors, and small entrepreneurs, a class with no use for socialism even as it resented the power of big business. It also sheds light on the speed and intensity of industrial concentration and provides a partial answer to the question of why Germany became less hospitable to the types of technological innovation associated with independent inventors than the contemporary United States.[41]

The fundamental organizing principle and most innovative aspect of the German Patent Code was the first-to-file principle. What were its consequences for the inventor? The new principle affected independents only marginally, as the inventor and the applicant for a patent typically were the same person. This was not true for employees, however, who lost the right to file under the new system. This, in turn, deprived them of authorship, professional recognition, and any claim to special rewards for their inventions. Much like today's technical writers and the authors of computer manuals, salaried inventors under the new Patent Code received no public credit or acknowledgment, remaining anonymous operatives behind the company label. The only name on the patent was that of the company or the employer.

Why the first-to-file system deprived employees of the right to apply for a patent is not immediately obvious. Through the workings of some crucial, intermediary steps, however, it ended up doing so with a degree of perfection unmatched anywhere else. Since the German Patent Code ignored authorship, the threshold question was not Who is the first inventor, but Who has the right to file—the employer or the employee? The statute was silent on this question, except for disallowing the claims, and voiding the rights, of filers who had acquired the invention illegally.[42] This point was critical but vague, since it was not clear what constituted illegal acquisition. The right to file therefore became the subject of frequent litigation between employers and employees and a matter of dispute among patent experts.[43]

Conservative legal scholars argued that employees owed the whole of their physical and mental powers to the company, which meant the company was entitled to file in all cases. This conclusion was based on labor law, according to which the product of an employee's work belonged to the employer. It was also justified by extending to salaried employees in the private sector the theory of the relationship between the state and the civil servant, according to which the civil servant merged his entire being with the state in return for a guarantee of lifetime employment and a pension.[44] The opposing, liberal school of thought continued to recognize inventor rights, even though the Patent Code did not mention them. Scholars in this tradition pointed out that labor law had only limited applicability to employees' intellectual products, the original rights to which always resided in the author. Analogously—and since only individuals as opposed to abstract legal persons such as companies could invent—the act of inventing created the inventing employee's right to the invention and therefore a right to file. Karl Gareis, future president of the University of Munich and the author of an early commentary on the Patent Code, argued precisely this point in an 1879 treatise on employee inventions: "a legal person or a corporation never discovers something itself; it is always the ... employees, etc. who do so, thereby creating a personal right, to which they themselves are entitled."[45]

The matter was never fully resolved, but the inventor's right remained firmly anchored in legal precedent and, despite the new Patent Code, survived many legal challenges.[46] In theory, therefore, the employee's right to the patent was quite broad. There were only two limiting conditions. The first derived from contractual obligations requiring employees to assign their invention to the employer. One possibility was that the employment contract specified certain technical and inventive duties, in which case the employer became automatically entitled to file patent applications for the relevant inventions. Just as in American law, however, an employer's right did not automatically extend to inventions unrelated to the employee's responsibilities or made at home, since an employment contract in itself did not suffice to assign all inventions to the company.[47] Another possibility was the separate patent agreement, which German employers in science-based industry after 1877 used just as often as their American counterparts. The Siemens company, for example, introduced its *Patentrevers* (literally, patent declaration) as early as six weeks after passage of the Patent Code. This document stated in part that "passage of the German Patent Code, according to which the first person to report an invention is entitled to the patent right so long as illegal acquisition cannot be shown, makes it necessary to establish certain principles for protecting the company's interests, ... the acceptance of which every technical employee must declare by his signature." Article I, which remained in effect until the end of 1928, stipulated that

> The company can claim as its exclusive property inventions or improvements of any kind made by the employees. Special evidence that the

employee has made the invention or the improvement in the course of his professional obligations or while using the means and experience of the company is not required.

The problem with such agreements from the employer's point of view (and an important difference from the United States) was that the courts not infrequently sided with employees, on the ground that patent agreements of this type were so broad as to violate the principle of "common decency," established in article 138 of the Civil Code.[48]

The second and more effective limitation of the employee's rights was constructed on the basis of the Patent Code's clauses pertaining to illegal acquisition of the invention. Inventing in employment typically involved collaboration among different employees or use of the company's equipment, in-house knowledge, or experience. This fact gave rise to the concept of the "establishment invention" (*Etablissementserfindung*), known after World War I as "company invention" (*Betriebserfindung*). The precise meaning and scope of the company invention were never clear. Liberals such as Gareis defined it narrowly, as an invention that resulted from so many different contributions by different collaborators and prior experiences that it was no longer possible to identify the individual who had made the decisive conceptualization.[49] Scholars in the industrial camp defined the company invention broadly, applying it to most inventions that resulted from the organization's division of labor and employer inputs. In either case, however, the company was said to be the original inventor and therefore the person entitled to file. If an employee filed instead, the employer could block granting of the patent or get the employee's patent voided in civil court, on the grounds that the invention had originated in collaborative work and was based on company know-how and materials. The invention was said to be original company property, which the employee had acquired illegally. By making the company the original inventor, the concept of the company invention eliminated the employee's exceptional, creative input and therefore his capacity as inventor altogether.[50]

The company invention was a direct consequence of the first-to-file system applied to inventing in employment. It became the employers' most effective weapon in their arsenal of legal tricks to acquire employees' inventions without financial compensation and moral recognition of authorship. The company invention proved more reliable than patent or employment contracts, although much depended on its precise legal definition. Industrialists therefore always strove to define the company invention as broadly as possible. The most aggressive employers defined essentially all inventions that way, while others extended it merely to include all ordinary professional achievements. Conversely, employees always attempted to get rid of the concept altogether, demanding fundamental reform and adoption of a Patent Code based on the first-to-invent system. Failing this, employed inventors constantly struggled to keep the definition of company invention as narrow as possible.

In time, the conflict gave rise to the emergence of informal conceptual refinements. On the eve of World War I, a tentative distinction had evolved between "company inventions," "service inventions," and "free inventions." Free inventions were mostly uncontroversial. Typically, they were defined as inventions unrelated to the employer's business and therefore the unrestricted property of the employee. Service inventions (*Diensterfindungen* or *dienstliche Einzelerfindungen*) originated with one or more individual employees, but the ownership rights went to the employer, in exchange for which the employee(s) received name recognition and perhaps a financial reward as well. Essentially, service inventions were all those employee inventions that were neither free nor company inventions, on whose definition they therefore depended. Company inventions originated in the collaborative research and development effort organized by the company. As collective inventions, they belonged to the employer outright and did not entitle the employee(s) to name recognition or financial compensation. As noted above, however, the exact definition of the company invention remained a matter of endless dispute, both before and after World War I.[51] Because the definitional question remained unsettled, the conceptual distinction among free inventions, service inventions, and company inventions also remained fuzzy and did little to mitigate the anti-employee thrust of the German patent system between 1877 and 1936. As a consequence, the company invention emerged as the single most contentious issue in management-labor disputes about inventing and patent rights in Germany—until the Nazi regime eliminated it with the 1936 Patent Code reforms.

Like high fees and compulsory working and licenses, the first-to-file system had wider ramifications. It became the source of bitter conflict between management and industrial scientists in German industry and developed into a political controversy that lasted more than half a century. Its thrust was to disenfranchise the professional employee, to reduce incentive, to undermine the sense of responsibility, and to weaken creative energy. Whether, and if so, to what extent this may have affected Germany's technological culture is explored more fully below. With regard to political significance, however, it is obvious that the first-to-file principle dovetailed neatly with the Patent Code's other abbreviations of the individual's rights.

Without changing any of the aspects discussed here, the 1877 patent statute was revised in 1891. Henceforth it was referred to as the Code of 1891. Fundamental reform of inventor rights had to wait until 1936. For almost six decades, therefore, Germany's Patent Code was a Janus-faced institution, in which sociopolitical reaction and economic modernity constituted two sides of the same coin.[52] In this respect it was very much like other institutions of Imperial Germany, such as its social insurance system, feudalized industrial elite, party system, and in general, the congruity of incongruities that has long fascinated students of German affairs as a source of political dynamism.[53]

Technological Culture and the Inventor's Reward

How did the anti-inventor thrust of Germany's Patent Code prior to 1936 affect the nation's prospects for technological progress? One hypothetical answer is that it retarded inventive energies and therefore harmed innovation. That conclusion played a major part in the inventor measures adopted by the Nazi regime and retained by the Federal Republic. Though plausible, the proposition is impossible to prove. Technological progress is multifaceted and depends on many different factors, some of which may cancel out one another. Any reduction in the inventive creativity of science-based industry caused by the Patent Code, for instance, may have been offset by improvements the statute made in the conditions for industry's commercial exploitation of new ideas, or by encouraging the pursuit of systematic, collaborative research. The law also had little relevance for important categories of inventors such as university-based scientists and researchers in institutions like the Physikalisch-Technische Reichsanstalt and the new Kaiser-Wilhelm Gesellschaft institutes, who constituted a crucial source of inventive talent in nineteenth- and twentieth-century Germany.[54] Nor do patent statistics—whatever their worth as a measure of technological progress—support a view of flagging inventiveness, showing instead a pattern of explosive growth from 1877 to 1930, when patent filings reached a peak that has not been equaled since.[55] Despite such difficulties and the appearance of unbroken inventive dynamism, it is worthwhile to inquire further into the problem of the 1877/91 Patent Code's impact on Germany's technological culture.

A first step is to consider more closely the concept of technological culture itself. Technological culture may be defined as the configuration of attitudes, practices, institutions, technical artifacts, and systems that set apart one society's or one period's approach to technology from that of another.[56] The leading cause of variation in this regard is the fact that technology is "socially constructed." Technology is shaped by the different natural environments and historical contexts of different societies. Because those conditions change over time, technological culture is not a static concept but is itself subject to change. As a consequence, one can delineate periods or epochs in the development of a society's technological culture.

The concept of technological culture is not merely descriptive of a society's various technological peculiarities but also serves analytical purposes. It seeks to explain the "why" of those peculiarities in terms of the reciprocal influences and the functional relationships between technology and other societal factors. Perhaps the most striking feature of Germany's technological culture in comparative historical perspective is its precocious and passionate professionalism and high degree of formalism. Since the first half of the nineteenth century, German engineers have shown a disproportionate interest in theory and abstraction, disdain for empiricism, and a hunger for credentialing. Another distinctive quality is the caution and conservatism that, relative to the United States, Germany has exhibited in its adoption of new technologies during the past two hundred years.[57] Both characteristics can be linked to Germany's bureaucratic tradition and to the early emphasis on school-based professionalization and specialized scientific training of the majority of its technologists. All of this generated a certain degree of intellectual compartmentalization and inventive conservatism. Organizational and social constraints, in other words, tended to produce inventors firmly anchored in one single "technological frame" (specialty) and therefore likely to channel their creativity into creating improvements or variants of established technologies.[58]

The pattern of inventing that strengthens and develops existing technologies has been described as "conservative inventing." It is associated with the work of industrial scientists in corporate research laboratories and the economic interests of the companies invested in those technologies. The counterpart to conservative inventing is "radical inventing," or "system-originating" advances that potentially threaten existing technologies. Radical inventions have been disproportionately associated with independent inventors. The great independents were not inventors narrowly defined or specialists, "but more like visionaries, individuals such as Henry Ford, Walt Disney, Thomas Edison, Werner Siemens, Rudolf Diesel, Carl Bosch, Emil Rathenau," and, one might add, entrepreneurs such as Steven Jobs or Bill Gates, pioneering whole new industries and technological systems and ways of life.[59] Conceptually, this view of the independent inventors is closely related to Joseph Schumpeter's concept of "creative destruction" and Max Weber's interpretation of charismatic religious leaders, innovators such as Jesus or Buddha, whose power is so great that they successfully challenge everyday reality and inspire new patterns of social organization. Sociologically, the inventive creativity and innovative power of independents can be linked to their being more marginal, less professionalized, and less committed to the traditions of a discipline than industrial scientists. Historically, independents have played a major role in the technological culture of the United States, whose most notable trait is probably the abandon with which it has been willing to innovate and discard older technological systems in favor of newer ones.[60]

The last quarter of the nineteenth century and the first two decades of the twentieth century were a golden era for the independents in the United States. After World War I, salaried professionals and specialized researchers began to move center stage, owing to the continuing growth of large corporations and aided by patent practices inimical to the independent inventor. The corporations bureaucratized inventing in the industrial research laboratory and created the modern system of scientific research and development in their quest for stable and controllable technological systems. The decline of independent inventors meant that radical inventions became less common, though they did not disappear. Technological progress became characterized by the predominance of conservative inventions, rationalization, consolidation, and the systematic elimination of problems that emerge as a consequence of advances in other parts of the system. Eventually, a world emerged that was dominated by huge technological systems possessing a great deal of autonomy and "momentum."

Momentum may be defined as the tendency of a technological system to continue to grow or persist far beyond the point at which the conditions that gave rise to it continue to operate. Technological systems acquire momentum because of the visions, the material interests, the discourse, the know-how—entire ways of life—of the people and institutions involved in them.[61] Momentum also is a leading cause of distrust and critique of large technological systems. In part, this is because momentum tends to suppress new and competing technology along with the advocates of change—a tendency that, in turn, may result in the loss of the system's own technological dynamism and in technological freezing, turning large systems with momentum into "dinosaurs."[62] Another source of apprehension is that adherence to the instrumental rationality of a large technological system easily becomes irrational or destructive from a human perspective, just as modern bureaucracy, in Max Weber's view, is both humankind's greatest organizational achievement and the attainment of complete dehumanization, the "iron cage of future serfdom."[63]

In the 1970s and 1980s, several highly publicized technological disasters and a steep decline in American competitiveness in industries such as automobiles, consumer electronics, and machine tools brought the problems of large technological systems to the attention of a wider public. Long before then, however, scholars had already begun to reexamine the technological culture created by the momentum of large technological systems. The questioning included a critical reexamination of the reigning ideology of inventing, which held that the large organization's scientific research laboratory and systematic R&D were vastly superior to the amateurish and unsystematic fumbling of the independent. This ideology counted many eloquent adherents, among them John Kenneth Galbraith, who in 1952 had written about a "benign Providence ..., [which] has made the modern industry of a few large firms an almost perfect instrument for inducing technical change ..." Galbraith continued:

> There is no more pleasant fiction than that technical change is the product of the matchless ingenuity of the small man forced by competition to employ his wits to better his neighbor. Unhappily, it is a fiction. Technological development has long since become the preserve of the scientist and the engineer. Most of the cheap and simple inventions have, to put it bluntly, been made.[64]

In a study first published in the late 1950s and reissued in an updated edition in the late 1960s, John Jewkes, David Sawers, and Richard Stillerman confronted the view that "originality can be organized." They investigated the origins and development of a large number of recent inventions and innovations, showing that independent inventors, individual genius, the "uncommitted mind," and the quality of being an outsider remain among the most important sources of invention. They also concluded that large, corporate research organizations, rather than being agents of change, frequently become centers of resistance to change. Captive research within the large company is less creative than that by independents, universities, and government research institutes. Within the organization it is usually the quasi-autonomous individual who is most creative.[65]

To counter stagnation and "offset the dangers of rigidity," Jewkes and his co-authors suggested the possibility of introducing measures to help independent inventors, such as changes in the patent system, public rewards, or tax credits.[66] The recommendation to assist independents clearly was a step in the right direction. But what about employed inventors? The greatest problem of American technological culture from the late 1960s to the 1980s probably was not a lack of original ideas and inventions, nor an inability to develop and introduce new products, but failure to generate the steady stream of improvements that sustain technological systems, conquer foreign markets, and keep products competitive in the long run. Why did this happen, and why do large technological systems and corporations in general have a tendency to ossify? There is no doubt that the answer has much to do with the nature of bureaucracy and the vested interests of all those who develop a stake in the technological status quo. But it seems quite plausible that any such inherent tendency toward stagnation is powerfully reinforced by corporate patent practices that deprive industrial scientists of the incentive to bring forth their best efforts and other labor policies that disregard the psychology of human motivation.[67] Reversing those policies by introducing award schemes for salaried inventors does not necessarily break momentum. But it might well reinvigorate conservative inventing and therefore resuscitate otherwise moribund technological systems.[68]

The plausibility of this contention is supported by the experience with employee-inventor schemes of nations such as Japan and Germany. These countries have managed to best the United States, not in the newest and most glamorous technologies, nor in inventive genius, but more typically in those industries—many of them older—whose success depends on the gradual perfection and ongoing accretion of many small changes, i.e., on the vitality of conservative inventing in the large corporation. It was largely this consideration that in the 1970s and 1980s prompted the reform initiatives by American politicians and others to remedy the incentive-killing effect of the employer's interposition between private inventor and public inducement.

Corporate management has framed its opposition to such efforts in terms of general philosophical considerations about restrictions on free enterprise, but it has also marshaled specific counterarguments about the nature of modern inventing and inventing in employment. A comparative study of employees' inventions groups those arguments into five different categories. The first centers on the idea that statutory awards for salaried inventors are unnecessary. The market is said to be a fair mechanism of determining the rewards for work performed. According to inventor and employer Jack Rabinow, speaking in the 1960s, "In this society, anyone who wants to change employers can, as I left the Government ... I think inventors should get all they can in a competitive society, such as ours. And if all they can get is a good salary, then that is all they deserve."[69] It is also said that inventors are a special breed of people who invent anyway, so that "there is no point in stimulating them to do what they already do"; and that "so few really important creative inventions are made that there is very little point in establishing the machinery of a statutory scheme to encourage them."[70]

A second argument maintains that rewarding salaried inventors is impracticable. "It is difficult, if not impossible in these days of teamwork and group invention to isolate the person who made the invention from those around him who assisted in it."[71] Such a thesis, which German employers also used frequently in the first part of this century, still has considerable scholarly support in the United States. In a 1985 study of the development of American radio one can read that invention is a "process ... essentially social and cooperative[,] ... with considerable duration in time, one to which many individuals contribute in a substantial way, and in which the conception of the thing invented or discovered changes." The "eureka moment," in this view, is largely an arbitrary decision and a romantic fiction resting on the "bias built into our patent laws."[72]

A third argument is that rewarding salaried inventors actually harms invention, because fear of losing credit discourages employees from sharing their ideas with colleagues, encourages them to consider only their own narrow self-interest rather than the well-being of the company, and disrupts teamwork.[73] The two final arguments rest on the notion that special awards are unfair to both employees and employers. Awards are said to be unfair to employees because they single out inventors when other employees in the organization are just as important to successful innovation or efficiency but get no special rewards. Awards are unfair to employers because they should be entitled to the employees' labor as a matter of course. Instead they will be flooded by worthless ideas and harassed by frustrated employees demanding the marketing of their inventions.[74]

The Inventor Debate in Germany

With a different chronology, different intensity, and different outcome, these issues were also debated in Germany. Conservative inventing by salaried employees and an emphasis on the importance of design as opposed to efficiency in production came to define Germany's technological culture somewhat earlier than in the United States. In the decade before World War I, the large corporations in electrical and mechanical engineering and chemistry already accounted for a great deal of German innovation, much of which consisted of customized systems and installations. The reasons centered on the synergies of the 1877 Patent Code, Germany's bureaucratic tradition, its highly developed and formalized system of technical education, relatively small markets, and the emergence of organized capitalism in the late nineteenth century.[75] Combined with the intense expectations of professional status and privilege among German engineers, those factors made the rights of salaried inventors and high patent fees into a political issue as early as the first decade of the twentieth century.[76]

In 1905 unionized engineers, progressive lawyers, and reformist politicians launched a campaign to revamp the Patent Code. The reformers, who attracted considerable public support, demanded that the first-to-file system be scrapped in favor of the Anglo-American first-to-invent system. They insisted that employers be prohibited from acquiring the patent rights of employees' inventions, except in exchange for steep royalty payments. They also pressed for abolition of the progressive patent fee schedule in favor of a low, one-time payment such as existed in the United States. The science-based industries put up a forceful defense of the status quo but were unable to hold the line. By 1913, the Imperial government had drafted a reform bill that steered a middle course between the two sides. The outbreak of World War I prevented passage of the bill.[77]

The details of the inventor conflict and struggle for patent reform in the decade before World War I have been described elsewhere and are not repeated here.[78] It is important to emphasize, however, that before 1914 the advocates of reform wrote and spoke exclusively in terms of social policy, fairness, and equity for inventors, and not yet with reference to problems such as technological

momentum or dynamism. The severe economic problems and manifestations of technological momentum during the 1920s brought a
new dimension to the struggle. The conflict expanded into a larger
problem, compounding the original social issue with fears about
technological stagnation. Critics of the status quo now argued that
the Patent Code and corporate inventor practices were not merely
unfair, but also harmed the vitality of Germany's technological culture and imperiled its destiny as a nation.[79]

The arguments of German industrialists opposing change were
variations of the positions one encounters among American management in the 1970s, 1980s, and earlier. There was, however, one
crucial difference. In most other countries the question of the initial
ownership of inventions was never contested. Typically, American or
British industrialists "swiftly conceded that, by virtue of the mere fact
of making an invention, the employee was entitled to be recognized
as the inventor, notwithstanding any part played by the employer in
financing or assisting in the making of that invention." Employers
merely contested the employee's right to compensation for the use of
his invention.[80] That was not true in pre-Hitler Germany, where
employers went one step further and relied on the company invention to also withhold inventor status from employees. This was the
case both before and after World War I. Consider, for instance, how
Carl Duisberg, the head of Bayer and the intellectual father of IG Farben, described inventing in the synthetic dye industry in 1909:

> A given scientific theory is simply put to the test, either at the instigation
> of the laboratory's supervisor or at the initiative of the respective labo-
> ratory chemist. The theory tells us that the product must possess dye-
> ing properties, but that matters less than finding out whether the new
> dye can do something new ... The chemist therefore simply sends every
> new product he has synthesized to the dye shop and awaits the verdict
> of the dyeing supervisor ... Not a trace of inventive genius: the inventor
> has done nothing more than routinely follow a path prescribed by the
> factory's method.[81]

Some scholars believe that Duisberg's observation accurately
reflected the realities of industrial exploitation of Azo-dyes chemistry before World War I, and that inventing in this field had indeed
changed to simple, industrial routine.[82] It is true that the underlying
"coupling process" had been known since the 1880s and the role of
individual inventive creativity reduced with respect to the basic
chemical principles. But it is also true that ongoing profitability in
the commercial production of Azo-dyes called for continual inven-

tiveness and rationalization in the manufacturing process.[83] More importantly, Duisberg made his remarks in the midst of the prewar struggle over inventor rights as a polemic against the inventor lobby. His words did not reflect management's private thinking or inventor policies, which held that routine inventing—even in dye making—was the exception.[84]

Nor did the alleged absence of inventiveness prevent the Bayer executive from making a complete about-face when it came to defending the grounds on which such "non-inventions" should be patented. In 1889 Duisberg had succeeded in preserving the patentability of "routine inventions" by convincing the *Reichsgericht* to widen the criterion of patentability in chemical invention from novelty of process to encompass novelty of usage as well. Subsequently this became known as the doctrine of the "new technical effect." Inventiveness and patentability, that is, were said to reside in having produced new and unexpected technical properties, a concept that figured prominently in the 1891 revision of the Patent Code and made possible continued patenting of inventions whose underlying process was no longer judged to be original.[85] In other words, the experts agreed that mere knowledge of the general paradigm did not eliminate inventiveness in producing specific results. Of course, this smuggled the human factor right back in.

In 1921, Ludwig Fischer, director of the patent division of the Siemens concern, published *Company Inventions*, a pamphlet that took the employers' position to its logical conclusion. Fischer defined inventing as "subjugating the forces of nature to a purpose," which he then compared to traveling an unfamiliar road strewn with "obstacles." The most important means to overcome those obstacles, or the actual process of inventing, according to Fischer, were the kind of scientific-technical training, experience, systematic activities, and rational planning that are within reach of "professional technicians with wholly average talents" and accounted for "ninety-nine per cent of the countless inventions made in the past twenty or thirty years."[86] The thing that

> fertilizes inventing more than anything else, however, and that makes it into rational activity, is the teamwork of many, who complement and stimulate each other; the sensitive use of the experience of others; time and freedom for specialization in one area, … ; and finally and most of all: possession of the necessary tools to do the work.[87]

Few if any of those conditions were within reach of the independent inventor, according to Fischer. It was the company that "create[d] the most important conditions for this rationalization" of inventing. It was the company that identified the thing to be invented; the company that provided the employee with the necessary tools, the laboratory, the time and money for experiments, the library, the division of labor, the experience, the state-of-the-art knowledge, the patent searches, and the overall climate of intellectual stimulation. The employee, therefore, was "exposed to a thousand influences, which are of the greatest significance for the gradual, step-by-step elimination of the obstacles," and which collectively may be designated as the "company spirit." According to Fischer, it was the "company spirit" that was the real inventor, the agency that "steers the activity of the employee in certain, rational channels and eliminates the obstacles, so that inventions must come about." Conversely, an "employee influenced by the company spirit who would never be able to find anything that might be patented as an invention ... would be a bungler."[88] Fischer concluded that the "employee invention as a rule is a company invention. The achievement of a company inventor as a rule does not exceed the measure of what can be expected from a competent full-time professional."[89]

Ironically, the individualistic dimension of inventing was assailed not merely from the right by big business, but also from the left by the working class, which for very different reasons articulated the same collectivist perspective. In 1923 the *Metallarbeiter-Zeitung* attacked the position that inventors ought to receive special compensation for their creativity. It did so by quoting the inventor and physicist Ernst Abbe, who had once turned down a financial interest in an invention he had made for the Zeiss firm on the grounds that accepting it would be an injustice. "'It is true that the invention originates with me,'" Abbe is quoted, "'but its reduction to practice and its development is the work of my numerous collaborators. I would not have been capable of marketing the invention profitably by myself. I owe this success to the collaboration of the firm's employees.'" "So it is with every collective endeavor," commented the metal workers' publication approvingly. "All those who are involved in it have a claim to it, and there is no possible way of determining how much the individual is entitled to ..."[90]

Similarly, in an article entitled "Invention or Development," the Social Democratic paper *Vorwärts* in 1921 argued that "in the currently prevailing conception of history the significance of the individual human being for historical development is vastly exaggerated. In the bourgeois conception of history, able field marshals figure as the motor of world history." It was the same with bourgeois interpretations of technological development, which looked only to the great inventors. This perspective was wrong because it ignored the more fundamental, economic forces that drove historical development. "Progress," concluded the paper, "does not depend on the genius and creative powers of individual human beings, but on the industrious collaboration, the enthusiasm of countless people."[91]

The convergence of the interpretations of inventing by big business and labor is a prime example of the threat to bourgeois culture and identity that middle-class professionals such as engineers and inventors detected in the Weimar Republic. It helps explain why, in defense of their world, such people could embrace extreme ideas and then the politics of extremism. This phenomenon is demonstrated with great clarity in the response of industrial scientists to the collectivist view of inventing, especially its employer version. Outraged by Fischer's orderly inventing machine, professional employees denounced *Company Inventions* as the work of a hired hand and attacked its author in other personal ways. They also countered with references to Goethe and Max Eyth, the well-known nineteenth-century engineer-poet, portraying the inventor as the genius whose flashes of insight propel human history, as the archetypal Promethean, made in the "image of the Creator, a being in which God has placed a spark of His own creative power."[92]

Many other Germans agreed that technological originality was not simply a matter of corporate organization. For instance, Carl Bosch, IG-Farben's chief executive and winner of the 1930 Nobel prize for inventing high-pressure catalytic hydrogenation, commented in 1921 on his engineering achievement by comparing it to the creativity of the artist. Neither the artist nor the engineer, Bosch maintained, "is in the final analysis master of his thoughts and ideas." It was "false to assume that everything has been calculated, everything figured out." The crucial idea comes to the inventor "at the right moment, just as it does for the artist in his creative urge."[93] Adolf Hitler, too, made his contribution to the debate. In *Mein Kampf* Hitler wrote that "the inventions we see around us are all the result of the creative forces and abilities of the individual person." It was "not the collectivity that invents and not the majority that organizes or thinks, but always, in all circumstances, only the individual human being, the person." As though this were not clear enough, Hitler added that "the most valuable part of the invention ... is first of all the inventor as a person."[94]

To be sure, Hitler's central concerns were not lower patent fees or royalties for salaried inventors. But this does not mean Hitler thought only of race and war. He also thought of himself as a revolutionary, a fighter for social mobility and equal opportunity within the racial community, and the historical agent who would restore charisma and individual dynamism to their rightful place in a nation he maintained was collapsing under the weight of Jewish conspiracy, socialist collectivism, and bureaucratic capitalism. Hitler was fascinated by technology and infatuated with the idea of the modern "Aryan" hero, the individual inventor as the source of vitality and dynamism in industrial society.[95] Given the general climate of despair, the manifestations of technological momentum and stagnation, and the particular frustrations generated by the Patent Code in Weimar, such views established an elective affinity with right-wing intellectuals and with many engineers and inventors.

Those affinities have been described as "reactionary modernism," a misguided blend of pro-technology and anti-Enlightenment values that was popular in the Weimar Republic and reappeared in subsequent Nazi policies. [96] The hyperbolic language used by engineers and inventors to describe their work does indicate that they thought in romantic categories, that they transferred the intellectual categories of Idealism and Romanticism to the world of twentieth-century business and technology. At the same time, their rhetoric about the spark of genius contained more than a grain of truth. More importantly, however, their language should be understood in context: it was above all a reaction against the systematic denigration of intellectual creativity by German employers using the modern but perverse concept of the company invention. "Reactionary modernism," therefore, should perhaps not be limited to the attitudes of discontented engineers and right-wing intellectuals after World War I. The term may be applied with equal if not greater justification to institutions such as the 1877/91 Patent Code, which the Weimar Republic inherited from Werner Siemens and the many other bourgeois authoritarians who helped make Imperial Germany.

To describe the company invention and the German Patent Code as examples of reactionary modernism is not to argue that employers' insistence on the reality of collective inventing in the corporate setting was a power grab only. Both sides of the debate were partially correct. The reality of modern inventing comprises both the organizational dimension and the individual one. Inventing in the industrial research laboratory is a combination of "individual act" and "collective process."[97] Moreover, the collective or teamwork dimensions today are "gaining ever greater significance in industrial research and development. Growing technical complexity as well as the general transition from 'Little Science' to Big Science' increasingly demand knowledge that integrates various specialties and is interdisciplinary." More than half of all German inventions in 1990 were so-called team inventions (two or more co-inventors), and in the chemical industry this figure climbs to about 80 percent.[98]

Teamwork, however, does not mean the absence of individual intellectual creativity or the disappearance of the independent inventor. Rather, research teams and the independent operate simultaneously and along a spectrum, working on different problems in different fields, in pulses of radical breakthrough and conservative consolidation of different technologies. Manfred von Ardenne, the scientific entrepreneur and enormously fertile independent who in 1937 at age thirty invented an electron-scanning microscope, refers to this fact in his autobiography. Reflecting on the story of his own creative spark and feverish capture of the essential conceptualization that led to the new microscope, Ardenne writes that "even under today's circumstances, in the pioneer stage of a young science the decisive impulse can come from a single researcher." According to Ardenne, who opted for life in the Soviet Union and the German Democratic Republic after World War II, this is true even though the independent stands on the shoulders of his predecessors. In the later stages, however, "especially when important 'developments' take place, it is rarely the case anymore that the decisive impulses stem from a sole researcher by himself. Usually, those impulses are vectors of components in which several partners, for example a collective, participate."[99]

In 1934, the chemist H. G. Grimm, director of IG Farben's Ammonia Laboratory in Oppau, made a similar observation. In a lecture on inventive creativity to the German Research Institute for Aeronautics that year, Grimm described inventing in large chemical research laboratories. Chemical "inventor factories," he pointed out, were typically organized in different research teams, some of which were as small as one or two members. The latter were often "specialists—for instance, mineralogists, bacteriologists, etc.—who serve the entire lab, or also solo inventors, 'one-horse carriages,' who have difficulty fitting in with team work, but who are frequently particularly valuable because of their original ideas." To get the large research laboratories to do the things for which they were designed, which was to "create new products and processes, i.e., economically usable inventions," argued Grimm, two conditions had to be met. First, there had to be a "good atmosphere inside the laboratory, and second, its internal organization had to function with so little friction that the individual inventor hardly notices it." Creating the requisite atmosphere consisted in establishing the "optimal mental and material conditions of work for the individual inventor, whose importance for the entire operation may be compared to that of the soldier at the front." The inventor must get the feeling that, "although he must fit into the larger whole and his research mission has been set by the needs of the factory, he can be free in his creativity and, within the given framework, his creative talent can roam freely and without hindrance." Citing Carl Bosch, Grimm concluded with the reminder "that for the technological inventor too, completely analogous to the artist, everything depends on the creative idea and the right mood. In neither case can one summon these by command or produce them artificially. All one can do is to establish the appropriate conditions for their appearance."[100]

Spoken in 1934, the above remarks might be interpreted as an attempt to cater to the pro-inventor sensibilities of the new regime. There may be some of that, but the chemical industry had a history of treading more gingerly than electrical or mechanical engineering when it came to industrial relations with inventors. As early as 1920, chemical employers and employees worked out a formula to differentiate among the various types of inventions and reward a majority of them with royalties and author recognition. Still, this was the exception, and after 1924 even the chemical industry sought to reduce costs and rationalize the innovation process by rolling back the financial gains industrial scientists had made earlier. Meanwhile, efforts to reform the Patent Code and pass labor legislation with compulsory rewards for employee inventors failed. On the whole, therefore, the inventor problem continued to defy solution in the Weimar Republic and contributed its small share to the growing desire for a different political system.

In the Third Reich, the debate about inventing continued, though the balance of power soon tipped toward the pro-inventor party. In the 1936 Patent Code, the Nazi government eliminated the company invention, introduced a version of the first-to-invent principle, and upgraded the inventor's position in a variety of other ways. Of course, the changes were couched in the typical Nazi rhetoric about subordinating the interests of the individual to the community. In fact, however, the new law was the precondition for what one chemist after 1945 called the inventors' "hour of liberation."[101] It accorded them the symbolic recognition they had sought for decades. It also made conceptual changes that prepared the ground for statutory compensation later. Adoption of the inventor's right principle and elimination of the company invention meant that employers could no longer deprive employees of authorship and block rewards by arguing that the company was the real inventor. With adoption of the 1936 Patent Code, Germany had reached the point where, in most other countries, the debate about employee inventions usually started.[102]

After 1936 the regime groped its way toward statutory compensation as part of a comprehensive system of inventor assistance, which was finally implemented in 1942 and 1943 and further limited management's autonomy. During the interval, the inventor conflict continued in different form. It became a paper war between the German Labor Front and Nazi party on the one hand and big business on the other over the details of the statutory compensation that was now inevitable. It was in the context of this bureaucratic conflict that the regime's inventor-policy experts, abandoning their sometimes naïve faith in the value of amateur and little inventors, developed the highly articulated and nuanced system for employee compensation that became the basis for the 1957 Act on Employee Inventions.

Historical Consequences

How did the politics of inventing relate to the evolution of Germany's technological culture? There is only sporadic evidence that problems of technological momentum and stagnation were the consequence of employers' inventor policies during the Weimar Republic. To the extent that the Patent Code had played a role in predisposing Germany's technological culture toward conservative inventing, it is plausible that the employers' hard line toward inventors made the problem worse. But it is impossible to say by how much. There is no doubt, however, that problems associated with momentum, bigness, rationalization, the treatment of inventors and engineers, and reluctance to embrace new technologies such as the automobile, telephone, and radio on a mass basis, created a climate in which the politics of inventing helped tear the fabric of Weimar society. The conflict over inventor rights and patent reform contributed to the polarization and political stalemate that led to Hitler.[103] Weimar was the crucible in which Nazi inventor policy was forged. Inventor assistance, statutory compensation, patent-fee reductions, and a host of other changes emphasizing the genius and creative energy of the individual inventor were a backlash against the inventive conservatism, corporate intransigence, and economic stagnation of the 1920s.

In the Third Reich, inventor policy became embedded in the comprehensive racism that was the regime's dominant organizing principle. Inventor assistance and financial rewards did not extend to
individuals categorized as *Ostarbeiter*, Poles, Jews, and "Gypsies."
The sizeable contingent of Jewish patent attorneys, consultants, and
participants in inventor discussions during Weimar disappears from
the sources virtually overnight after 1933. We do not have to ask
what happened to them. Nazi inventor policy was therefore an integral part of the Third Reich's descent into barbarism.[104] With regard
to the community of racial "citizens," however, the inventor measures represented a measure of social emancipation and progress,
which in turn appears to have had a positive effect on Germany's
technological culture.[105] This is true even though the motives of
reformers often were rooted in an exaggerated technological romanticism, and their expectations of a second wave of radical inventions
by independents remained mostly unfulfilled. But as a dialectical
counterpoint to the reactionary modernism of corporate inventor
policies in Weimar, the Nazi measures produced a synthesis that
energized conservative inventing among the pioneers of science-
based industry, both during the Third Reich and after. They created
incentives that helped prevent ossification in those industries—
machinery, electrical engineering, automobiles, pharmaceuticals,
and chemical engineering—that were no longer new but that
remained at the forefront of technological progress in their fields
and, after 1945, were at the center of West Germany's economic
success as an exporter.[106]

Notes

1. United States Constitution, Art. I, sec. 8, cl. 8; quoted in Bruce W. Bugbee, *Genesis of American Patent and Copyright Law* (Washington, D.C., 1967), 1.
2. Quoted in Noble, *America by Design*, 84.
3. Felix Wankel, "Patentamt, Erfinder, und Volkswirtschaft," in *Hundert Jahre Patentamt*, ed. Deutsches Patentamt (Munich, 1977), 324–32. On patent philosophy,
   see Fritz Machlup, *An Economic Review of the Patent System*, Study No. 15 of the
   Subcommittee on Patents, Trademarks, and Copyrights of the Committee on the
   Judiciary, U.S. Senate, 85th Congress, 2nd session (Washington, D.C., 1958);
   Fritz Machlup and Edith T. Penrose, "The Patent Controversy in the Nineteenth
   Century," *JEH* 10 (1950): 1–29. On economic aspects of patents: Jacob

Schmookler, *Invention and Economic Growth* (Cambridge, Mass., 1966); Klaus Grefermann, "Patentwesen und technischer Fortschritt," in *Hundert Jahre Patentamt*, 37–64.

4. The German Patent Code is governed by the principle of the "inventor's right," introduced in 1936. The "inventor's right" is a mixture of the first-to-file and first-to-invent principles. Under the inventor's right, "the inventor has the right to the patent." If more than one inventor has made the same invention, the patent is awarded to the one who files a patent application first. For reasons of practicality, the Patent Office assumes that the first filer is the inventor. If this is not so, the actual inventor can complain and get the patent reassigned (Paragraphs 6 and 7 of the Bundespatentgesetz).

5. The French patent statute dates from 1791, but the first-to-file principle was introduced only in 1978. On the history of the British patent system, see Christine McLeod, *Inventing the Industrial Revolution: The English Patent System, 1660–1800* (Cambridge, 1988). The new British Patents Act dates from 1977. The first United States patent law dates from 1790; it came of age with the reforms of 1836 and establishment of a regular patent office. Revisions in 1870 and 1952 did not affect the first-to-invent principle; Bugbee, *Genesis*, 149–58.

6. Friedrich-Karl Beier, "Wettbewerbsfreiheit und Patentschutz: Zur geschichtlichen Entwicklung des deutschen Patentrechts," *GRUR* 80 (Mar. 1978): 126.

7. Cf. the preamble of the French Patent Law of 1791: "Not to regard an industrial discovery as the property of its creator would be to attack the essence of the rights of man"; quoted in Ludwig Fischer, *Werner Siemens und der Schutz der Erfindungen* (Berlin, 1922), 7; Beier, "Wettbewerbsfreiheit," 126.

8. Noble, *America by Design*, 84–109. An excellent survey of this development: Catherine L. Fisk, "Removing the 'Fuel of Interest' from the 'Fire of Genius': Law and the Employee-Inventor, 1830–1930," *The University of Chicago Law Review* 65, no. 4 (fall 1998): 1127–98.

9. *American Law Reports, Annotated (ALR)* 153 (1944): 984.

10. Gerald P. Parsons, "U.S. Lags in Patent Law Reform," *IEEE Spectrum* 15 (Mar. 1978): 60–64.

11. Fisk, "Removing the 'Fuel'"; *ALR* 153, 995; Noble, *America by Design*, 90, 97–101; Jasper Silva Costa, *The Law of Inventing in Employment* (New York, 1953); Fredrik Neumeyer, *The Employed Inventor in the United States: R&D Policies, Law, and Practice* (Cambridge, Mass., 1971; Neal Orkin, "The Legal Rights of the Employed Inventor: New Approaches to Old Problems," *JPOS* 56, nos. 10 (Oct. 1974): 648–62, and 11 (Nov. 1974): 719–45; J. Rodman Steele, Jr., *Is This My Reward? An Employee's Struggle for Fairness in the Corporate Exploitation of His Inventions* (West Palm Beach, 1968), 7.

12. For university patent policies see Neumeyer, *Employed Inventor*, 425–96; state patent acts: Gerald S. Geren, "New Legislation Affecting Employee Patent Rights," *Research & Development*, Jan. 1984: 33.

13. Jeremy Phillips, ed., *Employees' Inventions: A Comparative Study* (Sunderland, 1981); Fredrik Neumeyer, *The Law of Employed Inventors in Europe*, Study No. 30 of the Subcommittee on Patents, Trademarks, and Copyrights of the Committee on the Judiciary, 87th Congr., 2nd sess. (Washington, D.C., 1963); Parsons, "U.S. Lags in Patent Law Reform."

14. Orkin, "Legal Rights"; Neal Orkin and Mathias Strohfeldt, "Arbn Erf G — the Answer or the Anathema?" *Managing Intellectual Property* 14 (Oct. 1992): 28–32;

Kenneth R. Allen, "Invention Pacts: Between the Lines," *IEEE Spectrum* 15 (Mar. 1978): 54–59; Parsons, "U.S. Lags."

15. Noble, *America by Design*, passim, but esp. 157–256; Layton, *Revolt of the Engineers*.

16. The IEEE, for instance, in 1979 withdrew support for a bill on employee inventions; Gerald P. Parsons to Robert W. Bradford, 4 Sep. 1979, SPEEA, File on Patent Agreement Negotiations. On American engineers: Peter Meiksins, "Professionalism and Conflict: The Case of the American Association of Engineers," *JSH* 19 (spring 1986): 403–21; Layton, *Revolt of the Engineers*; Robert Zussman, *Mechanics of the Middle Class: Work and Politics Among American Engineers* (Berkeley, 1985); Richard E. Walton, *The Impact of the Professional Engineering Union* (Cambridge, Mass., 1961); Robert Perrucci and Joel E. Gerstl, *Profession without Community: Engineers in American Society* (New York, 1969); Terry S. Reynolds, ed., *The Engineer in America: A Historical Anthology from* Technology and Culture (Chicago, 1991).

17. Neal Orkin, "Innovation; Motivation; and Orkinomics,"*Patent World* 1 (May 1987): 34; Steele, *Is This My Reward?* 109–13.

18. E.g., a 1922 request by industrial scientists at Hoechst to purchase stock options was denied; correspondence Vereinigung angestellter Chemiker und Ingenieure der Farbwerke vorm. Meister Lucius & Brüning, e.V., Höchst a/M with Hoechst management, Oct. 12 and 17, 1922, Hoechst, 12/35/4.

19. Noble, *America by Design*, chs. 6 and 7; U.S. figures: Orkin, "Legal Rights," 740; German figures: Klaus Prahl, *Patentschutz und Wettbewerb* (Göttingen, 1969), cited in Heggen, *Erfindungsschutz*, 124–25; Wolfgang Belz, *Die Arbeitnehmererfindung im Wandel der patentrechtlichen Auffassungen* (Munich, 1958), 4; Orkin and Strohfeldt, "Arbn Erf G," 28; Grefermann, "Patentwesen," 41–43; Erich Staudt et al., *Der Arbeitnehmererfinder im betrieblichen Innovationsprozess* (Bochum, 1990), Study No. 78, 4–5, which mentions a figure of below 20 percent for the late 1980s.

20. Leslie E. Simon, *German Research in World War II: An Analysis of the Conduct of Research* (New York, 1947), 43–44.

21. Walton Hamilton and Irene Till, quoted in John Jewkes, David Sawers, and Richard Stillerman, *The Sources of Invention*, 2d ed. (New York, 1969), 36, 34–39; Noble, *America by Design*, ch. 6.

22. Norbert Wiener, *Invention: The Care and Feeding of Ideas*, intr. Steve Joshua Heims (Cambridge, Mass., 1993), chs. 7–8, esp. 81–83, 89–91, 96, 145.

23. Cf. Noble, *America by Design*, 118–21.

24. In-house experience is defined conservatively, as the activities of the division or unit where the employee works. This is to prevent large corporations from claiming everything under the sun.

25. Eduard Reimer, Hans Schade, and Helmut Schippel, *Das Recht der Arbeitnehmererfindung: Kommentar zu dem Gesetz über Arbeitnehmererfindungen vom 25. Juli 1957 und deren Vergütungsrichtlinien*, 5th ed. (Berlin, 1975), 40–49; Matthias Ruete, "The German Employee-Invention Law: An Outline," in *Employees' Inventions*, 180–212; Robert Plaisant, "Employees' Inventions in Comparative Law," *IPQ* 5 (Jan. 1960): 31–55, esp. 37–44.

26. Economic value of an invention is usually determined according to the so-called license analogy: how much it would cost to obtain a license had the invention come from outside. The employee's share in the invention is a function of three factors: degree of initiative in formulating the *problem*, level of achievement in

finding the *solution*, and creative responsibilities and *position* in the company. The more articulation of the problem falls within the scope of professional responsibility, the smaller the factor. Similarly, the more conceptualization of the solution is a function of standard professional skills or company state of the art, the smaller the factor. Finally, the higher the rank and the greater the employee's responsibilities, the smaller the factor. The sum of the three factors represents the employee's share, which can range from zero to one hundred percent. The amount of financial compensation is the product of the employee's share and the invention's economic value, and ranges in practice from 2 to 7 percent of the profit from the invention. Reimer, Schade and Schippel, *Recht der Arbeitnehmererfindung*, 57–79; Orkin and Strohfeldt, "Arbn Erf G," 30. On early history of the formula, see ch. 12.

27. Orkin and Strohfeldt, "Arbn Erf G"; Orkin, "Innovation," 33; Grefermann, "Patentwesen," 41–42, 49, 63.

28. For an excellent discussion of this period with reference to the chemical industry, see Ernst Homburg, "The Emergence of Research Laboratories in the Dyestuffs Industry 1870–1900," *British Journal for the History of Science* 25 (1992): 91–111.

29. Hans Rosenberg, *Große Depression und Bismarckzeit: Wirtschaftsablauf, Gesellschaft und Politik in Mitteleuropa* (Berlin, 1967); Hans-Ulrich Wehler, *The German Empire 1871–1918*, trans. Kim Traynor (Leamington Spa/Dover, 1985); Ivo N. Lambi, *Free Trade and Protection in Germany, 1868–1879* (Wiesbaden, 1963); Mary Fulbrook, *A Concise History of Germany* (Cambridge, 1990).

30. Heggen, *Erfindungsschutz*, 19–68; Beier, "Wettbewerbsfreiheit," 126–29; Fischer, *Werner Siemens und der Schutz*, 9–19.

31. Two leading figures in the anti-patent movement were the economist John Prince-Smith and Rudolf Delbrück of the Prussian Ministry of Trade; Heggen, *Erfindungsschutz*, 69–85; Beier, "Wettbewerbsfreiheit," 129–30; Fischer, *Werner Siemens und der Schutz*, passim; Eric Schiff, *Industrialization Without National Patents: The Netherlands, 1869–1912; Switzerland, 1850–1907* (Princeton, 1971), 3–15.

32. On Werner Siemens: Wilfried Feldenkirchen, *Werner von Siemens: Inventor and International Entrepreneur* (Columbus, 1994); on the Siemens company: idem, *Siemens 1918–1945* (Columbus, 1999). On the ideal type of the inventor-entrepreneur: Wilhelm Treue, "Ingenieur und Erfinder: Zwei sozial- und technikgeschichtliche Probleme," *Vierteljahrsschrift für Sozial- und Wirtschaftsgeschichte* 54, no. 4 (Dec. 1967): 456–76; Homburg, "Emergence," 95–7.

33. Heggen, *Erfindungsschutz*, passim; Fischer, *Werner Siemens und der Schutz*, passim.

34. Cf. Noble, *America by Design*, chs. 1–3. On Siemens's goals: Alan Beyerchen, "On the Stimulation of Excellence in Wilhelmian Science," in *Another Germany: A Reconsideration of the Imperial Era*, ed. Jack R. Dukes and Joachim Remak (Boulder, 1988), 139–68; for similar, top-down motives in the organization of invention in the chemical industry after 1877: Homburg, "Emergence."

35. Speech of 10 Jun. 1883, quoted in Karl Hauser, "Das deutsche Sonderrecht für Erfinder in privaten und öffentlichen Diensten," *Die Betriebsverfassung* 5 (Sep. 1958): 168–75, 169.

36. Quoted in Hauser, "Das deutsche Sonderrecht," 169.

37. Initially, the only other early adherents of the first-to-file system were Belgium, Luxemburg, Lithuania, Greece, and Japan: Schiff, *Industrialization without National*

Patents, 3–4; Machlup, *Economic Review*, 19–22; 76–77; Fischer, *Werner Siemens und der Schutz*, 5; Ernst Heymann, "Der Erfinder im neuen deutschen Patentrecht," in *Das Recht des schöpferischen Menschen: Festschrift der Akademie für deutsches Recht anlässlich des Kongresses der Internationalen Vereinigung für gewerblichen Rechtsschutz in Berlin vom 1. bis 6. Juni 1936*, ed. Akademie für Deutsches Recht (Berlin, 1936), 104; Lothar Beckmann, *Erfinderbeteiligung: Versuch einer Systematik der Methoden der Erfinderbezahlung unter besonderer Berücksichtigung der chemischen Industrie* (Berlin, 1927), 26.

38.   Beyerchen, "Stimulation"; Heggen, *Erfindungsschutz*, 136–42; DVSGE, ed., *Vorläufige Entwürfe eines Patentgesetzes, eines Gebrauchsmustergesetzes und eines Warenzeichengesetzes*, spec. ed. of *GRUR* (Berlin, 1913), 7; Grefermann, "Patentwesen," 37–64; Schmookler, *Invention and Economic Growth*; Machlup, *Economic Review*; Homburg, "Emergence"; Henk van den Belt and Arie Rip, "The Nelson-Winter-Dos Model and Synthetic Dye Chemistry," in *The Social Construction of Technological Systems: New Directions in the Sociology and History of Technology*, ed. Wiebe E. Bijker, Thomas P. Hughes, and Trevor J. Pinch (Cambridge, Mass., 1989), 149–55.

39.   Heggen, *Erfindungsschutz*, 137–42. For illiberalism, see Fritz Stern, *The Failure of Illiberalism* (Chicago, 1975); Konrad H. Jarausch, "Illiberalism and Beyond: German History in Search of a Paradigm," *JMH* 55, 2 (Jun. 1983): 268–84. For the U.S. case: Noble, *America by Design*, chs. 6–7.

40.   A forced license could be obtained only if the invention was of compelling interest to society; it came into play after a patent had been in existence for at least three years; see *Patentgesetz vom 25. Mai 1877*, par. 11, reprinted in Heggen, *Erfindungsschutz*, 147–49. Compulsory working of patents (Article 11) was dropped in 1911. The change benefited big business far more than independents and small operators; cf. Robert Koch, "Forderungen an ein nationalsozialistisches Patentrecht," unpub. memorandum, 12 Oct. 1934, BAP 2343, Bl. 20, 2, 11–12.

41.   Gispen, *New Profession*, 223–336; Konrad H. Jarausch, *The Unfree Professions: German Lawyers, Teachers, and Engineers, 1900–1950* (New York, 1990), 27–114; inhospitality to independents: Wolfgang König and Wolfhard Weber, *Netzwerke Stahl und Strom 1840 bis 1914, Propyläen Technikgeschichte*, vol. 4 (Berlin, 1990), 271; Bernhard Volmer, "Entwurf eines Gutachtens über die Möglichkeiten einer wirksamen Förderung der Erfinder im Bundesgebiet," 31 Dec. 1951, pp. 10–12, in BA-Zwi, B141/2761; patent-fee issue: Richard Linde, Herbert d'Oleire, Felix Kaiser, and Joseph Loewe in *TgR* 4–6 (Oct.–Dec. 1928): 42–54.

42.   *Patentgesetz vom. 25. Mai 1877*, pars. 3 and 10.

43.   Konrad Engländer, *Die Angestelltenerfindung nach geltendem Recht: Vortrag vom 24. Februar 1925* (Leipzig, 1925), passim.

44.   B. Tolksdorf, "Das Recht der Angestellten an ihren Erfindungen," *ZfI* 3 (1908): 193–200; Max Weber, "Bureaucracy," in *From Max Weber: Essays in Sociology*, trans. and ed. H. H. Gerth and C. Wright Mills (New York, 1946), 196–244; Otto Hintze, "Der Beamtenstand," in *Soziologie und Geschichte: Gesammelte Abhandlungen zur Soziologie, Politik und Theorie der Geschichte*, ed. Gerhard Oestreich (Göttingen, 1964), 66–125; Jürgen Kocka, "Angestellter," in *Geschichtliche Grundbegriffe: Historisches Lexikon zur politisch-sozialen Sprache in Deutschland*, ed. Otto Brunner, Werner Conze, Reinhart Koselleck (Stuttgart, 1972), vol. 1, 110–128; idem, *White Collar Workers in America 1890–1940: A Social-Political History in International Perspective* (London, 1980).

45. Karl Gareis, *Über das Erfinderrecht von Beamten, Angestellten und Arbeitern: Eine patentrechtliche Abhandlung* (Berlin, 1879), 6.

46. Oscar Schanze, "Der rechtliche Schutz der Erfinderehre," *GRUR* 7 (Feb. 1902): 65ff.; *Denkschrift zum Erfinderschutz* (= *Sozialpolitische Schriften des Bundes angestellter Chemiker und Ingenieure, e.V.* First series, No. 6. Berlin; 1922), 6; correspondence between MAN patent experts Martin Offenbacher and Emil Guggenheimer (22 and 25 Jan., 2, 6, and 26 Feb. 1924) concerning *Reichsgericht* ruling of 1 Nov. 1922, siding with weapons designer Georg Luger against Deutsche Waffen- und Munitionsfabriken A.G., NGug; Ludwig Fischer, *Patentamt und Reichsgericht* (Berlin, 1934).

47. Paul Wiegand, "Die Erfindung von Gefolgsmännern unter besonderer Berücksichtigung ihrer wirtschaftlichen Auswirkungen auf Unternehmer und Gefolgsmänner" (*Dr.-Ing.* diss., Technische Hochschule Hannover, 1941), 2–9; Heinz Potthoff et al., eds., *Rechtsprechung des Arbeitsrechtes 1914–1927: 9000 Entscheidungen in 5000 Nummern in einem Band systematisch geordet*, ed. H. Potthoff et al. (Stuttgart, 1927), 185–86 (nos. 951–57).

48. Gareis, *Über das Erfinderrecht*, passim; Waldemar Meissner, "Unser Patentrevers," unpublished ms., 27 June 1899, in SAA 4/Lk78, 3–8; article 1 of 1889 Siemens patent agreement quoted in Engländer, *Angestelltenerfindung*, 46; different copies of Siemens patent agreements in SAA 4/Lk78. For the contemporary legal situation in the United States: Fisk, "Removing the 'Fuel.'"

49. Gareis, *Über das Erfinderrecht*, 7, 34; Wiegand, "Erfindung von Gefolgsmännern," 10; Reimer, Schade, and Schippel, *Recht der Arbeitnehmererfindung*, 80.

50. Gareis, *Ueber das Erfinderrecht*, passim; Gispen, *New Profession*, 264–87; Ludwig Fischer, *Betriebserfindungen* (Berlin, 1921), 3–32, 43–58; Wiegand, "Erfindung von Gefolgsmännern," 9–10, 14, 21–27.

51. Engländer, *Angestelltenerfindung*, 24–27, also 10–15; Müller-Pohle, *Erfindungen von Gefolgschaftsmitgliedern* (Berlin, 1943), 13–16; Belz, "Arbeitnehmererfindung," 18–23; background: DVSGE (Augsburger Kongress, 24–29 May 1914), *Vorschläge zu der Reform des Patentrechts: Denkschrift der Patentkommission und der Warenzeichenkommission* (Berlin, 1914), 1–18; idem, *Beschlüsse des Augsburger Kongress* (Berlin, 1914), 1–6; Bernhard Volmer and Dieter Gaul, *Arbeitnehmererfindungsgestz: Kommentar*, 2d ed. (Munich, 1983), 29–35; DVSGE, *Vorläufige Entwürfe eines Patentgesetzes* (1913), 10–17; Gispen, *New Profession*, 285–87; Wiegand, "Erfindung von Gefolgsmännern," 10–29.

52. Cf. Heggen, *Erfindungsschutz*, 121–24, 136–42; for a good survey: Rudolf Nirk, "100 Jahre Patentschutz in Deutschland," in *Hundert Jahre Patentamt*, 345–402.

53. The literature on this subject is huge. Some key texts: Wehler, *The German Empire 1871–1918*; Thorstein Veblen, *Imperial Germany and the Industrial Revolution* (Ann Arbor, 1968); Dahrendorf, *Society and Democracy*; Wolfgang Sauer, "Das Problem des deutschen Nationalstaates," in *Moderne deutsche Sozialgeschichte*, ed. Hans-Ulrich Wehler (Cologne, 1968), 407–36; Wolfgang J. Mommsen, *Der autoritäre Nationalstaat: Verfassung, Gesellschaft und Kultur des deutschen Kaiserreiches* (Frankfurt a. M., 1990); idem, ed., *The Emergence of the Welfare State in Britain and Germany, 1850–1950* (London, 1981); Gerhard A. Ritter, *Die deutschen Parteien 1830–1914* (Göttingen, 1985); Robert G. Moeller, "The Kaisserreich Recast? Continuity & Change in Modern German Historiography," *JSH* 17, no. 4 (1983): 655–83; Volker R. Berghahn, *Imperial Germany 1871–1914: Economy, Society, Culture and Politics* (Providence, 1996). Opposing interpretation: Blackbourn and Eley, *Peculiarities of German History*; Retallack, *Germany in the Age*.

54. On the significance of this sector and causes of its dynamism before World War I: Beyerchen, "Stimulation." Also: David Cahan, *An Institute For An Empire: The Physikalisch-Technische Reichsanstalt, 1871–1918* (Cambridge, 1989); Jeffrey Johnson, *The Kaiser's Chemists: Science and Modernization in Imperial Germany* (Chapel Hill, 1990).

55. For patent statistics, see ch. 10.

56. Cf. Thomas P. Hughes, "The Evolution of Large Technological Systems," in *The Social Construction of Technological Systems: New Directions in the Sociology and History of Technology*, ed. Wiebe E. Bijker, Thomas P. Hughes, and Trevor J. Pinch (Cambridge, Mass., 1989), 51–83; Radkau, *Technik*, 29–46, 171–76; Wolfgang König, *Ingenieurausbildung, Ingenieurberuf und Konstruktionstechnik in Großbrittanien, den USA, Frankreich und Deutschland seit der Industrialisierung: Ein vergleichender Essay* (Berlin, 1990); idem, *Künstler*; Stokes, "Wirtschaftswunder," 1–22, esp. 3.

57. Radkau, *Technik*, passim, esp. 323–26 ; Stokes, "Wirtschaftswunder," esp. 15–22.

58. Gispen, *New Profession*; Wiebe E. Bijker, "The Social Construction of Bakelite: Toward a Theory of Invention," in *The Social Construction of Technological Systems*, 159–90.

59. Hughes, "Evolution," 57–66; idem, *American Genesis*, 53–95, 180–83. The technological and cultural significance of independent inventors is systematically ignored in David Noble's *America by Design*, even as its author attacks corporate capitalism for having created a "world in which everything changes, yet nothing moves" (xvii). For an excellent discussion from an engineer's perspective of the relationships between invention, design, patenting, development, and innovation, see Henry Petroski, *Invention by Design: How Engineers Get from Thought to Thing* (Cambridge, Mass., 1996).

60. Joseph A. Schumpeter, *Capitalism, Socialism and Democracy*, 3d ed. (New York, 1975), 81–86; idem, *The Theory of Economic Development: An Inquiry into Profits, Capital, Credit, Interest, and the Business Cycle*, trans. Devers Opie (Cambridge, Mass., 1936), 57–94; Wolfgang J. Mommsen, *The Political and Social Theory of Max Weber: Collected Essays* (Chicago, 1989), 151–55.

61. Hughes, "Evolution," 71–80.

62. Hughes, *American Genesis*, 1–12, 462, 443–72.

63. For an imaginative application of some of the above (and other, related) concepts to the historiography of the Holocaust: Beyerchen, "Rational Means," 386–402.

64. John Kenneth Galbraith, *American Capitalism: The Concept of Countervailing Power* (Boston, 1952), 92, quoted in Jewkes, Sawers, and Stillerman, *Sources of Invention*, 35–36.

65. Ibid., 182. Wiener, *Invention*, passim.

66. Jewkes, Sawers, and Stillerman, *Sources of Invention*, 186–93.

67. Phillips, *Employees' Inventions*; Allen, "Invention Pacts," 54–59; Parsons, "U.S. Lags in Patent Law Reform"; Staudt, *Arbeitnehmererfinder*, 12–37; Grefermann, "Patentwesen," 41–43; Orkin and Strohfeldt, "Arbn Erf G," 28–32; Steele, *Is This My Reward?* 89–126.

68. Hughes, *American Genesis*, 459–72.

69. Quoted in Phillips, *Employees' Inventions*, 29–30.

70. Ibid., 31.

71. Ibid., 32.

72. Hugh G. J. Aitken, *The Continuous Wave: Technology and American Radio, 1900–1932* (Princeton, 1985), 548–49. I am grateful to Dr. Dill Hunley of NASA for bringing Aitken's position to my attention.

73. Phillips, *Employees' Inventions*, 32–34; Noble, *America by Design*, 100–101.
74. Phillips, *Employees' Inventions*, 35–36; Noble, *America by Design*, 100–101, 119–21.
75. On this configuration: Beyerchen, "Stimulation"; Homburg, "Emergence," which largely supersedes Georg Meyer-Thurow, "The Industrialization of Invention: A Case Study from the German Chemical Industry," *Isis* 73, 256 (Sep. 1982): 363–81; Jürgen Kocka, "The Rise of Modern Industrial Enterprise in Germany," in *Managerial Hierarchies: Comparative Perspectives on the Rise of the Modern Industrial Enterprise*, ed. Alfred D. Chandler and Herman Daems (Cambridge, Mass., 1980), 77–116; Gispen, *New Profession*, 113–59; and literature cited in note 56.
76. Gispen, *New Profession*, 264–87.
77. Ibid., 241–42, 264–87.
78. Ibid., 264–87.
79. See part II of this study.
80. Phillips, *Employees' Inventions*, 25.
81. Carl Duisberg at the Stettin Kongess für gewerblichen Rechtsschutz, 1909, quoted in *AC* 22, (Nov. 1909): 1667.
82. van den Belt and Rip, "Nelson-Winter-Dos Model."
83. Johann Walter, *Erfahrungen eines Betriebsleiters (Aus der Praxis der Anilinfarbenfabrikation)* (Leipzig, 1925), passim.
84. Cf. minutes of IG Farben patent commission meeting, 8 Mar. 1922, Hoechst 12/255/1.
85. Ibid. On the context of Duisberg's 1909 remarks see Gispen, *New Profession*, 230, 264–87.
86. Fischer, *Betriebserfindungen*, 4.
87. Ibid., 10.
88. Ibid., 11–13.
89. Ibid., 15; Beckmann, *Erfinderbeteiligung*, 14–35.
90. SAA, 11/Lf364.
91. Two copies in files of Siemens-Schuckert chief Carl Köttgen, SAA, 11/Lf352, and 11/Lf36. In 1929, the trade unions' principal expert on labor efficiency, Richard Woldt, maintained that "the human factor of production is no longer decisive; material forces are"; quoted in Nolan, *Visions of Modernity*, 102.
92. Engineer Herman Schmelzer, quoting Max Eyth, in "Erfinder oder Naturkraftbinder?" in *Der leitende Angestellte* 3, no. 23 (Dec. 1921): 178; BUDACI, ed., *Denkschrift zum Erfinderschutz, Sozialpolitische Schriften des Bundes Angestellter Chemiker u. Ingenieure e.V.*, 1st series, no. 6 (Sep., 1922), 45–55.
93. Quoted in Karl Holdermann, *Im Banne der Chemie: Carl Bosch Leben und Werk* (Düsseldorf, 1953), 95.
94. Adolf Hitler, *Mein Kampf* (Munich, 1933), 496–98, quoted in Robert Kahlert, *Erfindertaschenbuch* (Berlin, 1939), 1–3, trans. by the author.
95. Zitelmann, *Hitler*, passim, esp. 49, 83–84, 92–125, 259, 264, 321–25, 347, 401.
96. Jeffrey Herf, *Reactionary Modernism: Technology, Culture and Politics in Weimar and the Third Reich* (Cambridge, 1984), passim.
97. David A. Hounshell, "Invention in the Industrial Research Laboratory: Individual Act or Collective Process," in *Inventive Minds: Creativity in Technology*, ed. Robert J. Weber and David N. Perkins (New York, 1992), 285.
98. Staudt et al., "Arbeitnehmererfinder," 38–39.
99. Manfred von Ardenne, *Ein glückliches Leben für Technik und Forschung: Autobiographie* (Zurich, 1972), 139–40. Similar views: Walter Bruch (inventor of the PAL color-television system), "Ein Erfinder über das Erfinden," in *Hundert Jahre*

Patentamt, 317–24, 319; Petroski, *Invention by Design*, passim, esp. 8–42, 66–88, 120–40.

100. Professor Dr. H. G. Grimm to Director Dr. Brendel, 25 Oct. 1934, BASF B4/1978.
101. Dr. Stephan Deichsel to Bundestag deputy F. W. Wagner, 7 Jan. 1953, BA-Zwi, B141/2811, Bl. 9a–l; see ch. 13, below.
102. Phillips, *Employees' Inventions*, 25.
103. On stalemate specifically: Martin Broszat, *Hitler and the Collapse of Weimar Germany*, trans. and foreword by V. R. Berghahn (Hamburg, 1987); Ian Kershaw, ed., *Why Did German Democracy Fail?* (New York, 1990). The literature on the intertwining of Weimar disintegration and the rise of National Socialism is huge. Some selected titles: Thomas Childers, *The Nazi Voter: The Social Foundations of Fascism in Germany, 1919–1933* (Chapel Hill, 1983); idem, "The Social Language of Politics in Germany: The Sociology of Political Discourse in the Weimar Republic," *AHR* 95, no. 2 (Apr. 1990): 331–58; idem, "The Middle Classes and National Socialism," in *The German Bourgeoisie*, ed. David Blackbourn and Richard J. Evans (London, 1991), 318–37; Geoff Eley, "What Produces Fascism," 53–82; idem, "Conservatives and Radical Nationalists in Germany: The Production of Fascist Potentials, 1912–28," in *Fascists and Conservatives: The Radical Right and the Establishment in Twentieth-Century Europe*, ed. Martin Blinkhorn (London, 1990), 50–70; Juergen Baron von Kruedener, ed., *Economic Crisis and Political Collapse: The Weimar Republic 1924–1933* (New York, 1990); Charles S. Maier et al., eds., *The Rise of the Nazi Regime: Historical Assessments* (Boulder, 1986); Jane Caplan, "The Rise of National Socialism," in *Modern Germany Reconsidered 1870–1945*, ed. Gordon Martel (London, 1992), 117–39; Dick Geary, "The Industrial Bourgeoisie and Labor Relations in Germany 1871–1933," in *German Bourgeoisie*, 140–61; Bernd Weisbrod, "The Crisis of Bourgeois Society in Interwar Germany," in *Fascist Italy and Nazi Germany: Comparisons and Contrasts*, ed. Richard Bessel (Cambridge, 1996), 23–39; Peter Stachura, *The Nazi Machtergreifung* (London, 1983); Karl Dietrich Bracher, *The German Dictatorship* (New York, 1972); David Abraham, *The Collapse of the Weimar Republic: Political Economy and Crisis*, 2d ed. (New York, 1986); Henry A. Turner, Jr., *German Big Business and the Rise of Hitler* (New York, 1985); Harold James, *The German Slump: Politics and Economics 1924–1936* (Oxford, 1986).
104. One of the first priorities of the German Labor Front's inventor activists after 1933 was elimination of Jews from all patent- and invention-related occupations; cf. Erich Priemer, "Begrüßungsansprache und Rückblick auf 10 Jahre Erfinderbetreuung durch die NSDAP," in NSDAP, AftW, ed., *Tagung der Reichsarbeitsgemeinschaft Erfindungswesen im Hauptamt für Technik der Reichsleitung der NSDAP. Amt für technische Wissenschaften am 11. Januar 1944* (Munich, 1944), 3–5, BAK R131/22.
105. On "racial community": Burleigh and Wippermann, *Racial State*, 1–4, 8–22, 304–7; Ulrich Herbert, "Labor as Spoils of Conquest, 1933–1945, in *Nazism and German Society 1933–1945*, ed. David Crew (London, 1994), 219–74; Thomas Childers and Jane Caplan, "Introduction," in *Reevaluating the Third Reich*, ed. Thomas Childers and Jane Caplan (New York, 1993), 1–19.
106. Cf. Stokes, "Wirtschaftswunder," 2–4; Fischer, "Role of Science," 102–3; Christopher S. Allen, "Germany: Competing Communitarianisms," in *Ideology and National Competitiveness: An Analysis of Nine Countries*, ed. George C. Lodge and Ezra F. Vogel (Boston, 1987), 79–102, esp. 93–94.

Part II

THE WEIMAR REPUBLIC

Chapter 2

CHARTING SURVIVAL

The Chemists' Contract of 1920

The outbreak of World War I interrupted the drift toward a new Patent Code that had characterized the years 1913–14. A dearth of sources during the war makes it difficult to say what exactly killed patent reform. But the sudden disappearance of the documentary trail in September 1914 suggests that the government's 1913 bill fell victim to changed priorities. Concern with more urgent matters undoubtedly made inventor rights a problem that could wait for the return of normal times.

In the event, Patent Code reform did not reappear on the political agenda until 1928, when the Great Coalition government headed by Social Democrat Hermann Müller picked up where the Imperial government had left off in 1914. The long interval did not mean, however, that the social conflict at the heart of the patent question remained dormant as well. As early as December 1918 the politics of inventing resurfaced—in the new guise of an issue for collective bargaining between management and labor. Collective bargaining, regardless of its success or failure, became the Weimar Republic's signal method of dealing with industrial relations. Instead of continuing their prewar struggle over inventor rights in the arena of patent law, the contestants after 1918 redefined it as a question of labor relations, to be settled by collective bargaining, social legislation, labor law, or some combination of all three. The sources do not mention why this happened, but there is little doubt it had to do with the political gains and increased militancy of labor resulting

Notes for this section begin on page 75.

from World War I, discrediting of the authoritarian state, inflationary context, and, in general, the Revolution of 1918/19.[1]

The revolutionary government, to facilitate direct negotiations between the social partners, passed a special collective-bargaining ordinance as early as 23 December 1918. Following on the heels of the Stinnes-Legien Agreement of 15 November, which first established the collective bargaining principle, the December decree endowed private understandings between recognized organizations of management and labor with the status of law. This made such accords binding across entire sectors of industry.

With the chemical industry leading the way, management in that branch quickly moved out in front. In late 1918, chemical executives established their own employer association, the Arbeitgeberverband der chemischen Industrie Deutschlands (ACID), which took an immediate interest in employees' patent rights. The chemical industry's early focus on the inventor issue was not surprising. As the premier high-technology industry of its time, it depended on the morale of a relatively small cadre of chemists to a greater extent than mechanical and electrical engineering did on their larger and more variegated ranks of engineers. Finding a quick solution to the inventor problem was therefore deemed essential. Inevitably, the engineering industries also became involved in the issue, but they moved more hesitantly and were less inclined to compromise. Some intramanagement conflict over inventor tactics and strategy in Weimar was therefore inevitable.

In spite of their disagreements, the industrialists would prove better at maintaining cohesion than their "social partners," the salaried engineers, scientists, and chemists who hoped to strengthen their rights as inventors. The employees were fragmented from the outset, starting with divisiveness and internal conflict when forming the associations required for collective bargaining in the first place.[2] The ordinance of December 1918 prohibited organizations from concluding collective bargaining agreements if their membership included both employees and employers. Such mixed associations, however, were quite common in professional and white-collar circles. Especially among technically trained professionals, mixed professional organizations had remained popular until the end of the war. But the postwar climate made their survival more difficult and accelerated the trend toward unionization that had begun in the decade before 1914.[3]

The first trade union of technical workers and engineers had come into being in 1904. The Bund der technisch-industriellen

Beamten (Union of Technical-Industrial Salaried Employees, or BTIB) had initially sought to organize all technical employees in private industry above the rank of foreman. Unique for ignoring the educational divide between technical workers with and without higher education, the BTIB included many engineers and chemists with academic certification in addition to a majority of technicians with different and lower types of training. Led by a core group of academically certified engineers, the BTIB had been the driving force behind the movement to expand inventors' rights and improve other conditions of employment before World War I.

Since its organizational concept restricted potential membership to relatively small numbers, the BTIB even before World War I had experienced pressures to include foremen and merge with the Deutscher Techniker-Verband (Federation of German Technicians, or DTV), another association of lower-level technicians. Nothing came of those ideas before 1914, but the war radicalized the BTIB and changed its internal balance of power. Having lowered the barrier against foremen earlier, it merged in June of 1919 with the DTV and changed its name to Bund der technischen Angestellten und Beamten (Union of Technical Employees and Civil Servants, or BUTAB). As a consequence of the merger and the widespread enthusiasm for collective action after the war, the BUTAB's membership surged from approximately 25,000 for the old BTIB and DTV each in 1914 to more than 120,000 at the beginning of 1920.[4]

Closer in outlook and composition to Social Democracy and the blue-collar working class than its predecessors, the BUTAB nevertheless tried to hold on to its professional engineers and chemists and actively recruit new members in these circles. At the same time, it moved aggressively to demand concessions and higher wages from the employers. The two-pronged strategy did not prevent academically certified chemists and engineers from leaving in droves. The BUTAB's overtures to the left, its reaching out to the lower ranks of technical workers and its talk about the "unbridgeable difference between capital and labor" triggered a backlash among the higher ranks of salaried professionals. As one Ph.D. chemist put it in December of 1918, "penetration of our colleagues' ranks by the Bund endangers not only ... social peace but also threatens the social standing of salaried chemists with proletarization."[5] Such observations were not just an expression of abstract fears or ideological considerations. They also reflected anger over the BUTAB's failure to negotiate separate, higher pay scales for academically trained professionals in the labor pacts it concluded with employers.

Such tensions, in addition to mixed professional associations being unable to negotiate, prompted the formation of two new groupings of salaried professional engineers and chemists in the early days of the Weimar Republic.[6] The smaller but more successful of the new organizations was the Bund der angestellten Chemiker und Ingenieure (Union of Salaried Chemists and Engineers, or BUDACI). The BUDACI was born 10–11 May 1919 in Halle, from the merger of two short-lived predecessors that had sprung up in the preceding weeks and months to take the BTIB/BUTAB's place as a bargaining agency with the employers.[7] The BUDACI defined itself explicitly as a trade union whose sole purpose was to fight for the socioeconomic interests of its members. It did so by copying the blue-collar strategy of negotiating standardized salary rates and uniform labor contracts with employers, particularly in the chemical industry. This was a radical step for academically trained professionals, but there was little doubt among the BUDACI's ranks that it was the right decision.[8]

In the second half of 1919 the BUDACI joined the Arbeitsgemeinschaft freier Angestelltenverbände (Working Group of Free Employee Associations, or AfA). The AfA, also known as AfA-Bund, was the most left-leaning of Weimar's various umbrella groups of salaried-employee associations and was closely affiliated with the Social Democratic federation of blue-collar trade unions.[9] As early as 1920 the BUDACI switched to another new employee federation, the GEDAG (Gesamtverband Deutscher Angestellten-Gewerkschaften, Confederation of German Salaried Employee Trade Unions). To the right of center, the GEDAG was affiliated with the Christian trade unions and hostile to the Social Democratic platform.[10] Feuding between the BUDACI and the BUTAB, an AfA core group, was the main reason for the realignment.

Membership in the BUDACI was open to all academically certified salaried chemists of both sexes employed in any rank and branch of industry, the universities, commercial, drug, agricultural, or food laboratories. In addition, it welcomed physicists, academically certified engineers, architects, physiologists, pharmacologists, pharmacists, bacteriologists, and anyone else whose education could be considered the equivalent of academic certification and who was not a civil servant or a senior manager.[11] Initially, industrial chemists made up the vast majority of the BUDACI's membership, which numbered less than 2,000 at the time of its founding.[12] Though small, this figure represented as many as 50 percent of Germany's chemists in industry and was therefore remarkably inclusive in the chemical sector.[13] In fact, the broad overlap of membership in the BUDACI and the Verein deutscher Chemiker (VDC), the older professional association of chemists that included both employers and employees, resulted in an agreement in the fall of 1919 spelling out the relationship and division of labor between them.[14]

In subsequent years academically trained technicians from other disciplines and industries—especially electrical engineering—contributed disproportionately to the BUDACI's membership growth. Eventually there were about as many of them as there were chemists. Even so, the BUDACI always remained small and in 1929 had between 7,000 and 8,000 members.[15] Still, it did better against the employers than other salaried-employee organizations many times its size. The BUDACI owed its accomplishments, limited though they were, to two factors. One was a complete lack of sentimentality about the employment conditions of modern salaried professionals in industry. The other was the power of knowledge: it represented a big part of Germany's intellectual capital in the high-technology industries.

The BUDACI was especially successful in the chemical industry. Its troubles in other sectors were directly related to the existence of a competing organization. The Vereinigung der leitenden Angestellten in Handel und Industrie (Association of Managerial Employees in Commerce and Industry, or VELA) was the second new organization of salaried professionals to emerge at the end of World War I. Formed in Berlin in late 1918, also by former BTIB members, the VELA had its home base in the city's numerous mechanical and electrical engineering firms, whose professional employees joined in large numbers. But the VELA also accepted chemists and, like the BUDACI, soon expanded into a nationwide organization.

While it aimed to do for its members what the BUDACI was achieving in the chemical sector, the VELA was not nearly as effective. Unlike the BUDACI, the VELA limited its membership to the higher—i.e., middle managerial or professionally more advanced—ranks of salaried employees. Whereas the BUDACI represented all technically trained employees so long as they possessed higher education, the VELA admitted only those who had climbed beyond the bottom ranks, placing less emphasis on academic certification and occupational specialization. As a consequence, the VELA did not organize engineers alone but also included substantial numbers of nontechnical professionals such as accountants, lawyers, and economists. Theodor Heuss, for instance, the future president of the Federal Republic and a political economist by training, was a member of the VELA during Weimar.[16]

As an interest group of senior employees—men who interpreted industry-wide contracts and standardized pay scales as an affront to their professional dignity—the VELA refused to engage in collective bargaining over salaries. Its chosen mission was rather "to erect a bulwark against the leveling tendencies" unleashed by the Revolution of 1918/19. The greatest threat to the elite of the salaried employees, according to the VELA, was the "danger that they will be deprived of the position to which education and superior understanding of business affairs entitle them, and that they will gradually be reduced to the level of the proletariat." The VELA's highest priority in the summer of 1919, therefore, was to pursue the "struggle against these equalizing tendencies," which it considered "even more important than vigorous defense of its members' purely economic interests."[17]

The struggle against leveling and egalitarianism did not mean the VELA was ready to join antidemocratic umbrella groups of salaried employees such as the GEDAG, or, for that matter, any of the other salaried-employee federations. It remained unaffiliated until forced to dissolve by the Nazi government in 1934. Neither the refusal to affiliate nor the decision to exclude junior salaried professionals appears to have hurt the VELA's numerical strength. From just over 9,000 in 1919, its membership grew to 31,350 in 1931.[18] Several times larger than the BUDACI, the VELA was also much more heterogeneous. This, and the wishful notion of being above collective bargaining, constituted its greatest weaknesses. To mask its ineffectiveness, the VELA adopted the stratagem of portraying the BUDACI as excessively radical but quietly following in its footsteps to take advantage of whatever pact the latter managed to sign with industry.

Even when taken together, the BUTAB, BUDACI, and VELA, with their approximately 150,000 members, constituted only a small minority of the many other salaried-employee organizations, whose combined membership was about 1.5 million in the Weimar Republic. Salaried employees, in turn, were only a junior partner in the labor movement as a whole, which was dominated by the organizations of blue-collar workers with a total strength fluctuating between five and nine million members.[19] It is therefore not surprising that inventor rights and patent reform only rarely made the headlines and for most observers remained a marginal issue.

For those immediately affected it was, of course, a very different matter. Management's undiluted power in the matter of patents and inventions was a source of considerable frustration for all three technical employee groups. This is not to say that they judged inventor rights equally important or collaborated or put them at the top of their agendas. The BUTAB, for instance, initially treated inventor rights almost as an afterthought. This was especially true immediately after the war, when it had many other and higher priorities. In 1919 the BUTAB was at the center of white-collar strikes in Berlin, Bavaria, and Saxony to force employers to the bargaining table. It pushed for the eight-hour day, co-determination, and the socialization of industry. It lobbied for a Works Councils Act with broader powers than the government's 1919 draft recommended. And it constantly accused its rivals of cowardly and antidemocratic policies.[20] In this hectic atmosphere the BUTAB devoted relatively little attention to inventor rights. Its widely copied contract with the Bavarian metal industrialists of 27 June 1919, for instance, included only five brief paragraphs about inventions and did little to expand inventor rights.[21]

The most important inventor feature of the BUTAB's Bavarian contract was its recognition of the threefold division into company inventions, service inventions, and free inventions. In theory, this was a big change from the prewar situation and a major step forward. In practice, the contract reclaimed with one hand what it granted with the other. It had a sweeping definition of the company invention that assigned all rights to the employer. Uncompensated, service inventions also belonged to the employer, though the employee did gain the right of being mentioned as the inventor. Free inventions belonged to the employee, subject only to the stipulation that, if they fell within the scope of the company's product line, the employer had a right of first refusal.[22] This agreement gave the inventor virtually nothing, but there is no evidence the BUTAB tried to do much about it.

Like its more plebeian rival, the BUDACI also began by putting salaries first. As one of its members wrote in March of 1919, "the social question is first of all a salary question." The author went on to explain that chemists' salaries had failed to keep up with the large war profits of the chemical industry, with the disproportionate increases in blue-collar incomes, and with inflation in general. The consequence was both relative and absolute deprivation of those who deserved most of the credit for the industry's exceptional achievements.[23] There can be little doubt that many chemists and engineers shared such sentiments and harbored a similar sense of social dislocation and injustice. Since the employers at first turned a deaf ear, uncoordinated work stoppages, demonstrations, and protests by professionals in the chemical industry became increasingly common in the first half of 1919.[24]

Even during those initial postwar months, inventor rights were already an integral part of professional employees' demands in the chemical industry. In the April 1919 issue of the *Zeitschrift für angewandte Chemie*, for instance, a Dr. Stettbacher argued that employees' patent rights would have to be interpreted more liberally if the chemical industry were to succeed in the future. Stettbacher contended that if inventors and laboratory scientists were to put forth their best efforts, they would have to be freed from corporate anonymity and get the same name recognition for their intellectual products as academics, writers, and other intellectuals.[25] Other chemists wanted more than just name recognition, demanding material rewards in the form of premiums, royalties, or co-ownership in patents as well. Inventor rights were one of two principal reasons that chemists and engineers at Hoechst decided to join what employers called the "terrorism" of a salaried-employee strike in early April of 1919.[26] Similar irregularities and unrest were widespread in chemical factories and laboratories all over Germany.

The employers' initial reaction was one of shock over the outrageous demands by a group from which they recruited their own. The relatively progressive Carl Duisberg, for instance, had quickly accepted the need for salary increases among professionals. But when the Bayer chief learned that chemists and engineers at Hoechst, another large chemical firm, had struck because "they also want the right to dispose over all the inventions made during their employment," Duisberg thought "any possibility of negotiation ceases to exist." If the employees succeeded with their demand it would be better to close up shop altogether. Duisberg urged Hoechst director Adolf Haeuser not to make any patent concessions that might become a precedent for employee demands elsewhere.[27]

Duisberg's fears proved well founded. In June 1919 the chemical employers in Berlin reluctantly sat down with a joint AfA/BUDACI team to negotiate a local contract that included concessions on inventor rights. Elsewhere, the employers attempted holding out and resisted collective bargaining with academically trained employees—particularly over inventor rights. But salaried professionals everywhere were reconstituting themselves as BUDACI locals, staging wildcat strikes and other actions, until the chemical industrialists finally agreed in August 1919 to recognize the BUDACI and start negotiations for a nationwide contract.[28] Even then, senior executives at companies such as Bayer and Agfa fought rear-guard actions to block collective bargaining or keep patents and no-compete agreements off the agenda for those talks. But this, too, failed, and in a meeting in Frankfurt a. M. on 5 September 1919 the chemical industry's employer association voted by a small margin to "acknowledge the employee's legal right to special compensation for individual service inventions." The form and size of compensation would be determined on an individual basis by the employer.[29]

The employers' Frankfurt vote in favor of financial compensation for service inventions turned out to be a crucial event. On the one hand, it paved the way for successful negotiations and labor peace between the BUDACI and the chemical industry. This resulted in what was probably the longest-lived of all the collective bargaining agreements concluded after World War I. First signed in 1920, the so-called Chemists' Contract remained in effect throughout the Weimar Republic as well as the Third Reich and was still operative almost forty years later, in 1959.[30] It became a model for all subsequent reforms, including Nazi inventor policies and Germany's current inventor-rights legislation.

On the other hand, the chemical employers' September vote also set the stage for a great deal of trouble. Once the principle of financial compensation had been conceded in one industry to one group of employees, demands for similar concessions in other industries acquired new urgency and became a source of discontent that would not die down until the contenders were satisfied—which never happened during Weimar.[31] Other employers refused to follow the chemical industry's example of making timely concessions that averted worse. The mechanical and electrical equipment manufacturers—and especially the heavy iron and steel industrialists—opposed inventor compensation to the bitter end, regardless of the political costs. Not surprisingly, this had a polarizing effect and hardened attitudes on both sides of the issue. Growing rigidity, in turn, wrecked the climate in which collective bargaining could survive. This led employees to seek redress by mandatory labor legislation, which the employers on their part opted to fight tooth and nail. In this way, the inventor-rights issue contributed to the wider gridlock that destroyed the Weimar Republic.

Considering the consequences, one might ask whether the chemical employers realized what they were doing when they voted for inventor compensation in September 1919. In one sense they obviously did not, being no better at predicting the future than anyone else. But does this mean their concession was the result only of temporary disarray, weakness, or calculation to use the inflation as a way of buying social peace?[32] To some extent, perhaps, yes, but their compromise, rooted in the prewar past, also had longer-term causes. Specifically, the chemical employers' 1919 vote reflected extensive prewar discussions, which had culminated in 1914 in the so-called Bonn Resolutions of the Association of German Chemists. Adopted at the VDC's annual convention in June of 1914, the Bonn Resolutions centered on the effort to develop a distinction between company inventions and service inventions. The basic approach was to classify a new idea as a company invention when the contribution of the company was greater than that of the individual employee, while with service inventions it was the other way around.[33] The owner of the company would be entitled to all the rights in the company invention, including name recognition. Service inventions would transfer to the owner of the business with two provisos: that the employee had a right to be named as the inventor and was entitled to receive special monetary compensation.[34]

It was these still somewhat vague but already accepted principles that the chemical employers voted to adopt at their meeting in Frankfurt in September 1919. The decision helped smooth relations with the BUDACI, which during the summer and early fall of 1919 continued to press the industrialists, signing a number of regional settlements favorable to professional employees.[35] By October the employers signaled they were ready to start talks for a comprehensive, nationwide agreement. Broadened at the last minute by the inclusion of delegates from the VELA, discussions got underway toward the end of November and were successfully concluded in February 1920. The definitive contract was signed 27 April 1920. In August of the same year the government declared the agreement binding for the country's entire chemical industry, retroactive to June.[36]

A broad "framework agreement" that emphasized procedures and minimum conditions rather than actual amounts and numbers, the Chemists' Contract covered terms of hiring and firing, hours and days of work, salaries, bonuses, seniority, overtime, vacation, arbitration, no-compete agreements, and inventor rights. Overall, employers made enough concessions on those points to reassure industrial scientists of their status as middle-class professionals and achieve labor peace. Compared with most other labor pacts in Weimar, the agreement was unusually progressive and attractive for those it covered.[37] This was especially evident in the area of patent and inventor rights.

In August 1919, the BUDACI had published "Inventor Protection," a position paper that made demands far greater than those agreed to in the 1914 Bonn Resolutions.[38] The BUDACI's postwar paper started with the threshold assumption of Anglo-American patent law that all inventions originated as the intellectual property of the inventor. It therefore did not recognize the legitimacy of the familiar triad of company inventions, service inventions, and free inventions, accepting only service inventions and free inventions. In the prophetic words of one of the union's patent specialists, "a company invention is an absurdity."[39] The employee had an obligation to offer service inventions to the employer, who would be entitled to acquire them under conditions highly favorable to the inventor. The invention would become free if the employer did not decide to claim it within two weeks; the inventor was to receive name recognition; the employer would pay for reassigning the invention to the employee should the former decide not to renew the patent; the employee would be entitled to profit-sharing terms of up to 15 percent for all inventions.

The BUDACI arrived at its figure of 15 percent by way of several new concepts. The union began with a fundamental critique of the employer position. It heaped scorn on the notion that modern inventing was strictly a function of industrial organization, which could succeed without individual creativity. To do justice to the inventor in the conditions of organized inventing in the industrial research laboratory, the BUDACI divided the inventive process into three parts: definition of the problem to be solved, conceptualization of the solution, and reduction to practice. An inventor who could take credit for all three components was considered a "full inventor" and should get 15 percent of the invention's net profit. Partial inventors, the BUDACI said, were responsible for one or two components and should get 5 or 10 percent respectively. In the case of multiple inventors, the same fractions would be divided among them. In case the company made an exceptional contribution to the invention (e.g., large infusions of capital for a high-risk venture), the inventor's share would drop by a third, while conversely the inventor's share might go up by the same fraction if there was little or no company input.[40]

Dividing the invention into components of inventive responsibility was a creative and pioneering step that became the basis for future compromise. If at first this idea received little attention it was because the BUDACI's aggressive figures for compensation overshadowed everything else. The union's claim that 15 percent ought to constitute the norm of an inventor's share of profits rested on tenuous grounds. It asserted that under normal circumstances an inde-

pendent inventor was likely to share equally the profits of an invention with a financier, and that the inventor-financier team would be
able to sell the invention to a third party for royalties amounting to
30 per cent of the net profit.

The BUDACI also demanded 50 percent of the sale price for an
invention should the employer sell it to a third party; 50 percent of
any royalties derived from licensing arrangements; and a guarantee
that the employer assume liability for any third-party defaults.
Unpatentable improvements in the production process, too, would
have to be compensated, as well as inventions and improvements
the employer failed to use. Compensation was to run for the entire
duration of the inventor's employment, and, upon termination, at a
reduced rate of 75 percent for the first twenty years and 50 percent
during the next ten years.[41]

Clearly, these were extraordinary demands by any standard, and
one can hardly fault the employers for rejecting them out of hand. In
the political climate of 1919/20, however, the chemical industrialists
proved remarkably flexible. To be sure, the final agreement gave the
employees far less than they had originally demanded.[42] None of
their more extreme demands, such as elimination of the company
invention and their 15 percent solution, survived. In three crucial
respects, however, the Chemists' Contract did go beyond the Bonn
Resolutions to produce significant gains.

First, there was the BUDACI's innovative deconstruction of
employee inventing into its component parts. Dividing the invention
into three parts was the first serious attempt to assess an individual
inventor's contribution to the inventive process in industry and then
link the amount of compensation to that contribution. This
approach resulted in a more precise conceptualization of the difference between company and service inventions and mitigated the
undesirable consequences for employees of the former.

Limited to patented and patentable inventions, the contract defined
a company invention as one in which the "suggestions, experiences,
preparation, and auxiliary services by the company were such" that
the employee's inventive achievement did "not exceed the level of
routine activity (or ... of ordinary professional competence)." In
other words, the company's share in this type of invention was considered so large that the employee's inventive contribution did not
merit any special recognition or reward. The employer could therefore claim it without further ado. A service invention was an inventive idea that (1) came within the scope of the company's business,
or (2) resulted from the employee's professional obligations to

invent.[43] The crucial difference from a company invention was that the service invention depended on an employee's powers of intuition, combination, and observation to such an extent that it surpassed ordinary technical proficiency. It was an invention in which the employee's share exceeded that of the company, regardless of whether the latter contributed "suggestions, experiences, preparation," etc. This was why a service invention entitled employees to name recognition and special, "appropriate compensation."

Without further interpretation, these definitions did not eliminate all ambiguity. Much still depended on the meaning given to vague concepts such as "routine professional competence," inventiveness, or company input. For that reason, rival organizations such as the BUTAB denounced the agreement as a cowardly surrender by the BUDACI. The BUTAB insisted that a company invention be defined narrowly and unambiguously, as one whose author could no longer be identified—a situation so rare that in effect it would have eliminated the concept altogether.[44] While the BUDACI did not like the company invention any better than the BUTAB, its compromise with the chemical industrialists solved the problem in practice. By informal arrangement, both sides agreed that the criterion for separating service inventions from company inventions would be patentability. In most cases the awarding of a patent (or a ruling that the invention could be patented) was construed as evidence of exceptional job performance and signified a service invention. This was the crucial step that made compromise possible. Without surrendering the company-invention principle, the chemical employers silently accepted that patented company inventions henceforth would be the exception in their industry. As the BUDACI's 1920 commentary on the Chemists' Contract correctly put it, "the great majority of inventions by the employees under the jurisdiction of this agreement are service inventions."[45]

The second benefit of the Chemists' Contract for employees was financial. Henceforth, service inventions entitled the inventor to special compensation. Of course, the entitlement was conceded only in principle, since the employer retained the right to determine what it meant in practice. Other than mentioning it would be "appropriate," the agreement did not stipulate the amounts or forms of compensation, nor methods for calculating it. Still, the contract moved in that direction by specifying the duration, conditions for cancellation, contract breach, and various other details relevant to the size of rewards. Subject to various escape clauses, for instance, inventor compensation ran for fifteen years from the day a patent was filed, regardless of whether the inventor remained an employee or not. And while employers remained free to determine the size and method of inventor rewards, in practice they settled on a system of royalties that went as high as 5 percent of the net income produced by patented service inventions.[46]

Since employees were not entitled to examine the company's books, it was impossible for them to evaluate the accuracy or honesty of industrialists' profit-sharing formulas. There is evidence that in spite of the agreement inventors were regularly treated arbitrarily and unfairly. Individual complaints and litigation became a common occurrence, in part because the pact referred disagreements about inventions to the courts.[47] As a wider socioeconomic question, however, the issue of inventor rights subsided in the chemical industry—at least until the second half of the 1920s. Eager for peace in the laboratories, employers on the whole stayed within the bounds of the contract—and they paid to prove it. The employees, too, recognized they had gained an important measure of inventor protection—far better, at any rate, than any other organization or rival group of employees in the Weimar Republic. Despite occasional grumbling, therefore, neither side in subsequent years was willing to challenge the agreement in public, and both held it up as the type of sensible compromise the Weimar Republic needed to function.[48]

Finally, the inventor agreement gave industrial scientists in the chemical sector something on which it was impossible to put a price: formal recognition and tangible evidence of their separate social standing as professionals, above the mass of ordinary white-collar technicians represented by organizations such as the BUTAB. The Chemists' Contract in general, and its highly articulated inventor clauses in particular, lifted academically trained professionals up from BUTAB-negotiated labor contracts, which most of them considered inadequate and demeaning. The 1920 patent agreement therefore counteracted the socioeconomic leveling that professionals in Weimar so intensely feared and that would eventually drive so many of them into Hitler's arms.

Notes

1. For broader context, see Gerald D. Feldman, *The Great Disorder: Politics, Economics and Society in the German Inflation, 1914–1924* (New York, 1993), 99–130, esp. 106–8 on revolution and emergence of collective bargaining; idem, "The Social and Economic Policies of German Big Business, 1918–1929," *AHR* 75, no. 1 (1969): 47–55; idem, "German Business Between War and Revolution: The Origins of the Stinnes-Legien Agreement," in *Entstehung und Wandel der modernen Gesellschaft: Festschrift für Hans Rosenberg zum 65. Geburtstag*, ed. Gerhard A. Ritter (Berlin, 1970), 312–41; Eberhard Kolb, *The Weimar Republic* (London, 1988), 3–22, 138–47.

2. On German white-collar workers, see Fritz Croner, *Die Angestellten in der modernen Gesellschaft* (Vienna, 1954); idem: *Soziologie der Angestellten* (Cologne, 1962); Kocka, "Angestellter"; idem, *White Collar Workers in America*; idem, "White-Collar Employees and Industrial Society in Imperial Germany," in *The Social History of Politics: Critical Perspectives in West German Historical Writing since 1945*, ed. Georg Iggers (Leamington Spa, 1985), 113–36; Siegfried Kracauer, *The Salaried Masses: Duty and Distraction in Weimar Germany*, intr. Inka Mülder Bach, trans. Quintin Horare (London, 1998); Dieter Langewiesche, "Die Angestellten in industriekapitalistischen Systemen," *Geschichte und Gesellschaft* 6, no. 1 (1980): 283–96; Emil Lederer, *The New Middle Class: (Der neue Mittelstand) in Grundriss der Sozialökonomik, IX. Abteilung, I. Teil. Tübingen, 1926* (New York, 1937); Prinz, *Vom neuen Mittelstand*; Hans Speier, *German White-Collar Workers and the Rise of Hitler* (New Haven, 1986).

3. Mixed associations included the VDI, VDC, VDDI, and VDE.

4. Gispen, *New Profession*, chs. 8–12; *DTZ* 1, no. 1 (4 Jul. 1919): 1–5, and 2, no. 1 (9 Jan. 1920): 1.

5. Dr. F. Jander at a VDC meeting, 17 Dec. 1918, quoted in *AC*, pt. 2, *chemisch-wirtschaftliche Nachrichten*, 1919: 96; *Bundesblätter* 3, no. 20 (1 Oct. 1921): 210–15. Cf. Feldman, *Great Disorder*, 83–85.

6. Dr. Karl Höfchen, 12 Aug. 1919, in *AC*, pt. 2, *chemisch-wirtschaftliche Nachrichten*, 1919: 525; *Dipl.-Ing.* A. Schmidt, 11 Mar. 1919, ibid., 215; Dr. Mittelstenscheid, ibid., 285. Pay scales: *DTZ*, 1919–1920.

7. *AC*, pt. 2, *chemisch-wirtschaftliche Nachrichten* 1919: 269, 356.

8. Contrast BUDACI-VDDI: *VDDIZ* 11 (Sep. 1920): 96. On the VDDI: Gispen, *New Profession*, chs. 8, 12.

9. Voigtlander-Tetzner, *Der Mensch in der BASF*, vol. 3, 618, unpubl. ms., BASF; Gispen, *New Profession*, 263.

10. The dominant organization in GEDAG was the German National Commercial Assistants Federation (Deutschnationaler Handlungsgehilfen-Verband, DHV). Fascist tendencies of DHV: *DTZ* 2, no. 2 (23 Jan. 1920): 23.

11. *AC*, pt. 2, *chemisch-wirtschaftliche Nachrichten* 1919: 269. The BUDACI's 1920 contract, which applied equally to men and women, covered "architects, chemists, engineers, physicists and other salaried workers with completed higher technical or natural scientific education and pharmacists and natural-scientific or technical employees whose professional responsibilities and achievements contractually rated them as equals of the former." *Prokuristen* and laboratory and other supervisors were included. BUDACI, ed., *Kommentar zum Reichstarifvertrag für die akademisch gebildeten Angestellten der chemischen Industrie* (Berlin, 1920), 5.

12. Robert Berliner, "10 Jahre Verband angestellter Akademiker der chemischen Industrie e.V." *Der Leitende Angestellte* 9 (Apr. 1959): 58–60.

13. In 1919 the VDC had 5,500 members. Two-thirds were employees: *AC*, pt. 2, *chemisch-wirtschaftliche Nachrichten* 1919: 96, 185.

14. Agreement of 25 Oct. and 22 Nov. 1919, in *AC*, pt. 2, *chemisch-wirtschaftliche Nachrichten*, 1919: 792.

15. BUDACI membership: BUTAB, ed., *25 Jahre Technikergewerkschaft, 10 Jahre BUTAB: Festschrift zum 25jährigen Jubiläum des Bundes der technisch-industriellen Beamten (Butib) und zum 10-jährigen Jubiläum des Bundes der technischen Angestellten und Beamten (BUTAB) im Mai 1929*, (Berlin, 1929), 170; Berliner, "10 Jahre Verband angestellter Akademiker," mentions 6,000 to 7,000. The BUDACI in 1925 changed its name to Bund angestellter Akademiker technisch-naturwissenschaftlicher Berufe, e.V. (Union of Salaried Academicians with Technical-Natural-Scientific Occupations, registered association); the acronym BUDACI survived.

16. G. Kleine, "Zum 75. Geburtstag von Theodor Heuss," *Der leitende Angestellte, Monatsschrift der ULA* 9 (Feb. 1959): 21.

17. *AC*, pt. 2, *chemisch-wirtschaftliche Nachrichten*, 1919: 461.

18. D. Petzina, W. Abelshauser, and A. Faust, *Sozialgeschichtliches Arbeitsbuch III: Materialien zur Statistik des Deutschen Reiches 1914–1945* (Munich, 1978), 112.

19. Ibid.

20. *DTZ*, 1919–20, passim.

21. *Zwanglose Mitteilungen für die Mitglieder des Vereines deutscher Maschinenbau-Anstalten* (1 Dec. 1919): 710–11.

22. *DTZ* 1,1 (July 4, 1919): 13; VDMA, *Zwanglose Mitteilungen* (1 Dec. 1919): 710-11; correspondence between Dr. Weidlich and Dr. Frank, 4 Oct. 1920, Hoechst, 12/255/1.

23. *Dipl.-Ing.* A. Schmidt in *AC*, pt. 2, *chemisch-wirtschaftliche Nachrichten*, 1919: 215. Cf. Feldman, *Great Disorder*, 83–95.

24. Strikes and unrest: correspondence between Hoechst's Adolf Haeuser and Bayer's Carl Duisberg, 4 and 11 Apr. 1919, in AS Duisberg; Hoechst 12/255/1, 12/35/4, and 12/11/2; grievances: meeting of 4 May 1919, Hoechst, 12/35/4.

25. Dr. Alfred Stettbacher, "Über einige notwendige Freiheiten für den Aufstieg der Begabten im Industrieleben," *AC*, pt. 2, *chemisch-wirtschaftliche Nachrichten*, 1919: 217.

26. Haeuser to Duisberg, 9 Apr. 1919, AS Duisberg.

27. Duisberg to Haeuser, 11 Apr. 1919, AS Duisberg.

28. Three of seven issues in Series I of the BUDACI's *Sozialpolitische Schriften* published between 1919 and 1922 dealt with inventor rights.

29. See file, "Arbeits- und Sozialverhältnisse 1923–39. Akademiker Reichstarif Verhandlungen," Hoechst, 12/11/2.

30. Berliner, "10 Jahre Verband angestellter Akademiker," 58–60.

31. This was why the VDA in 1921–22 preferred a legislative solution to collective bargaining; ACID memo, 28 Nov. 1921, Hoechst, 12/255/1.

32. For inflation before 1923 as facilitating labor peace: Feldman, *Great Disorder*; idem, "The Historian and the German Inflation," in *Inflation through the Ages: Economic, Social Psychological and Historical Aspects*, ed. Nathan Schmukler and Edward Marcus (New York, 1983), 386–99; Charles S. Maier, *Recasting Bourgeois Europe: Stabilization in France, Germany, and Italy in the Decade After World War I* (Princeton, 1975), passim, esp. 3–15, 579–94; Carl-Ludwig Holtfrerich, *The Ger-*

man *Inflation 1914–1923* (Berlin, 1986); Detlev J. K. Peukert, *The Weimar Republic: The Crisis of Classical Modernity*, trans. Richard Deveson (New York, 1992), 109–12.

33. The Bonn Resolutions defined company inventions as "inventions that are made in industrial firms and whose realization is in essential ways conditioned by the suggestions, experiences, preparatory activities and auxiliary facilities of the firm." Service inventions were inventions that did not fit the definition of company inventions but (a) resulted from the employee's responsibility to invent or (b) could be exploited within the framework of the company's business. Minutes of ACID meeting, 5 Sep. 1919, at Casella's in Frankfurt, Hoechst, 12/11/2.

34. Minutes of ACID meeting, 5 Sep. 1919, Hoechst, 12/11/2.

35. BUDACI settlements with employers at Hoechst on 2 Aug.; Agfa, 26–28 Aug.; Berlin, 1 Sep.; Griesheim, 4 Nov.; Bavaria, 6 Dec. 1919, Hoechst 12/11/2.

36. *RAB* (1922) (unofficial part): 721. VELA participation: *Bericht über die Verhandlungen zwecks Abschlusses eines Reichstarifvertrag für die in der chemischen Industrie angestellten Chemiker und Ingenieure am 27. und 28. Nov. 1919 im Hause der Berufsgenossenschaft der chemischen Industrie zu Berlin*, Hoechst 12/11/2; memo by Richard Weidlich, 12 Nov. 1924, Hoechst, 12/255/1.

37. Men and women got equal pay for equal work; seniority and pay scales were calculated in terms of years of professional experience rather than company service; *Bericht über die Verhandlungen zwecks Abschlusses*; BUDACI, *Kommentar*, 5, 22–23.

38. "Erfinderschutz," *Bundesblätter* 1, no. 5 (15 Aug. 1919): 17–18; ibid., nos. 6/7 (10 Sep. 1919): 21–26.

39. Dr. Kerkovius, 24 Nov. 1919, meeting of DVSGE, quoted in *GRUR* 24 (Dec. 1919): 244–51.

40. "Erfinderschutz," *Bundesblätter* 1, no. 5 (15 Aug. 1919): 17–18; ibid., nos. 6/7 (10 Sep. 1919): 21–26.

41. Ibid.

42. BUDACI, *Kommentar*, 10–16.

43. RTV, article 9 (*Erfinderrecht*), section II (*Einteilung der Erfindungen*).

44. BUTAB's Karl Sohlich at a DVSGE meeting of 24 Nov. 1919, in *GRUR* 24 (Dec. 1919): 244–51; BUTAB, ed., *Die Neuregelung des Erfinderrechtes der Dienstnehmer: Gegenentwurf des Bundes der technischen Angestellten und Beamten zu den pars. 121 bis 131 des vom Arbeitsrechtsausschuß aufgestellten Entwurfes eines Allgemeinen Arbeitsvertragsgesetzes (28. Sonderheft zum Reichsarbeitsblatt, 2. Stück) nebst Begründung. Sonderabdruck aus dem Reichsarbeitsblatt 1926 Nr. 12 und 13* (Berlin, 1926), in SAA 11/Lf391. BUTAB critique of BUDACI compromise with employers: Fritz Pfirrmann, "Ein neuer Erfinderrevers," *DTZ*, 4 Jan. 1929.

45. BUDACI, *Kommentar* 1920, 11. Management accepted this interpretation, and company inventions became the exception; minutes of IG patent committee meeting, 8 Mar. 1922, Hoechst, 12/255/1.

46. IG patent committee meeting, 31 Aug. 1920, Frankfurt a.M., Hoechst, 12/255/2 and 12/255/1.

47. BUDACI, *Kommentar*, 27.

48. Weidlich to Engländer, 15 Jan. 1925; ACID correspondence among Doermer, Weidlich, Curschmann, Staubach, 8–29 Jun. 1926, Hoechst 12/255/1.

Chapter 3

STRUGGLES AND SETBACKS, 1920–1924

The ability to compromise, which characterized relations between management and salaried inventors in the chemical industry, was lacking in mechanical and electrical engineering. When engineering industrialists got wind of what their counterparts in chemistry were up to in late 1919 and early 1920, they became very nervous. Among the first to sound the alarm was Reinhard Poensgen, a spokesman for the mechanical engineering firms in the Ruhr region. Reviewing an early draft of the Chemists' Contract, Poensgen in November 1919 criticized its inventor clauses as "an extraordinary danger," which would do "the most serious harm" in his industry. Engineering inventions were invariably the result of collective effort, wrote Poensgen, insisting that "no distinction is to be made between company inventions and individual service inventions ..."[1]

Poensgen summed up the position of management in the engineering industries: Everything new was a company invention, service inventions and inventor rewards were a nightmare, and the best that employees could hope for they had already got in the June 1919 agreement between the Bavarian metal industrialists and the BUTAB. In fact, the engineering industrialists touted the Bavarian pact as evidence of their flexibility and goodwill and held it up as a model for collective bargaining with other employee organizations. The Association of German Machine-Building Firms (Verein Deutscher Maschinenbau-Anstalten, VDMA) portrayed the Bavarian inventor clauses as a point of harmony between employers and employees.[2] The Federation of German Metal Industrialists (Gesamtverband Deutscher Metallindustrieller, GDM) recommended them for patent agreements in individual employment contracts. Even this was too much for the Association of German Iron and Steel Industrialists. In June of 1921 this powerful trade organization instituted fines for any member firm signing a collective inventor agreement that "accommodate[d] the salaried employees any further than the law," thus forcing adherence to the authoritarian and increasingly outdated standards of the 1891 Patent Code.[3]

Why did the engineering industrialists reject the consensus model embraced by their counterparts in the chemical sector? Considering the space for accommodation opened up by the inflation, which promoted rapid economic reconstruction, expansion, and full employment in the immediate postwar years, one might have expected engineering management to be more flexible in its dealings with the salaried employees as well. Accommodation and labor peace as cemented in the Central Working Community (Zentralarbeitsgemeinschaft, ZAG), after all, were the hallmark of management-labor relations during the first few years of the Weimar Republic.[4] As the engineering employers' rigidity in the inventor issue shows, however, the broader pattern did not necessarily extend to white-collar and professional workers. True, the chemical employers had struck a deal with their scientists, but it is unclear what role the inflationary context played in bringing it about. What is clear is that factors such as the prior state of the discourse and the structure and organization of the chemical industry had been important. The same appears to be true for the engineering industries. Inflation did not seem to make much difference, while engineering's underlying, structural and organizational conditions weighed heavily, indeed. A brief discussion of some of the pertinent differences between the engineering and chemical sectors will serve to make the point.

To begin with, there were important cultural differences between the two branches. Engineering management exhibited an arrogance of power and a confidence in its ability to impose autocratic solutions that is absent in the documentary evidence left by the chemical industrialists, who could flatter their industrial scientists when they had to. Engineering's combativeness stemmed in part from its cultural proximity to the heavy iron and steel barons, who were notorious for their authoritarianism and insistence on being the complete "masters of their own house."[5]

Such attitudes were reinforced by the social divide between senior management in the engineering industries and its technical staff—a gap considerably larger than in the chemical sector. Unlike most chemists, who constituted a fairly homogeneous educational and functional elite, the majority of mechanical engineers did not hold doctorates and occupied a much greater variety of functions within the firm. Mechanical engineering lacked a clear functional demarcation between university trained engineers and engineers educated at lesser technical institutes. Employers therefore opposed as unwarranted and divisive the effort of academically certified engineers to professionalize by seeking special privileges and different employment contracts. The electrical industry had somewhat greater internal career differentiation. Employing a greater percentage of university graduates than mechanical engineering, it occupied a middling position between the latter and chemistry. But long-term oversupply and underemployment of all engineers in Weimar limited any tendency of the two specializations to diverge too far and counteracted their social differentiation.[6] As a consequence, engineers in industry found climbing the corporate ladder and ascending to the ranks of higher management even more difficult than the chemists. Being addressed by their employers as "fellow Ph.D.s," "friends," and "equals" was simply unimaginable in the engineering industry.[7]

A second reason for rejecting the example of the chemical industry was the large number of engineers, which contrasted with the more concentrated form of intellectual capital represented by chemists. Of the approximately 150,000 members of the BUTAB, VELA, and BUDACI in the early 1920s, there were only about 4,000 chemists. This facilitated employer compromise with the latter, but made inventor rewards in the engineering industry a very different and much more costly proposition.

The different magnitudes of the numbers were a function of the different organizations of labor and the different systems of producing innovation in chemistry and engineering. Innovation in the chemical industry centered on the discovery of new processes, which contrasted with the creation of new products and systems in mechanical and electrical engineering. Typically the chemical industry patented processes, which could ordinarily be associated with the laboratory work of specific individuals or teams. The calculation of special inventor rewards in the form of royalties from patent-based product sales, therefore, remained a manageable proposition in the chemical industry.[8]

Engineering tended to patent physical products, usually components or subsystems as well as numerous trivial improvements, whose relationship with a given innovation or a marketable system was indeterminate. Engineering innovations, as management never tired pointing out, typically incorporated numerous inventions, patents, and improvements. These came from many different individuals and groups along the line that led from laboratory to shop floor to point of sale. A relatively simple radio transmitter marketed in 1914 by Telefunken, for example, incorporated some twenty-seven patents, ranging from major discoveries to minor design details and based on inventions by twelve separate inventors and one two-person team.[9] In general, the engineering industries filed many more patents than the chemical industry. The chemical firm BASF applied for 2,341 patents between 1877 and 1922, while the electrical company Siemens during the same period filed 107,000 applications in Germany alone and another 20,000 abroad.[10] With such mass patenting, calculation of the value added to the final product by a specific inventor became extraordinarily complicated and expensive.

Given the importance of collaboration in the production of engineering innovation, management also feared that inventor rewards for service inventions would harm progress. The lure of special financial gain, managers said, would encourage employees to invent outside the area of their immediate duties and expertise, discourage them from asking for advice or using the company's laboratories, promote internal secrecy, and in general wreck the collaborative process at the heart of generating innovation in assigned areas of responsibility. Royalties, once in place, were an incentive to disparage or suppress new ideas by colleagues and subordinates that might jeopardize the employee's additional source of income.

The best way to avoid inventor problems in their industry, engineering managers believed, was to define just about everything as a company invention. Poensgen essentially denied there was any technological progress in engineering that was the achievement of an individual person. "There are only a very few cases in machine building factories … where it can be proved without any doubt that [an] 'individual service invention' has not been influenced by the experiences and other support facilities that are at the inventor's disposal by virtue of his capacity as an employee." Inventions in specialized branches "in ninety-nine percent of all cases" were made possible only by company experience gathered over many decades, company laboratories, etc. The inventor, "therefore, may be said to be using capital in whose formation he has had no part."[11] Carl Köttgen, general director of the Siemens-Schuckert Works in Berlin and the engineering industry's central figure in the struggle over employee inventor rights during Weimar, asserted time and again that, "ninety-five to ninety-eight per cent of all inventions are company inventions."[12]

It was attitudes such as these—and employees' consequent inability to negotiate inventor terms comparable to the Chemists' Contract—that explain why the engineering unions became disillusioned with collective bargaining. Getting nowhere, they changed strategy and began to pursue a legislative solution. As early as November of 1919, engineer Karl Sohlich, the BUTAB's leading patent expert, demanded a uniform, legislative solution instead of collective bargaining agreements, which varied from industry to industry and left most salaried engineers out in the cold. Sohlich called for curtailment or outright elimination of the company invention and for statutory inventor rewards. Two years later the BUDACI, with many more engineering members than before, adopted the same position. The VELA, too, by 1921 demanded legislation to eliminate the company invention and pay statutory compensation for all service inventions.[13]

The bourgeois and Social Democratic coalition governments of the early 1920s proved receptive to the employees' appeals for help. The Reich Labor Ministry, which in October 1920 had begun work on a uniform labor code, shortly afterward appointed a study group for salaried inventor issues. Made up of civil servants, academics, patent experts, and labor lawyers, the inventor-rights committee quickly made the employees' cause its own. Its dominant figure was the economist and labor lawyer Heinz Potthoff. Before World War I, Potthoff had been a Reichstag deputy for the Progressives and served briefly as syndic for the BTIB. Together with Frankfurt law professor Hugo Sinzheimer, a friend of labor and member of the inventor-rights committee, Potthoff edited the periodical *Arbeitsrecht* (*Labor Law*). Other committee members included Göttingen law professor Peter Oertmann, who had supported employee-inventor rights since the days of the Empire, and Berlin lawyer Stephan Oppenheimer, business manager of the Federation of Berlin Metal Industrialists (Verband Berliner Metallindustrieller, or VBM). Oppenheimer was the committee's chief conduit to the engineering employers.[14]

The Potthoff group soon began circulating legislative proposals that greatly alarmed the industrialists. Chemical employers were afraid that a uniform and nationwide solution across all sectors of industry would upset the delicate balance of the Chemists' Contract, which was precisely tuned to conditions in their industry and allowed for a compromise that might otherwise not be possible.[15] The engineering industrialists also spoke about the need for branch-specific solutions via collective bargaining agreements, instead of a standardized law that would do justice to no one. But their chief worry centered on the specifics of the government's proposals, which they considered "utterly one-sided in [their] representation of the employees' interests." To coordinate effective opposition, the engineering employers in January 1921 formed an intellectual-property-law committee in the Reichsverband der Deutschen Industrie (Federation of German Industry, RDI). Siemens's Köttgen quickly emerged as the group's leader. At the very first meeting Köttgen articulated the fundamental objective: "the far-going rights of employees in the chemical industry cannot be extended to their counterparts in the mechanical industry."[16]

The best way to persuade the government that it was on the wrong track, the industrialists decided, was to fight fire with fire, or rather, paper with paper. The upshot was the outbreak of a minor war—a war of position papers, memoranda, meetings, hearings, lobbying, and stratagems—which raged for several years but eventually fizzled in 1924, in the rising tide of unemployment created by currency stabilization and deflation from late 1923.[17] The first shot, after the Labor Ministry's initial trial balloons, was a detailed report the RDI's intellectual property committee sent to the government in February 1921, explaining why its plans for inventor legislation in general and its specific pro-inventor proposals in particular were a big mistake.[18]

If the industrialists thought that the Labor Ministry's plans went too far in one direction, the employees insisted they did not go far enough. In April 1921 the AfA-Bund came out with a bill of its own that trumped the government's proposal. The AfA-Bund's draft coincided with the latter on many points, but it went further in two crucial regards, both of which prefigured the solution eventually imposed by the Nazi regime. Whereas the government's plan retained a domesticated version of the company invention, the AfA-Bund wanted it eliminated outright. Second, the employee federation called for statutory inventor rewards for all inventions claimed by the employer. The principle of employee compensation, in other words, should no longer be tied to the invention's profitability or to some other attribute the employer could manipulate. As in the Göring-Speer ordinances of 1942/43 and in current law, the employee would be entitled to compensation the moment the employer declared his intention to claim the invention for the firm.[19]

Not surprisingly, the industrialists heaped scorn on the AfA-Bund's bill, which would lead to "unimaginable conditions."[20] It is not clear what the response was in the Labor Ministry, but considering that its own, less radical plans were already under attack—not merely from big business, but also from other government quarters—the AfA-Bund plan is unlikely to have been welcome there either. In the fall of 1921 the Ministry of Labor lost its jurisdiction over the employment conditions of government scientists and engineers to the Finance Ministry, which immediately joined the engineering industrialists in the effort to defeat inventor legislation and save money.[21]

The new allies explored possibilities such as transferring Labor's remaining jurisdiction in the matter to the Justice Ministry, on the grounds that inventor rights were really an aspect of patent law. But this raised the old specter of a new Patent Code based on the first-to-invent principle. The idea was dropped almost as quickly as it was raised.[22] Another problem they faced was intra-industry discord. Chemical executives such as Hoechst's Richard Weidlich continually pointed out the danger of a one-size-fits-all approach in legislation. Time and again Weidlich and his colleagues urged their counterparts in engineering to make enough concessions to give meaningful collective bargaining a chance. This would be the only way to stop the government from proceeding with legislation.[23] The engineering industrialists interpreted the situation differently. Unwilling to make the kind of concessions that might result in a working relationship with the employee unions, they reckoned with the inevitability of inventor legislation and concentrated most of their effort on changing the content of whatever law the government would propose.[24] Despite temporary setbacks, this approach succeeded beyond expectations. The engineering industrialists gained control of the RDI's inventor policy and ended up playing a crucial role in shaping the Labor Ministry's 1923 bill. There were costs, of course: alienation of their colleagues in the chemical sector—which they managed to repair—and opposition from the majority of Weimar's engineers and industrial scientists—which they did not. But the engineering industrialists did not worry about that problem.

From early 1921 to the summer of 1923, when the government finally published its bill for a unified labor code, engineering employers badgered the Labor Ministry with a constant stream of objections, explanations, demands, and recommendations. Led by Köttgen and MAN director Emil Guggenheimer, they objected to almost every proposal or suggestion by the Labor Ministry. Inventor legislation, the engineering industrialists argued, would jeopardize technological progress, harm corporate finances, and hurt competitiveness with nations such as the United States, which carried no comparable burden. In Köttgen's words, "other countries do not have such impediments; America will protect its maneuverability."[25]

Another employer tactic was to say there was no reasonable way of calculating an individual inventor's contribution to the typical mechanical or electrical invention.[26] To prove this assertion, Köttgen presented the ministry with detailed case studies of two typical Siemens products: an overhead power supply for streetcars and a radio transmitter. The radio transmitter incorporated twenty-seven patents for inventions by fourteen different scientists and engineers. How could the company possibly calculate each inventor's share in the final product and come up with accurate royalties for every one of them? What about those employees whose role in the project had been to exclude certain technological possibilities? Should they be penalized for doing their job?[27] Whatever approach one took, a fair solution was impossible. The whole idea was a nightmare and should be dropped. If the government did not believe him, Köttgen wrote, its inventor-law experts were welcome to inspect the company's laboratories, so they could see with their own eyes he had spoken the truth.[28] Protests such as this were not without results. The Labor Ministry's experts began arguing among themselves and in late 1921 withdrew their initial draft bill. Potthoff immediately began work on a new plan, for which he solicited input from both the engineering unions and the employers.[29]

Defining Invention: Siemens versus Hitler

The government's decision to go back to the drawing board was only partially the consequence of direct lobbying by the employers. It also resulted from their attempts to influence the climate of opinion in more indirect ways. Industry veterans of the prewar battles over Patent Code reform such as Guggenheimer had sensed early on they would be fighting an uphill battle. They realized that bombarding the government with protests and memoranda and bills of their own was good, but not good enough. They knew that the lay public and most politicians believed in the individualist paradigm and the genius of the heroic inventor. There was no doubt in their minds that those ideas were an anachronism and irrelevant nonsense, which had nothing to do anymore with systematic problem solving in the modern, corporate research laboratory. But determining how to prevent the employees from exploiting the stereotype, how to energize their own ranks, and how to change the inherited concept of invent-ing—that was the real challenge. The answer, Guggenheimer and Köttgen managed to persuade their colleagues, was a campaign of propaganda and reeducation, which would reach wider circles and weaken the enemy from within.[30]

The industrialists' basic idea was to produce a literature targeted at inventors and their allies that portrayed the process of inventing so the reader could not help but conclude that organization, division of labor, system, plan, and order—in short, the company—was everything, and the employee's individual creativity nothing. One early product of this strategy was an article in the April 1921 issue of the *Journal for Industrial Law* by Georg Meyer, managing director of the electrical engineering firm Paul Meyer AG in Berlin and a respected inventor in his own right.

Meyer assumed a posture of impartial analysis and benevolent neutrality. From that vantage point he could not help but note a deplorable German tendency to overestimate the inventor—to con-sider every invention a "flash of genius," every inventor a "bene-factor to humanity," and every employee invention a "source of gold" for the employer. Meyer gave several detailed illustrations from his professional experience to show how wrong such ideas really were. In the world as he knew it, "great pioneer inventions are extremely rare, and the object of a normal patent is no gigantic mental achievement but the result of craftsmanship in design, just like most everything else that is not patented." Instead of linking employee incentives to inventor rewards, it would be much fairer and much more rational to reward productive researchers in other ways—for instance, with profit sharing based on the company's annual financial results.[31]

Meyer's article, a copy of which the industrialists sent to the Min-istry of Labor, had appeared in a specialized professional journal that reached a narrow audience only. The same was true for most other articles and writings of this type, which reinforced the opin-ions of those who did not need convincing but did little to convert a wider audience. Stronger medicine was needed. The cure was administered in the form of Ludwig Fischer's little book *Company Inventions*, which appeared with Carl Heymann's Verlag in Septem-ber 1921 and provoked more public controversy than any other work written on the subject before or since.[32]

Fischer was Köttgen's right-hand man in all intellectual property matters. A member of Germany's small and exclusive patent bar, he had been hired in 1899 by Wilhelm von Siemens to centralize all of the concern's patent matters into a single division, which he created and ruled with an iron hand until his retirement in 1934. Fischer was also a prolific writer in the cause of employer patent rights. He cast many of his uncompromising views as scholarly treatises of the Patent Code's original intent and that of its chief author, Werner Siemens.[33]

Fischer appears to have written *Company Inventions* on his own initiative, but also on company time and with the active encouragement and guidance of his immediate superior. There can be no doubt that Köttgen was behind the project from the outset and was, in that sense, its true instigator. Köttgen was also in charge of the book's publication and its marketing plan. His basic strategy was to aim for the widest possible distribution to responsible engineers and scientists in all enterprises, public and private, across the country. The publisher would make the book available to corporate buyers at a sizeable discount when ordered in sufficiently large quantities. As for content, the work was designed to give the appearance of a straightforward factual exposition, professional but without the usual patent-law jargon, concerning an issue about which misconception and error abounded. If all went as planned, even casual reading in the volume would persuade the more thoughtful employees that most inventions were indeed company inventions. Discussions with colleagues would encourage further reading and more discussion, and so the conviction would gradually spread among the ranks of salaried inventors themselves that their demands had been excessive and unreasonable.

The details of this plan were outlined in printed, confidential guidelines and a cover letter that Köttgen sent along with the book to Guggenheimer in September of 1921. Acknowledging that the industrialists "still had a great deal of further polemicizing to do," he explained that the work's tone was intentionally kept "not too polemical," because it aimed to "lay the foundation for a serious discussion regarding this very important subject." Siemens and AEG had agreed to give copies to all the "better technical employees," and Guggenheimer was asked to do the same at MAN.

The accompanying confidential guidelines mentioned that the "extraordinarily far-reaching demands of the employees" were the product of such fundamental misconceptions and such profound ignorance of the truth that they threatened to be anchored in the law imminently. An educational campaign was essential, since even the most senior scientists, laboratory supervisors, and in some cases top managers, business leaders, and employers themselves were "applauding the [government's] draft." Everywhere, "people are groping in the dark; enlightenment is urgently needed." Fortunately, the instructions continued, enlightenment was precisely what Fischer's *Company Inventions* offered, for it did not deal in empty "slogans" but appealed to the predilection for "thinking along" and the appreciation of "professionalism and comprehensiveness" among scientists and engineers with higher education."[34]

Whether Fischer's book actually did what its purveyors claimed and hoped for may be doubted. Many of Germany's leading papers and professional journals, as well as the entire community of patent and inventor-rights experts, received copies and gave the work substantial coverage in reviews and commentary. Almost all the reviews were positive, in the sense that they appreciated the persuasiveness of the author's arguments and the strength of his case for modern industrial inventing as systematic and organized problem solving. Many accepted its conclusions as well, conceding the need for reassessing the standard pro-inventor view and praising Fischer for daring to swim against the tide of public opinion. But others, while acknowledging the work's intellectual achievements, remarked on its highly partisan thrust. These commentators regretted Fischer's refusal to make any concessions to the other side and criticized his "extreme radicalism on behalf of the employers." One critic thought that the evidence Fischer had attempted to marshal was "completely erroneous." Even the "most moderate contemporary opinions" placed the "ethically higher justification" on the side of the employees. According to the same writer, the Imperial government's 1913 patent bill and the 1920 Chemists' Contract proved that the employers' traditional hard line had lost its credibility in recent times.[35]

Critiques such as these represented a minority view among Germany's tightly knit community of patent lawyers, which not surprisingly tended to side with management rather than labor. More significant, therefore, was the resistance Köttgen's scheme encountered among representatives of the managerial class itself. While Guggenheimer and other MAN executives such as Richard Buz and director Gertung enthusiastically supported it, Imanuel Lauster, the company's top manager at the main Augsburg plant, thought that distributing the book to all the engineers would be a terrible mistake. Fischer's work was so blatantly pro-employer, Lauster argued, that it would provoke rather than persuade most employees. The engineers would not bother to study it closely, and the book would end up having the opposite effect from that intended. Köttgen's plan might work with politicians and legislators, Lauster thought, but the company should acquire just a few copies for the library and circulate nothing more than an announcement of the work's availability. After further discussion and reflection, the other MAN executives reluctantly came around to Lauster's view, and in the end the company's employees were spared this particular experiment in corporate reeducation.[36]

Other engineers and industrial scientists were not so lucky, since their superiors mainly decided it would be a good thing for them to learn that the "company spirit" rather than the individual person's insight or conceptualization was the critical factor in generating invention. It would also help if the employees came to understand that inventing was nothing more than "subjugating the forces of nature to a purpose." This phrase of Fischer's indicated that "inventions and technological innovations in general are wholly rational creations that lack individual characteristics." Inventions, therefore, were "fundamentally different from artistic creation, whose essence is individuality and whose uniqueness consists precisely in being created only once and by one single person only." In other words, argued Fischer, the inventor's activity was either the same as doing workmanlike design such as applying known techniques and principles, or it was analogous to discovering natural phenomena. In either case, the difference from artistic creation was crucial, he maintained, and it explained why the principles governing patent law differed from those of copyright law and why salaried inventors were not entitled to ownership rights or special compensation. Copyright law, according to Fischer, protected the individual author or artist precisely because the uniqueness of the creation constituted the basis of true intellectual property. In contrast, patent law was not about an individual's intellectual property. The reason was that the invention was a rational, universal creation that lacked in individuality and uniqueness and, like the discovery of a law of nature, belonged in principle to all mankind. A patent for such a discovery was merely an instrument of economic policy: an incentive to rapidly divulge it so as to spur technological progress and benefit society at large. Naturally, this incentive should go to the entrepreneur who organized and produced such progress. All in all, it was clear from this and other arguments that the employees' "demands on many points, and precisely on the most important ones, are on objective grounds both unwarranted and impracticable."[37]

As Lauster had predicted, *Company Inventions* did not sway employees but outraged them instead. The VELA responded with an immediate, point-by-point refutation of Fischer's arguments spread across four issues of its biweekly, *The Managerial Employee*, in late 1921 and early 1922. Entitled "Inventor or Subjugator of Nature?" it was written by Hermann Schmelzer, an engineer and patent consultant in Kassel. The thrust of Schmelzer's essay was to emphasize the irreducible element of personal creativity in the act of inventing, and to posit the latter as a manifestation of the same ideal forces that made for artistic genius and for the progress of human civilization in general.[38]

Schmelzer developed this argument in two directions. He played up the affinity between Fischer's materialistic and collectivist theory of invention and Marxism, in order to discredit the Siemens patent chief as an unwitting agent of the Bolsheviks and a promoter of "intellectual-property Bolshevism" (*Urheber-Bolschewismus*). At the same time, Schmelzer anchored his own position in what was both Germany's proudest cultural legacy and its hackneyed legitimizer of moral superiority over other countries—the civil religion of *Bildung*, Romanticism, and Idealism.

Schmelzer prefaced his article with a quotation from Johann Peter Eckermann's *Conversations with Goethe*:

> All production of a higher order, every significant observation, *every invention*, every great thought that bears fruit and has consequences, stands in thrall to no one and exists high above all earthly power. Man has to realize these are undeserved gifts from up high, pure children of God, which he should accept and increase in happy gratitude.

In keeping with this lofty vision, Schmelzer relied heavily on the definitions of inventing advanced by outside, mostly older authorities to prove Fischer's fundamental error. He cited Joseph Kohler, dean of German patent scholars, who defined inventing as the "technical manifestation of a person's intellectual creativity" (*"eine zum technischen Ausdruck gebrachte Ideenschöpfung des Menschen"*). He also quoted Alois Riedler, one of Germany's most famous engineering professors and a widely recognized authority on social aspects of technology, to show that Fischer had not probed deeply enough to locate the true sources of inventing. Testifying as an expert witness in a patent case before the *Reichsgericht*, Riedler in 1898 had debunked the exaggerated faith in scientific methods and scientific principles as a means to inventing. Science, methods, fundamentals, and calculation, Riedler had observed, should not be confused with, nor taken for the whole of, inventing. The original conceptualization was always prior and always more important, even though after the fact the theoreticians could often hardly recognize this anymore.

"The tools of science can only be used critically, not creatively," according to Riedler, and "hegemony resides not in the tool, but in its use by the person who works with it." Once the mental effort that produced the invention's conceptualization was complete, however, it was easy for the experts to figure out the inventor's results and reproduce them at will.[39] So it was with Fischer, argued Schmelzer. "Fischer's subjugations of nature are not inventions, and his nature-subjugators are designers, but not inventors."

Schmelzer also quoted at length from Max Eyth (1836–1906), an engineer, world traveler, author, romantic, and technology enthusiast whose writings frequently served German engineers as a source of moral authority.[40] Eyth's lectures on the philosophy of inventing made it clear that perspiration alone did not suffice, that inventing transcended hard work and hard thinking. Rather, Eyth had written in 1904, the crucial moment of conceptualization was akin to poetic inspiration, a "mental flash, completely divorced from the environment and even the intellectual efforts of the moment, which suddenly seizes the entire soul as though lit up in happiness." Many people had such momentary flashes of insight, according to Eyth, but only "exceptionally talented people" were "capable of capturing the fleeting, shadowy image and recognizing its meaning." Eyth's observations, commented Schmelzer, gave a much better idea about how inventions originated than Fischer's misleading assertion that ninety-nine percent of all inventions could be made by average professionals. If Fischer really believed that, Schmelzer added, it merely proved he was not very good at his job. But if Fischer ever had an opportunity to study real inventions and real inventors up close, he would perhaps come to grasp the truth of Eyth's observation that

> Even in the remote future we can only hint at, the same human spirit that in prehistoric times invented a method of starting fires by rubbing two sticks together will test its powers against greater problems. From the bottom of man's soul, the flash of insight will light up time and again, to illuminate yet another part of his progress through space and time. In this earthly existence, the inventor will never come to rest so long as man remains what he is: an image of the Creator, a being in which God has placed a spark of His own creative power.[41]

Having thus disproved Fischer's theory of invention, Schmelzer had little difficulty showing that the Siemens patent chief was wrong to make a radical distinction between patent law and copyright law. Rather, the two were merely slightly different manifestations of the same underlying principle. Conceding that this point was sometimes disputed in the literature by others besides Fischer, Schmelzer noted that most legal scholars and theorists shared his interpretation and that, therefore, "patent right is an author right." If, however, the invention belonged to the inventor, then Fischer's argument—that society in general rather than the private individual was entitled to it—really amounted to a form of Bolshevism. Fischer's argument on this score would "cause the great happiness of all Bolsheviks," and his position was even more outrageous than that of the "communists and socialists who maintain that the soil and the land, the riches of the earth and machinery, should belong to society at large." Of course, Schmelzer continued, Fischer did not follow the logic of his argument to this, its final conclusion. Rather, he said that for pragmatic, economic-policy reasons the right to a patent of invention went to the person who divulged and filed it first with the Patent Office. But if someone could actually interpret the law this way, and "if the author of an invention should play the minor role and the divulger the lead role," this was the "best proof of the necessity to clarify the situation in a fundamental way and, in the new legislation, anchor the first-to-invent right firmly in the Patent Code." In sum, the only conclusion one could draw concerning Fischer was that in "authoring such theories, he ma[de] himself into the vanguard of Bolshevism in the field of inventing."[42]

A potent counterattack in its own right, Schmelzer's critique of Fischer gained added significance when it became the rhetorical centerpiece of the BUDACI's response to *Company Inventions*. The union replied with a ninety-six-page *Study Concerning Inventor Protection*, which appeared in the spring of 1922 in its regular series of *Sociopolitical Reports*.[43] Reproducing large segments of Schmelzer's article and repeating all known pro-inventor arguments and proposals for legislation it had articulated earlier, the BUDACI report was notable more for its comprehensiveness than its originality. At the same time—and this makes the report particularly interesting—certain of its passages exhibit a striking similarity to Hitler's subsequent observations concerning inventing in *Mein Kampf*.

Taking its cue from Schmelzer, the BUDACI study began by analyzing the essence of invention. In the absence of a commonly accepted definition of the concept, the union's anonymous author(s) cited a mass of authorities—including Goethe, Ernst Mach, the 1913 patent reform bill, *Reichsgericht* rulings, and numerous patent experts—to prove that an invention was neither scientific discovery nor engineering design but creative spirit objectified. Inventing, it maintained, could never be replaced by systematic research and development purporting to guarantee success merely on the basis of "honest work." Technological creativity could never be generated on demand to solve problems in the same way that money, time, or effort could. The BUDACI supported this position with a 1919 Patent Office decision, which had ruled "untenable" an employer's argument that "the mental activity of inventing [was] ... a psychic condition of creative synthesis" akin to effort. Inventive creativity, according to the Patent Office, was not simply a factor of production "subject to an arbitrary act of will" and comparable to the way in which "one 'spends' money, time or effort for something—as though it were a permanent state of readiness to be deployed on demand, more or less like an organist pulling the registers of his organ." The BUDACI concluded that "the invention is therefore a creation of the spirit." To be sure, the creation might be small, but "one thing [was] certain. Without the personal inventor, in whose psyche at least this last obstacle to the invention is overcome, it never comes off."[44]

From this, and the fact that employers could always find the inventor when they filed for a patent in the United States, the BUDACI concluded that, "the theory of impersonal company inventions has been proved false." This was as true for the various collectivist theories underpinning the company invention as it was for the denial of productive intellectual creation by exponents of the Siemens Company. Arguments by men such as Wilhelm Siemens and Siemens director Emil Budde in their 1908 pamphlet, the *Inventor Right of Employees*, were as groundless as "Fischer's 'Company Spirit,'" a concept the union dismissed as a "bad joke." Amplifying their point, the BUDACI authors subjected Fischer's work to a devastating critique by adding their own running commentary to that of Schmelzer. They attacked Fischer as the hired hand of a company known for its "hostility to inventors" and the "mouthpiece of one-sided business interests." His clever "dialectics were aimed at preventing a just solution of the problem" and a "slap in the face of the creative engineer's sense of justice." It was to be hoped that everyone who had anything to do with the inventor issue would avoid contamination by the "unhealthy spirit that breathes through Fischer's pamphlet."[45]

The problem with all collectivist theories of inventing, including Fischer's, according to the BUDACI, was their failure to recognize "that the inventive deed presupposes the overcoming of impeding, material and individual blockages [and] that it is essentially spiritual (*geistiger Art*)."

> The company as such cannot think; what thinks within it are the employees of the company. It is not the company, therefore, but its employees who invent.[46]

Both the thrust and the phrasing of this particular passage are very close to Hitler's language on inventing in *Mein Kampf*, which in the Third Reich would become the ultimate source of legitimacy for inventor policy. According to the future dictator,

> It is not the mass that invents and the majority that organizes and thinks, but in everything always only the individual human being, the person.[47]

This sentence was the rhetorical climax of the approximately one-page passage in *Mein Kampf* Hitler devoted to the place of the inventor in a future and better society.

Not surprisingly, Hitler embedded his admiration for the inventor in Social Darwinism, *völkisch* rhetoric, and the myth of heroic, "Aryan" leadership. This was a crucial mutation from the liberal individualism and romanticism that constituted the BUDACI's point of departure, even as the union instrumentalized those ideas for purposes of strengthening the position of salaried professionals in the industrial bureaucracies of modern capitalist society.[48] Still, the emphasis on the role of the individual in Hitler's inventor language was striking. The entire passage reads as though it were directly inspired by the public campaign of the VELA and BUDACI to destroy the credibility of Fischer and his industrial backers, and to help pass pro-inventor legislation. Much like Max Eyth and other pro-inventor ideologues, Hitler cast the inventor as the hero of modern industrial society. He maintained that, "all inventions are the result of the creative effort of a person."

> All these persons themselves are, consciously or unconsciously, greater or lesser benefactors of all mankind. Their activity later gives millions, even billions of human beings the means to facilitate the conduct of their struggle for existence.

Hitler explained the origins of contemporary material civilization in terms of the reciprocities among inventors and the mutual interdependencies of their inventions. "All production processes, in turn, are to be equated in their origin with inventions and thereby depend on the person." It followed that

> a human community is organized correctly only if and when it eases the work of its creative forces in the most accommodating way possible and

uses them to benefit society as a whole. The most valuable aspect of the invention, regardless of whether it resides in the material world or that of ideas, is first and foremost the inventor as person.[49]

It is impossible to tell whether Hitler actually read Eyth, Schmelzer, or the BUDACI's *Study*. But it is not altogether implausible. The wave of publishing activity and controversy generated by Fischer's *Company Inventions* became a very public affair in the early 1920s. The anger, the exaggerations, and the overstatements that employer rigidity incited among engineers produced natural and elective affinities with Hitler's cosmology. Even if Hitler himself might have failed to notice this, the engineers in his entourage surely would have pointed it out to him. Quite apart from it being an issue close to their hearts, the inventor conflict was valuable propaganda material. It provided an opportunity to exploit the resentments of industrial scientists—and salaried employees in general—against predatory capital and proletarian labor alike. Whatever the exact mechanism by which Hitler became aware of the conflict, it is quite clear that his remarks about inventors and inventing were not simply a random broadside or a general manifestation of reactionary modernism. They reflected direct acquaintance with the inventor issue, which appeared in the news as the propaganda and labor relations struggle over Fischer's *Company Inventions* and remained a constant source of irritation thereafter.

Throughout the 1920s, of course, Hitler and the Nazis were a marginal phenomenon, so even if Köttgen and his colleagues had known they helped write sections of *Mein Kampf*, they would probably not have done anything differently. Hitler's latching on to the inventor issue may have proved to be significant in the long run, but in the 1920s it was of no consequence. Even so, with hindsight it is clear that the episode was a microcosm of Weimar's larger political tragedy. The existence and survival of the Chemists' Contract clearly indicates that some people recognized the need and scope for compromise. But refusal to compromise and polarization from both sides constituted the dominant pattern. It was stronger, for instance, than any "inflation consensus," as is demonstrated by the radicalization of concepts of inventing on the part of both engineering employers and employees in 1921 and 1922.[50] The stalemate, the frustration and the hostility this bred in both camps, in turn, created Hitler's opportunity and made converts to National Socialism. To be sure, this did not happen overnight, but the uncanny similarity in the language of *Mein Kampf* and the BUDACI's critique of Fischer suggests that early on Hitler laid the foundations of support from those who could have been management's allies or the Republic's supporters but instead became Weimar's "dying middle." They subsequently voted for the Nazis, and in the Third Reich, "working towards the Führer," hammered the nails in the coffin of the employers' inventor arguments.[51]

Potthoff's Conversion and the 1923 Labor Code Bill

If the long run augured ill for the employers, their immediate prospects in 1922 did not at first seem much more auspicious. After the setbacks of 1921, the Labor Ministry designed its new plans once more on the basis of employee thinking rather than the views of the industrialists. Even so, Potthoff was careful to solicit input from the engineering employers as well, in the hope of bringing about a compromise. In May 1922 he invited Köttgen to submit a plan for inventor legislation that would be acceptable to all of German industry.

With the assistance of Fischer and MAN directors Guggenheimer and Martin Offenbacher, Köttgen developed proposals that dovetailed with the conditions of invention in the engineering industry but took no account of the chemical industry. The distinction between company and service inventions used in chemistry, for instance, was too imprecise for their taste, while the concession to reward all service inventions was considered too generous.[52] The engineering managers instead came up with a complicated plan that would have been quite generous when it came to name recognition but rewarded practically no employee inventions with money. To create the appearance of making at least some material concessions, though, Köttgen and his associates also relaxed the language specifying which inventions the employer could claim outright and without compensation. The RDI's committee for intellectual property rights approved the plan at its June 1922 meeting.[53]

When Hoechst's Weidlich and the other chemical industrialists learned what had happened they immediately protested. On the one hand, they said, Köttgen's changes cutting back on inventor royalties were "dangerous" and would lead to unrest in the laboratories. On the other hand, the concessions Köttgen had introduced jeopardized management's grip on inventions that lay outside the employee's responsibilities but inside the scope of the company's business interests. Weidlich felt betrayed and urged his colleagues in the chemical sector to "decisively reject" the Köttgen plan.[54]

The two branches of industry soon made up. Weidlich and Köttgen both realized that internal divisiveness of industry would doom all chances of influencing the Labor Ministry and worked out a compromise during a meeting in Berlin in September 1922. Attended by top business executives such as Carl Friedrich von Siemens, as well as senior officials from the RDI and the Ministry of Finance, the meeting forged its compromise at the expense of employee interests. Industry's new plan combined the worst features—from the employee's perspective—of the Chemists' Contract and the Köttgen plan: the employer could claim almost everything, while the employee would be entitled to a reward only in rare cases. Even then the industrialists were reluctant to show their plan to the Ministry of Labor, for fear it might be interpreted as an opening bid rather than the limit they were prepared to accept. So they decided to keep it in reserve as a final, deadlock-breaking solution. Oberregierungsrat Dr. Schilling of the Reich Finance Ministry would introduce it during negotiations with the Ministry of Labor, but only after the industrialists had first trashed the government's plans and offered nothing.[55]

The industrialists executed their game plan at a meeting with the Labor Ministry's inventor-rights committee in October 1922. The two sides started far apart. The ministry's proposals closely resembled the most recent recommendations of the BUDACI, which granted employees far-going inventor rights.[56] Weidlich and Köttgen opened with extreme positions of their own. Weidlich portrayed chemical inventing as a regular business routine that had nothing to do with romantic notions of genius or creativity and reiterated how inherently unfair any inventor reward scheme really was. Köttgen stressed that industry's right to dispose over all employee inventions was essential. Industry could not possibly assume the financial and administrative burdens sought by employees and the ministry's inventor experts. He contended that the overwhelming majority of patents were all but worthless, proving the point with statistics showing that after fifteen years a mere 5 percent of patents were still in effect.[57]

After the two industrialists had made their presentations and left the meeting, Schilling introduced industry's plan as his own, confident that the inventor-rights committee would take the bait. But the effect was not what he and his companions had expected. According to committee member Oppenheimer, who reported a few days later to Köttgen what happened, the meeting disintegrated into "pandemonium." Industry's proposal went nowhere, and the entire matter was referred back to a subcommittee.[58]

When the Labor Ministry's experts recovered and were ready for the next round in January 1923, they produced a draft bill that differed only slightly from their earlier, pro-employee design.[59] Not surprisingly, industry reacted with anger and dismay.[60] In Köttgen's words, "the social tendency under the leadership of Herr Potthoff has triumphed after all." The government's plan, if it were to become law, would create an impossible situation:

> the company undertakes years of research, builds laboratories or experimental installations and spends vast sums to develop technology and bring about progress. And then the result of these labors is first going to be the property of the employee, from whom it then has to be purchased.

This should never come to pass. Rather, "the law should unambiguously state that the employee *must assign* those inventions to the company. Without that, smooth functioning is not possible." Even the most socially inclined supporters of the employees, Köttgen wrote to Oppenheimer, would "have to realize one cannot require the company to undertake research without being absolutely sure it can dispose over the result." If this principle were not anchored in the law, "progress will suffer, which is the most dangerous thing that could possibly happen to us in the German economy just now." If the government did not modify its proposal, Köttgen warned, industry would launch a barrage of the "very sharpest criticism on all fronts" the moment the bill was revealed in public.[61]

The sources do not reveal what Oppenheimer did with Köttgen's letter, but it appears to have struck a nerve. Before a week had passed, Potthoff had arranged to visit the Siemens research laboratories, to learn first hand what inventing in industry was really like. During an all-day tour that included breakfast, lunch, and dinner, he received briefings from various technical executives and laboratory directors. Potthoff was accompanied everywhere by Fischer, who served as his personal guide and explained everything. Fischer described the experience in a report for Köttgen, mentioning how he had taken Potthoff to look at products that incorporated numerous different patents. He had explained the differences between real inventions and those that were so in name only (because they met the legal criteria of patentability but otherwise represented nothing special). He had made sure that when they visited the central research laboratory, Potthoff would recognize "that in the environment of a laboratory with such tremendous means of support, an employee acquires rich suggestions and experiences from his interaction with others and that such an employee unquestionably *must*

arrive at inventions, if he is not exactly inferior." According to Fischer's report, "Dr. Potthoff confirmed without reservation" that the
desired effect had been achieved. "Everything went as planned," he
wrote, and "it was clearly noticeable that Herr Dr. Potthoff experienced a change of heart. In the end, we were in almost complete
agreement." After spending the entire day with Potthoff, it was Fischer's "overall impression that *on very important points he has changed
his opinion.*"[62]

Perhaps Fischer exaggerated his own powers of persuasion, but
there can be no doubt that the visit made a difference. Potthoff
underwent a conversion of sorts and came away with the idea that
corporate inventing was indeed a collective undertaking, in which
the division of labor, company know-how and the company's materials and facilities played the decisive role. Company inventions, he
decided, did exist after all. This did not mean Potthoff suddenly
sided with the employers. On the contrary, the collectivism to be
observed in the Siemens laboratories, Potthoff believed, argued in
favor of reforming the laws of incorporation and restricting the
employer's powers over the company in a variety of other ways.[63]

Fortunately for the industrialists, those "very dangerous" ideas of
the Labor Ministry's principal inventor-rights expert went nowhere.[64]
Potthoff's new appreciation for the realities of corporate inventing,
on the other hand, weakened his immunity to industry's arguments.
The engineering managers immediately pressed their new advantage. Through Oppenheimer, Köttgen and Fischer made numerous
suggestions for seemingly minor editorial changes in the language of
the government's inventor-protection proposal. The effect of their
recommendations, many of which were accepted, was to render the
bill harmless. As Köttgen wrote to Weidlich after having studied a
confidential draft in early May 1923, "on balance, it must be said that
the bill does now more justice to the needs of the employers than in
the past." A few days later, Oppenheimer in the final editorial meeting managed to win approval for "even more improvements along the
lines suggested by Herr Dr. Fischer." Fischer, who ordinarily saw
dangers everywhere, concluded that, "almost all the objections we
raised have been eliminated." Köttgen congratulated Oppenheimer
on his invaluable assistance, thanks to which "many of our wishes
have been accepted." The RDI, too, concluded that the "language of
the bill in its current version is on the whole tolerable."[65]

The subject of all this celebrating was the section on inventor
rights in the Labor Ministry's "Draft of a General Law on Employment Contracts," published in the *Reichsarbeitsblatt* of 1 August 1923.

The bill's inventor language followed the design of the Chemists' Contract, although its provisions for inventor rewards were considerably less generous. It also categorized the various types of employee inventions differently from the familiar triad of company inventions, service inventions, and free inventions. While it retained those three concepts, it introduced an additional, fourth type: "company-related free inventions." These were inventions that the Chemists' Contract included in the service invention. They lay outside the employee's responsibilities but inside the orbit of the company's business. The employer, in return for special compensation, was entitled to the "company-related free invention," but that entitlement was not quite as strong as in the Chemists' Contract.[66]

As far as the engineering industrialists were concerned, the bill's inventor terms were very good: they limited special compensation to a minute fraction of the total number of employee inventions while still securing adequate company control over them. The responsibilities of engineers and scientists in the engineering industries were normally defined broadly, and their numerous small and collective inventions did not commonly yield radically new technologies or processes—as might more readily happen in chemistry. Company inventions and uncompensated service inventions would therefore be the norm, while the compensated "company-related free inventions" would be relatively rare. Even if the latter did occur, the employer had a way to get his hands on them. No wonder Köttgen thought the bill was "by and large tolerable."[67]

The chemical industrialists had a very different reaction. They were deeply disturbed by the threat they detected to the company's control over employee inventions. Directors Weidlich, Frank, Doermer, Koebner, and other chemical executives attacked the bill as "intolerable." In their view, it failed completely to take into account the conditions of intellectual production in their branch. The central problem—a "question of life and death"—was the "company-related free invention."[68] Unlike in engineering, inventions outside the employee's professional duties but inside the orbit of the company's business were common in chemistry. The idea that the employee might have an *a priori* ownership right in such inventions, which the employer was merely entitled to acquire but which the employee might not even have to report and could patent under his own name, was absurd.

Compared to the vital question of undiluted corporate control over all the inventor's results, the cost of special inventor compensation was of secondary importance in the oligopolistic and highly profitable chemical industry. This was why the Chemists' Contract used a broad definition of the service invention and rewarded all of them. As the chemical employer association's Dr. Frank pointed out, "it is of fundamental importance for us that the concepts of service invention, company invention and free invention are preserved exactly as in the language of our collective bargaining contract. The introduction of a new concept of 'company-related free invention' is absolutely impossible for us."[69] Other executives in the chemical industry shared these sentiments. The bill's concept of "company-related free inventions" was a "confused and linguistic monstrosity."[70]

Apart from these comments by the two industries most affected, there was practically no other reaction—not by the BUTAB, the VELA, the BUDACI, the broader community of patent experts, nor the public at large. It was as though the government's plans for inventor legislation disappeared into a black hole, just as had happened in 1914. The Labor Ministry itself did not press the issue either, ostensibly to gather more input, but in truth because the opportunity for enacting major social legislation—if it ever existed—had already passed. Once again, the timing of the bill's announcement could not have been worse—or better, depending on one's point of view. In the summer of 1923 the country was in the grip of hyperinflation and political chaos. The Hitler putsch and other, left-wing coup attempts and currency stabilization were just around the corner. In this environment, almost no one had the time or patience to concentrate on the arcana of patent and inventor rights. With the sociopolitical sea change of late 1923 and early 1924, labor was thrown on the defensive, management grew more intransigent, and party politics came to center on bourgeois coalitions.[71] In this context, any hope for the passage of a comprehensive labor code was out of the question. And so, in late 1923 the politics of inventing temporarily went into remission. As Offenbacher noted with palpable relief in a 1925 letter to Guggenheimer, the inventor conflict was "currently dormant."[72]

The new power relations between management and industrial scientists that came into being with the transition to Weimar's middle phase are reflected in a small but telling incident, which will serve to conclude this chapter. More than a year after the Labor Ministry first announced its bill for a uniform labor code, the BUDACI published a critique of the inventor-rights paragraphs in the *Reichsarbeitsblatt*. Registering its "utmost disappointment" with the government's design, the union criticized it on all fronts and presented an alternative bill favoring the employee. What was needed, the BUDACI argued, was fundamental change, including an overhaul of the Patent Code. Company inventions were to be eliminated. All employee inventions should be free, unless a patent agreement described the employee's duties in detail and clarified exactly when the employer might claim the employee's intellectual achievements as service inventions. The latter would have to be rewarded liberally, and special boards of arbitration or labor courts were to adjudicate disagreements.[73]

As soon as the chemical industrialists read the BUDACI's demands they cried foul, accusing the union of unilaterally abandoning the spirit of the Chemists' Contract, which they had so valiantly defended against the "far more extensive wishes of the machinery industry." Now the BUDACI had violated that trust. To make matters worse, management had also got hold of a BUDACI report for the International Labor Office in Geneva, in which the union accused the employers of taking unfair advantage of the glutted labor market for industrial scientists. "It will be the industrialists' duty," Weidlich informed the industry's employer association, "to resume—and carry through—the battle that the Bund apparently wants to renew again." Every single clause in the Chemists' Contract was being attacked and every single issue raised again, resulting in "totally unacceptable" demands. "The spirit that animates the Bund's proposals," he concluded, "is not a good one."[74]

Preparing for unilateral cancellation of the Chemists' Contract, Weidlich and his colleagues decided to give discussion with the union one last chance.[75] The two sides met in Hoechst in November 1924. The employers played hardball, their brusqueness a striking contrast to the kid gloves they had worn during the negotiations of 1919 and 1920. It was madness, Weidlich told the BUDACI's delegates, to contend that employee inventions belonged to the inventor. It was duplicity to go public without first talking to their contractual partner. It was time, apparently, to go back to war. Dr. Knorr, the BUDACI's business manager, beat a hasty retreat and apologized profusely for the radicalism of the recent demands. He assured Weidlich and the other industrialists that the chemists in reality had no quarrel whatsoever with their employers and were eager to abide by the terms of the Chemists' Contract. The counter draft, he explained, was the product of the "massive resistance" to an acceptable collective-bargaining contract the industrial scientists had encountered from the leaders of other industries. It was not at all meant to wreck the good relationship with the chemical industry. Knorr promised he would urge his colleagues to publish a disavowal—which he would first show the employers—rejecting a legislative solution altogether.[76] In the event, no such disavowal ever appeared. But the BUDACI had effectively been silenced and it simply dropped the matter. Although the Chemists' Contract survived, this phase of the politics of inventing had ended with an unequivocal victory for big business.

Notes

1. Poensgen to Arbeitnordwest, 14 Nov. 1919, Hoechst, 12/255/1.
2. VDMA, *Zwangslose Mitteilungen*, 1920: 186; *GRUR* 24 (Dec. 1919): 244–51.
3. GDM, suppl. 1 to circular 291/M, 24 Nov. 1920, Hoechst 12/255/1; Dr. E. Hoff to Phoenix steel works, 14 Jun. 1921, Mannesmann, P2.25.45.
4. On the history and historiography of the inflation, including the ZAG, see Feldman, *Great Disorder*; idem, "Historian and the German Inflation"; idem, *Iron and Steel in the German Inflation 1916–1923* (Princeton, 1977); Holtfrerich, *German Inflation*; Maier, *Recasting Bourgeois Europe*.
5. On the politics of heavy industry before World War I: Hartmut Kaelble, *Industrielle Interessenpolitik in der Wilhelminischen Gesellschaft* (Berlin, 1967); after the war: Feldman: *Iron and Steel*.
6. Gispen, *New Profession*; BUDACI 1924 report to ILO in Geneva, minutes ACID Akademiker Ausschuß, 18–19 Nov. 1924, Hoechst, 12/255/1.
7. "Bericht über die Verhandlungen zwecks Abschlusses eines Reichstarifvertrages für die in der chemischen Industrie angestellten Chemiker und Ingenieure am 27. und 28. Nov. 1919 im Hause der Berufsgenossenschaft der chemischen Industrie zu Berlin," Hoechst, 12/11/2; Harald Mediger, "Gedanken zur Gestaltung des Rechts des nicht-selbständigen Erfinders im Lichte der Betriebsverknüpftheit seiner Erfindung" (Munich, 1948), pp. 35–38, BA-Zwi, ver. B141/2793.
8. Beckmann, *Erfinderbeteiligung*, 9–10; Walter, *Erfahrungen eines Betriebsleiters*; Köttgen to Arbeitnordwest, 29 Oct. 1928, p. 5, SAA, 11/Lf237/238.
9. Prof. Franke (Siemens & Halske) and Dr. Schapira (Telefunken) to Köttgen, 27 Jul. 1921, SAA, Lf352/11; the Osram lightbulb was covered by more than one hundred patents, Beckmann, *Erfinderbeteiligung*, 9; Köttgen to Arbeitnordwest, 29 Oct. 1928, p. 5.
10. BASF figures: Karl Holdermann, *Geschichte der Patentabteilung. Zusammenstellung von 1877 bis 1968*, unpub. ms., BASF, B4/E 01/1; Siemens figures: "Besprechung über das Erfinderrecht der Arbeitnehmer mit Vertretern des Reichsverband der Deutschen Industrie und der Vereinigung Deutscher Arbeitgeberverbände im Reichsarbeitsministerium am 19.1.1922," SAA, Lf352/11.
11. Poensgen memorandum, 14 Nov. 1919, Hoechst, 12/255/1.
12. Minutes of meeting, 19 Jan. 1922, SAA, Lf352/11. On Köttgen, a leader of Weimar's principal efficiency lobby, the National Productivity Board (Reichskuratorium für Wirtschaftlichkeit, RKW), see Nolan, *Visions of Modernity*; J. Ronald

Shearer, "Talking About Efficiency: Politics and the Industrial Rationalization Movement in the Weimar Republic," *CEH* 28, no. 4 (1995): 483–506; James, *German Slump*, 166.

13. Sohlich: minutes of meeting, *GRUR* 24 (Dec. 1919): 244–51; BUDACI: *Bundesblätter* 3, no. 20 (1 Oct. 1921): 208–10; VELA: Dr. Zellien, "Zum Recht der Angestelltenerfindung," *Der leitende Angestellte* 3, no. 21 (1 Nov. 1921): 161–63.

14. Members of Arbeitsrechtausschuß, list of 4 Oct. 1920, BAP, RAM 3451; Oppenheimer to Köttgen, 25 Nov. 1922, SAA, 11/Lf364; Gispen, *New Profession*, ch. 10.

15. Dr. Plieninger to HK Frankfurt, 3 Jan. 1921; Dr. Doermer to HK Frankfurt, 15 Jan. 1921, Hoechst, 12/255/1.

16. "Bericht über die konstituierenden Sitzung des Ausschusses für gewerblichen Rechtsschutz im RDI am 25. Januar 1921," NGug.

17. On the postinflationary era of recession, underemployment, and slow growth, see James, *German Slump*; Peukert, *Weimar Republic*, 107–28; Knut Borchardt, "Constraints and Room for Maneuver in the Great Depression of the Early Thirties: Towards a Revision of the Received Historical Picture," in his *Perspectives on Modern German Economic History and Policy*, trans. Peter Lambert (Cambridge, 1991), 143–60, esp. 152–60; Thomas Childers, "Inflation, Stabilization, and Political Realignment in Germany 1924 to 1928," in *The German Inflation Reconsidered: A Preliminary Balance*, ed. Gerald D. Feldman et al. (Berlin, 1982), 409–31.

18. RDI (Dr. Herle) to intellectual property committee, 10 Feb. 1921, Hoechst, 12/255/1; RDI, *Geschäftliche Mitteilungen* 3, no. 6 (Feb. 1921); Guggenheimer to RDI, 18 Feb. 1921; correspondence Köttgen and Herle, 2–3 Mar. 1921, NGug.

19. Labor Ministry's plan: Herle and Schweighoffer to Guggenheimer, 7 Apr. 1921, NGug; AfA-Bund plan: AFA-Bund to RAM, 13 Apr. 1921, Hoechst, 12/255/1; text of RAM, AfA-Bund, and RDI proposals in RDI, *Geschäftliche Mitteilungen* 3, no. 16 (May 1921), supplement 3.

20. RDI, *Geschäftliche Mitteilungen* 3, no. 16 (May 1921), enclosure 3.

21. RFM invitation of 18 Oct. 1921 for conference on "Entwurf zur Regelung des Rechtes an Erfindungen der behördlichen Angestellten," SAA Lf352/11; Finance Ministry collaboration with industry Oct 1921–Nov. 1922: Hoechst, 12/255/1; SAA Lf352/11, 11/Lf364; extensive correspondence in NGug; Engländer, *Angestelltenerfindung*, 45.

22. Meeting of DVSGE on 23 Feb. 1921; Guggenheimer to Köttgen, 3 Mar. 1921; RDI to intellectual property committee, 9 Mar. 1921 (Leitsätze zur Frage, "Gehört die Angestelltenerfindung ins Arbeitsrecht oder ins Patentrecht?); minutes of RDI intellectual property committee meeting, 16 Mar. 1921, NGug; VDMA, *Zwangslose Mitteilungen* 21 (Jun. 1921): 224.

23. Weidlich to Herle, 13 May 1921; minutes of ACID Akademikerkommission meeting, 16 Jun. 1921, Hoechst, 12/255/1.

24. Correspondence between Weidlich and Herle, 10 and 23 May, 2 Jun. 1921, Hoechst, 12/255/1.

25. RDI, *Geschäftliche Mitteilungen* 3, no. 36 (Dec. 1921): 391, item 686; Köttgen to RAM working committee, 17 Nov. 1922, SAA, 11/Lf364.

26. Confidential memo, RDI headquarters to intellectual property committee and branch units, 31 May 1921, NGug.

27. Franke and Schapira to Köttgen, 27 Jul. 1921; Hochschild to Köttgen, 2 Jun. 1921, SAA, Lf352/11.

28. RDI to Köttgen, minutes and report of "Besprechung über das Erfinderrecht der Arbeitnehmer mit Vertretern des RDI und der VDA im RAM am 19.1.1922," 19 Jan. 1922, SAA, Lf352/11; ACID to Akademiker Ausschuß, 20 Jan. 1922, Hoechst, 12/255/1.

29. "Vermerk über ein Besuch des Herrn Dr. Potthoff beim Herrn Staatssekretär," 13 Dec. 1921, BAP, 10375, Bl. 138–39; RDI to Köttgen, 19 Jan. 1922, SAA, Lf352/11.

30. RDI (Dr. Herle) to intellectual property committee, 10 Feb. 1921, Hoechst, 12/255/1; Guggenheimer to RDI, 18 Feb. 1921; RDI, *Geschäftliche Mitteilungen* 3, no. 6 (Feb. 1921); correspondence of Köttgen and Herle, 2–3 Mar. 1921, NGug.

31. Georg J. Meyer, "Praktische Beispiele aus der Elektrotechnik zum Problem der Angestelltenerfindung," *Zfl* 15, no. 1/2 (15 Apr. 1921): 1–8; RDI to industrial property committee, 31 May 1921; RDI, *Geschäftliche Mitteilungen* 3, no. 16 (May 1921), supplement 3.

32. For summary of Fischer's *Betriebserfindungen*, see ch. 1.

33. Ludwig Fischer, *Bericht über meine dienstliche Tätigkeit 1899–1934* (unpublished, 225 pp. typescript in the patent offices of the Siemens Company, Franziskaner-str. 14, Munich; photocopy in possession of the author); for Fischer's publications, see Works Cited.

34. Köttgen to Guggenheimer, 16 Sep. 1921, NGug.

35. Compilations of reviews and reactions in Fischer to Köttgen, 6 and 21–22 Oct. 1922, SAA, 11/Lf364.

36. Correspondence of Lauster, Gertung, Offenbacher, Buz, and Guggenheimer, 14 Nov. and 6 Dec. 1921, NGug.

37. Fischer, *Betriebserfindungen*, 4, 9, 12–13, 24–27, 59.

38. All quotations from Hermann Schmelzer, "Erfinder oder Naturkraftbinder?" *Der leitende Angestellte, Zeitschrift der Vereinigung der leitenden Angestellten in Handel und Industrie e.V.* 3 (1921), nos. 22–24 (15 Nov–30 Dec. 1921): 170–72, 177–81, 184–85, and 4 (1922), no. 1 (2 Jan. 1922): 5–6.

39. Alois Riedler, "Das deutsche Patentgesetz und die wissenschaftliche Hilfsmittel des Ingenieurs," *VDIZ* (1898), no. 48: 1319, quoted in Schmelzer, "Erfinder oder Naturkraftbinder," 178–9. Riedler argued the opposite case in a paid expert opinion for MAN in 1927: "For about a generation, and with ever increasing speed and in the midst of technologies highly developed with respect to both means and effects, inventions can only be achieved systematically and with scientific means—only on the basis of systematic planning, by experiments, usually large-scale experiments in company laboratories; and the result of innovations must be calculated with production-scale machines before they can be shipped to the customer. In the past the conceptualization of the invention as idea was the most important thing; today it is only a simple beginning and the easiest part of the achievement. The difficult, the laborious, and the expensive thing today is always the inventive idea's development. By itself, a 'good hunch,' a 'lucky throw of the dice,' is nowadays practically worthless without the ability and the experience to develop the invention technologically and economically until it is ready for use." A. Riedler, "Gutachten zum Patentstreit der Germania Werft A.G. in Kiel gegen die MAN in Augsburg betreffend die Nichtigkeit des DRP 367393," MAN (Augsburg), Nachlaß Lauster.

40. On Eyth: Conrad Matschoss, "Max Eyth zum hundertsten Geburtstag," *Abhandlungen und Berichte des Deutschen Museums* 8, no. 2 (1936): 29–41; Carl Weihe, *Max Eyth: Ein kurzgefasstes Lebensbild mit Auszügen aus seinen Schriften* (Berlin, 1922).

41. Max Eyth, "Zur Philosophie des Erfindens" (orig. pub. 1903), in his *Lebendige Kräfte, Sieben Vorträge aus dem Gebiete der Technik,* 4th ed. (Berlin, 1924), 240, 262ff., quoted in Schmelzer, "Erfinder oder Naturkraftbinder?" 179. Almost identical language by Adolf Hitler writing about the "culture-founding Aryan": "He is the Prometheus of mankind from whose bright forehead the divine spark of genius has sprung at all times, forever kindling anew that fire of knowledge which illumined the night of silent mysteries and thus caused man to climb the path to mastery over the other beings of this earth"; *Mein Kampf,* trans. Ralph Manheim (Boston, 1943), 290. For a modern version of Eyth's definition and theory of invention, consider the remarks by Walter Bruch, the 1962 inventor of the PAL color television system: "Inventions frequently are conceived as an unconscious flash of insight caused by nothing. Earlier I already spoke of the great leap—by which I mean the leap from one approach that prior logic said ought to work to another one. We are only capable of this creative intuition, which is frequently triggered by accidental circumstances, if we concentrate our focus on solving the problem." Bruch, "Erfinder," 317–24, esp. 323.

42. Schmelzer, "Erfinder oder Naturkraftentbinder?" 180–81. In 1925 Professor Konrad Engländer, a respected patent expert, made the same point about Fischer's argument with less inflammatory language. "It is not without provocative stimulation that one sees how ... the Marxist interpretation is represented, not by the economically weaker party of the employees, but rather by big business. For it is the best heritage of Karl Marx and Friedrich Engels that the entrepreneurs have incorporated into their conceptual treasures"; Engländer, *Angestelltenerfindung,* 20.

43. *Denkschrift zum Erfinderschutz, Sozialpolitische Schriften des Bundes Angestellter Chemiker und Ingenieure E.V.,* first series, no. 6 (1922).

44. *Denkschrift zum Erfinderschutz,* 10–11.

45. Ibid., 12–13, 43, 56.

46. "Das Betrieb als solcher kann nicht denken; was in ihm denkt, sind die Betriebsangehörigen. Also nicht der Betrieb, sondern seine Angehörigen erfinden," ibid., 13.

47. "Nicht die Masse erfindet und die Majorität organisiert und denkt, sondern in allem immer nur der einzelne Mensch, die Person." Quoted in Hans Frank, ed. *Nationalsozialistisches Handbuch für Recht und Gesetzgebung,* 2d ed. (Munich, 1935), 1032; cf. Hitler, *Mein Kampf,* 444–46. For similar parallels between Hitler's language and that of Schmelzer and BUDACI, see note 41.

48. On the numerous mutations of liberal, scientific, and romantic ideas into *völkisch* thought and Nazi ideology see Uwe Puschner, Walter Schmitz and Justus H. Ulbricht, eds., *Handbuch zur "Völkischen Bewegung" 1871–1918* (Munich, 1996), ix–xxvii; Rüdiger vom Bruch, "Wilhelminismus: Zum Wandel von Milieu und politischer Kultur," ibid., 2–21; Günter Hartung, "Völkischer Ideologie," ibid., 22–41; see also Beyerchen's discussion of Aryan physics in *Scientists under Hitler,* 123–40.

49. Hitler, *Mein Kampf,* 444–46.

50. "Inflation consensus," see Feldman, "Historian and the German Inflation," 386–99.

51. Cf. Larry Eugene Jones, "The Dying Middle: Weimar Germany and the Fragmentation of Bourgeois Politics," *CEH* 5, no. 1 (March 1972): 23–71; idem, *German Liberalism and the Dissolution of the Weimar Party System, 1918–1933* (Chapel Hill, 1988); Childers, *Nazi Voter.* On the concept of "working towards the

Führer," see Ian Kershaw, *Hitler 1889-1936: Hubris* (New York, 1999), 527-60, and idem, *Hitler 1936-1945: Nemesis* (New York, 2000), passim.

52. Offenbacher to AEG patent division, 27 May 1922, NGug.

53. Inventions not based on the experiences, suggestions, or means of the company and not made in the course of professional responsibilities but falling within the orbit of the company's business activities would belong to the employee, subject to the employer's right to claim them in exchange for special compensation. Offenbacher to Guggenheimer, 16 and 27 May, 9 Jun. 1922, NGug; Bericht über die Sitzung des Ausschusses für gewerblichen Rechtsschutz des RDI, 16 Jun. 1922, SAA, 11/Lf364.

54. ACID to Akademiker Ausschuß, 27 Jun., 5 Jul., 4 Aug. 1922, Weidlich to ACID, 8 Aug. 1922, Hoechst, 12/255/1.

55. Offenbacher to Guggenheimer, 13 Sep. 1922, confidential report of joint RDI-VDA meeting in Berlin, 13-15 Sep. 1922, 19 Sep. 1922, NGug.

56. RAM's Dr. Feig, 4 Oct. 1922, SAA, 11/Lf364.

57. Joint RDI-VDA position paper, 1 Dec. 1922, SAA, Lf11/391; drafts of prepared remarks by Weidlich and Köttgen, 28 and 31 Oct. at RAM; Köttgen to Arbeitsausschuß beim Reichsarbeitsministerium zur Bearbeitung des Arbeitsrechts, 17 Nov. 1922, SAA, 11/Lf364; Köttgen to Guggenheimer, 17-18 Nov. 1922, NGug.

58. Oppenheimer to Köttgen, 1 Nov. 1922, SAA, Lf11/364.

59. Company and service inventions were defined narrowly, free inventions broadly. The employee was entitled to a reward for a service invention if the profit it generated was disproportionate to the salary. The employer had a right to acquire "company-related free inventions" in exchange for "unalienable and appropriate compensation." Ergebnisse Besprechung Entwurfs (Dr. Potthoff) eines Allgemeinen Arbeitsvertragsgesetzes, 3. Fortsetzung der 3. Lesung im RAM durch die Unterausschüsse 1 und 2 in den Tagen vom 3. bis 6. Januar 1923, BAP, RAM 10378, Bl. 121-25.

60. Exchange between Köttgen and Weidlich, 13 and 15 Jan. 1923; Guggenheimer to Köttgen, 17 Jan. 1923; Fischer to Köttgen, 25 Jan. 1923; SAA, Lf11/364.

61. Köttgen to Oppenheimer, 7 Feb. 1923, italics underlined in original, SAA, Lf11/364.

62. Fischer to Köttgen, 15 Feb., 13 Apr. 1923, italics underlined in original, SAA, Lf11/364.

63. Heinz Potthoff, "Grundsätzliches zur Angestelltenerfindung," *Bundesblätter* 5, no. 3 (15 Mar. 1923): 17-18; "Nachwort zur Angestellten-Erfindung," ibid., no. 4 (15 Apr. 1923): 27-28. See also *Arbeitsrecht*, 1923, no. 2, 105-10, and no. 4, 233-35. Charles Steinmetz (1865-1923), chief engineer at General Electric, had a similar notion of corporate inventing being compatible with socialism; cf. Noble, *America by Design*, 42.

64. Correspondence of Guggenheimer and Offenbacher, 9, 16, 25 Jul. 1923, 25 Jan. 1924, 14 May 1925, NGug; the BUDACI also rejected Potthoff's proposals: "Das Sonderheft 'Angestelltenerfindung' der Zeitschrift 'Arbeitsrecht' vom November 1925," *Der angestellte Akademiker* 7, no. 12 (15 Dec. 1925), 84-88.

65. Köttgen to Weidlich, 4 May 1923, Oppenheimer to Köttgen, 14 May 1923, Fischer to Köttgen, 19 Jun. 1923, Köttgen to Oppenheimer, 20 Jun. 1923, SAA, 11/Lf364; RDI *Geschäftliche Mitteilungen* 6, no. 23 (1924): 6.

66. "Entwurf eines Allgemeinen Arbeitsvertragsgesetzes" (Paragraphs 122-31), *RAB*, 1923, no. 15 (1 Aug.), official part, 498-507.

67. Meeting 27 Nov. 1923, RDI, *Geschäftliche Mitteilungen* 6, no. 1 (Jan. 1924): 6.
68. Ibid.
69. Dr. Frank, meeting of RDI intellectual property law committee, 23 Nov. 1923, Hoechst, 12/255/1.
70. Boehringer's Dr. Koebner to ACID, 28 Dec. 1923, Bayer's Dr. Doermer to ACID, 5 Jan. 1924, Hoechst, 12/255/1.
71. On economic and political changes associated with currency stabilization, dismantling of the eight-hour day, and the collapse of the ZAG, in addition to literature cited in notes 4 and 17: Peukert, *Weimar Republic*, 107–46, 207–21; Gerald D. Feldman and Irmgard Steinisch, "Die Weimarer Republik zwischen Sozial- und Wirtschaftsstaat: Die Entscheidung gegen den Achtstundentag," *Archiv für Sozialgeschichte* 18 (1978): 353–439; Bernd Weisbrod, *Schwerindustrie in der Weimarer Republik* (Wuppertal, 1978).
72. Offenbacher to Guggenheimer, 11 May 1925, NGug; Fischer to Engländer, 22 Jan. 1925, SAA, 11/Lg709.
73. Numerous additional demands and conditions: *RAB*, 1924, no. 20 (8 Oct.), 495–99.
74. Weidlich to ACID, 12 Nov. 1924; Hoechst, 12/255/1; ACID to Akademiker Ausschuß, 16 Oct. 1924.
75. Minutes of meeting, ACID's Akademiker Ausschuß, 18 Nov. 1924, Hoechst, 12/255/1.
76. Minutes of meeting of ACID and BUDACI, 19 Nov. 1924, Hoechst, 12/255/1.

Chapter 4

COMPROMISE FOUND AND LOST, 1925–1929

*F*rom the end of 1923 until late 1925 the inventor issue largely disappears from the sources. The "regime change" of 1923/24 associated with the transition from inflation to deflation apparently caused a hiatus, as engineering labor was thrown on the defensive and the bill for a uniform labor code gathered dust in the archives of the Ministry of Labor.[1] Still, for reasons to be discussed shortly, inventor rights were back on the agenda by fall 1925.

During the interval, the stage on which the conflict took place had undergone several important changes. One of the leading hard-liners in industry's camp had departed forever. In June 1925, MAN director Emil Guggenheimer, chairman of the RDI's intellectual property law committee, died from heart failure. Guggenheimer had battled employee-inventor rights for over twenty years, and twice his efforts had paid off—in 1914 and again in 1923/24.[2] Guggenheimer's counterpart in the electrical industry, meanwhile, had begun to soften his stand and become interested in some kind of accommodation with the industrial scientists. Köttgen's change of heart came in the early phases of the "rationalization boom," in the wake of a 1924 visit to the United States to study American industrial efficiency. Considering its timing, Köttgen's new course is likely to have been inspired by the close relationship between engineers and management observable in the United States and a desire to emulate it for purposes of industrial recovery and productivity gains in Germany. As director of the National Productivity Board, Köttgen was a pacesetter in the postinflationary rationalization movement, and rationalization among other things held out the promise of more productive relations with labor. Another, related reason for working with the engineers was Köttgen's hope of forestalling separate inventor legislation, which the government was likely to introduce as it stepped into the void left by a weakened labor movement and increasingly intervened in management-labor relations.[3] Whatever the exact mix of his motives, Köttgen in the mid-1920s became an advocate of accommodation with industrial scientists and for the next several years worked hard to bring about an informal understanding in the inventor-rights issue.

The chemical industry witnessed a change as well. In December 1925, after years of increasingly close cooperation among the leading chemical firms, IG Farben came into being. Receiving a final impetus from the overcapacity and excess inventory built up during the inflation, the formation of IG Farben was itself an instance of rationalization as well as the cause of further rationalization inside the new firm. Intensified by the difficult business climate and "sick economy" of Weimar's middle period, the streamlining had important consequences for the inventor issue. It included a concerted effort to cut costs by simplifying the inventor-reward procedures spelled out in the Chemists' Contract, whose hollowing-out and subversion during the years 1926–31 fostered growing resentment among professional employees.[4]

Both Köttgen's new course and IG Farben's policies in the second half of the 1920s were responses to the altered economic climate after 1923. For all its appearances of stability, the economy entered a phase of troubles that lasted right up to the Great Depression and was a major factor in the collapse of the Weimar Republic.[5] In this "crisis before the crisis" rationalization became the shibboleth of German industrialists as well as the labor movement. But rationalization, a "catchword ... tantalizing in its ambiguity," meant different things to different people.[6] For some it was the road to a modern, American-style consumer society based on the latest technologies and economic well-being for broad layers of the population. For others it was primarily a way to increase productivity and efficiency, to get more from labor without having to pay for it and without investing much in new technological systems (or anything else, for that matter). Rationalization could also serve as a political strategy of obfuscation, a language of technocracy, modernity, and progressiveness, which made it possible to seize the moral high ground even as the claims of technological innovation far outdistanced the reality.[7]

The true extent of technological innovation that went on behind all the smoke remains something of a mystery, but there is little doubt that management tended to de-emphasize invention and the launching of new technologies in favor of cost cutting and efficiency in the manufacturing and expansion of existing systems. Köttgen's prescription for recovery as set forth in his 1925 study, *Economic America*, for instance, is mostly harder work and longer hours and investment in machinery to make existing technologies more efficient.[8] Important exceptions notwithstanding, Weimar's industrial leaders exhibited a lack of technological imagination—inadequate appreciation for the economic importance of what Thomas Hughes calls radical inventions or qualitative change.[9] Even when this was not the case, much of the new investment that did take place appears to have been misdirected and backfired.[10]

Organizational innovation, on the other hand, especially with regard to the use of labor, was extensive. The evidence from the politics of inventing, at any rate, indicates that management spent a great deal of energy trying to reduce the expense and friction caused by white-collar workers such as engineers, inventors, and industrial scientists. Not surprisingly, the latter felt threatened by this development. At best they were suspicious. Many complained bitterly about the neglect and abuse they suffered during this time and about their bosses' failure to pursue new technologies—technologies that might have stimulated economic growth and brought not only American efficiency, but also American affluence. Those complaints, in turn, and the dashed hopes of reaching the New World, became a key component in the larger discontent with Weimar, the insidious seductiveness of National Socialism, and the pro-inventor policies of the Third Reich.[11]

Demoralization, disillusionment, and susceptibility to Nazi solutions were the outcome in both the chemical industry and mechanical and electrical engineering. It is ironic that this should have been the case, for the two industries pursued very different versions of "rationalizing" the relationships with their technical staffs. Under Köttgen's leadership the electrical and mechanical industry pursued an integrationist strategy, while the chemical industry, led by IG Farben, adopted a more confrontational model. How those different routes converged on the same point is explored in this and the next chapter.

The 1928 Patent Agreement Between the RDI and the BUDACI

The inventor issue resurfaced in the fall of 1925, in the wake of Austria's 1925 patent reform, which included the "most extensive legislation concerning the salaried inventor question in the entire world."[12] The principle of inventor protection in Austria went back to the country's original, 1897 Patent Code, which adhered to the first-to-invent theory, did not recognize company inventions, and expressly prohibited patent agreements depriving employees of the "appropriate reward" for their inventions. The Austrian government after World War I had undertaken to amend the code for reasons unrelated to employee inventor rights. But the social-democratic League of Industrial Employees (Bund der Industrieangestellten), a sister organization of Germany's BUTAB, hoped to exploit the occasion to strengthen the rules against predatory patent agreements. This was necessary, according to the union, because the Austrian subsidiaries of German firms such as Siemens and AEG "enforced policies that deprived salaried inventors of every right." Following repeated breakdowns of management-employee negotiations and parliamentary blockage by the Social Democrats of any bill that did not include better inventor protection, the Austrian industrialists finally gave in.[13] The new legislation included almost everything that was on the German employees' wish list, from name recognition and statutory rewards to favorable definitions of free inventions and service inventions.[14]

The Austrian reforms hit German industry like a bolt from the blue. The chemical industrialists were caught completely off guard. Weidlich found it "astonishing there were no reports of any resistance by the Austrian employers." Not even the Soviet Union's 1924 Patent Code was as generous to employees. It was extremely unsettling, Weidlich observed, how "the most extreme demands concerning salaried-employee inventions had been realized in the new law, without regard for the question of whether they could be accommodated or realized in practice." It was imperative to prevent a similar outcome in Germany.[15] Managers at Siemens, too, worried about the threat posed by the Austrian law, which raised the "great danger that the German government, prompted by the Austrian arrangement, will introduce a bill in the Reichstag" that was inspired by the BUDACI.[16] The union's proposals, wrote Ludwig Fischer, "are so outrageous that any serious discussion about them is impossible. They are simply the demands of puerile brains, which do not know what they are doing."[17]

Not surprisingly, the pro-inventor side had a very different response. The Austrian legislation served as a call to arms, to which labor eagerly responded. The BUDACI dusted off its 1924 bill and returned to the offensive.[18] The Ministry of Labor began work on stand-alone employee-inventor legislation, independent of the languishing Labor Code bill of 1923.[19] Potthoff in November 1925 devoted a special issue of *Arbeitsrecht* to employee inventions, with a contribution of his own and others by Köttgen, Fischer, Dr. Knorr (the business manager and a former chairman of the BUDACI), and Karl Sohlich (BUTAB). The inventor problem was back.[20]

The renewed specter of legislation once again prompted the chemical industrialists to urge their counterparts in engineering to sit down with the employees. A collective-bargaining agreement in the engineering sector, they hoped, would obviate the need for legislation.[21] As IG Farben's Weidlich put it, he and his colleagues would do everything in their power "to work on the metal industrialists, to get them to extend their hand to the employees about a broad framework contract."[22] The effort seems to have paid off. In December 1925 the Federation of German Employer Associations authorized Köttgen to explore the feasibility of collective bargaining with employee-inventors in the engineering industries.[23]

Informal discussions between management and the industrial scientists got underway in early January 1926. Köttgen and Knorr were both on the board of the Green Society, the professional-scholarly society of patent experts that was preparing to write a report on employee-inventor legislation. It seems that at one of the meetings the two men talked briefly about exploring common ground. Knorr followed up the initial contact with a letter informing Köttgen that the BUDACI was ready to sit down with the employers.[24]

The engineering industrialists had meanwhile decided against a real collective-bargaining agreement.[25] True, a contract might forestall legislation, but the price was too high. A collective-bargaining agreement, "which would inevitably go beyond the subject of inventions, would have to be avoided at all costs," Köttgen pointed out.[26] His reasoning was the same as that in the early 1920s. The structural differences between the engineering sector and the chemical industry were too great, the number of employees to be covered too large and too diverse, and the volume of patented inventions too great. The financial cost of special compensation would be too high, and the overall management-labor relationship under a collective-bargaining arrangement too rigid.[27] Instead, a small RDI committee—a few key people well versed in patent matters and chaired by

Köttgen—should try for a more limited, informal, and voluntary patent agreement with the BUDACI. The two sides should try to find compromise language concerning inventor rights, which the RDI would then recommend for adoption by engineering companies in employees' individual contracts. The arrangement would have none of the disadvantages of a collective-bargaining agreement but still preempt the need for legislation.[28]

Köttgen outlined this plan to Knorr at a meeting of the Green Society in March 1926. Not surprisingly, the BUDACI leader was disappointed. His initial reaction was to reject the offer as meaningless, for reasons he spelled out in a long letter to Köttgen. Knorr's letter merits close attention. It opens a revealing window on the mentality of Weimar's industrial scientists and contains clues as to why they succumbed so easily to Hitler just a few years later.

Knorr began with assurances that his views represented those of other BUDACI leaders as well. It was "out of the question," he wrote, "that the employee side ... would want to have anything to do" with Köttgen's proposal. The RDI was not an authorized employer organization and therefore unable to enforce a voluntary patent understanding in industry. The agreement would remain a "scrap of paper" that might forestall legislation but would not help inventors. There was nothing in it for the employees. A similar agreement with an authorized employer association such as the Federation of Berlin Metal Industrialists, on the other hand, might work. It would carry sufficient weight that other employer groups would soon follow suit. A nationwide contract about inventor rights, in turn, could serve as the nucleus of a much broader collective-bargaining agreement, and the end result would be a nationwide metal contract (*Reichsmetalltarif*).

A nationwide metal contract for industrial scientists and academically certified engineers—that was the true goal and "crux of the entire question," Knorr explained. The higher-level employees, he wrote, had been fighting for years "to liberate themselves from the shackles of the labor contracts geared toward the lower categories of technicians that the employer associations habitually sign with the BUTAB." Academically certified professionals had "always considered it a degradation of their social standing that they fall under the contracts of an organization whose *Weltanschauung* offends them in the vast majority, and with which they have nothing in common." The idea of a single category of technical employees might have made sense years ago, but "in the course of time salaried professionals with academic certification developed a greater sense of their separate social identity and solidarity." They were entitled to separate negotiations with the employers just as much "as the federation of physicians, the association of attorneys, and other professional organizations." And if this was possible in the chemical industry, it could be done in the mechanical and electrical engineering industry as well. A "collaborative arrangement (*Arbeitsgemeinschaft*), which may bear fruit in all sorts of ways"—that was "the core idea of the academic professionals' desire for a collective bargaining partnership."

The fundamental question, Knorr continued, was "whether the profession of academically trained employees can maintain its independence as a vital link in the economy or is condemned to be assimilated by the mass of all workers and employees." His colleagues were "struggling to preserve their cultural and ideological autonomy and fighting against the leveling tendencies of the free trade unions, which seek to undermine their social standing and professional consciousness." If they did not prevail, Knorr warned, the professionals could easily "turn away in resignation." Defeatism and resentment, in turn, "would be bitter, not merely for the profession but also for industry." Possibly the industrialists could "not quite assess the danger as clearly as someone who has an opportunity to look more deeply into the mentality of colleagues who are naturally reticent when face to face with their supervisor and paymaster." Without specifying what the danger might be, Knorr once again pleaded for help in preserving the "profession's good traditions." The solution was for the industrialists to support "liberation from the influences of the free trade unions and, as visible proof, creation of the professionals' own labor contract."[29]

There is no record of Köttgen's reaction to Knorr's letter, but whatever it was, it did not cause him to change course. The chemical industrialists who read Knorr's letter all saw its logic, and Weidlich "left nothing untried" in talking the engineering employers into going forward with collective bargaining. The latter remained unconvinced.[30] Köttgen gave all sorts of reasons why a real contract was impossible, but the cost of inventor compensation was, "of course, the most important ... point."[31] If nonetheless discussion between the two sides did not collapse, culminating instead in a voluntary patent agreement in 1928, it was because a common threat unexpectedly brought them together.

In April 1926 the BUTAB published a plan for inventor-rights legislation that was more radical than anything seen thus far. Trumping the recent Austrian reforms, the BUTAB plan sought changes in both labor legislation and the Patent Code that would have given the employee extensive rights and reduced those of the company to the bare minimum.[32] The fact that the plan appeared in the *Reichsarbeitsblatt* suggested it might have the imprimatur of the Reich Labor Ministry. Suddenly, it looked to the employers, the risk of truly dangerous inventor legislation was very real again. In Köttgen's words, "we realized that discussion of the matter in the Reichstag ... might bring the most unpleasant surprises." Most lay people—which Köttgen thought included "the majority of the members of all the parties in the Reichstag"—had such primitive and stereotypical notions about inventing that a political settlement would likely "result in especially far-going considerations to the inventors." This was all the more probable because the subject was being "exploited as a most welcome source of agitation by the employee associations, especially those that otherwise do not care about employees who make inventions."[33]

The BUDACI felt threatened as well. Accustomed to being the dominant player in the inventor-rights issue, the industrial scientists suddenly faced competition from a group they considered their ideological enemy and social nemesis. If the BUTAB succeeded in reviving inventor legislation, it would inevitably doom the BUDACI's project of professional emancipation outlined in Knorr's letter. Keeping alive the chances of compromise with the employers was therefore essential. As Knorr put it during one of the meetings in the Green Society, "we [i.e., labor and management] must find a common front, and there is a common front, for both sides have a stake in inventing and its encouragement. If the employee side makes excessive demands, however, it saws off the branch on which it sits."[34]

Such talk and collaboration in demolishing the BUTAB's proposals in the meetings of the Green Society brought the BUDACI and management closer together.[35] In the second half of 1926 both sides looked with fresh eyes at the idea of a voluntary patent agreement, which might be mutually advantageous after all. Following a period of extensive preparation, exploratory talks began in March 1927. Since neither side had a firm mandate, however, progress was slow. Bargaining continued for many months and into the fall of 1928. Finally, in October 1928 the RDI and the BUDACI signed a voluntary agreement, "Regulations concerning Inventions and Rights of Protection" (*"Bestimmungen über Erfindungen und Schutzrechte"*).[36]

An elaborate and imperfect compromise, but quite possibly also a workable one, the 1928 Patent Agreement was a small example of how the Weimar Republic might have survived. The new understanding closely resembled the Chemists' Contract, although its provisions for inventor rewards—geared to engineering—were both less generous and more complicated. The main innovation was that special compensation should be a function of multiple factors, such as employee salary, company contribution, market potential, patent strength, and entrepreneurial risk, all of which were spelled out and took the place of the undifferentiated and elastic concept "appropriate compensation" in the Chemists' Contract. The new approach still left room for disagreement. But it moved in the direction of creating objective criteria for measuring the relative weight of the contributions by the employee and the company, which becomes the central issue in all employee inventions once it is acknowledged that such inventions typically are not about either-or but about both-and. The new approach was also what would eventually make possible elimination of the concept company invention altogether, without thereby destroying the basis of company control over employee inventions.[37]

If the 1928 Patent Agreement had been implemented as intended, it might have succeeded in lowering the tensions surrounding the inventor issue. The RDI cast the agreement in the best possible light and gave it the highest possible recommendation. The BUDACI published an upbeat editorial by Dr. Georg Baum, the union's top legal adviser, who praised the pact as an important step forward.[38] Not surprisingly, the BUTAB criticized it as a despicable sellout of employee rights and a "stab in the back."[39] But the VELA, which at first was critical also, soon came around and asked the RDI for a parallel understanding, which in fact was signed a year later, in November 1929.[40]

Acceptance of the 1928 agreement by the BUDACI and the VELA suggests it was not primarily the fault of these professional unions that inventor rights would remain a burning issue in the last phase of the Weimar Republic and the Third Reich. To be sure, the BUTAB remained unreconciled, but the principal reason the agreement failed to live up to its expectations was obstructionism from within the ranks of industry. Köttgen and his allies in the RDI desperately tried to save the pact, but other industrialists associated with heavy industry deliberately torpedoed it.

Problems developed almost immediately. The pact had been concluded by the RDI, an organization with no jurisdiction over labor relations and not powerful enough to make even a voluntary inventor agreement stick. Of course, there were good reasons that the RDI, rather than some other industry group, had negotiated the agreement on the employer side. The federation's lack of authority to negotiate a binding labor agreement was a good excuse for concluding a voluntary and informal agreement rather than a real labor contract, which had no chance to begin with. A nonbinding agreement also might appear harmless in the eyes of those opposed to any kind of accommodation with any part of labor.[41]

Another consideration was the subject's inherent difficulty. The inventor issue was so arcane that, apart from the chemical employers, most industry groups lacked the expertise and the inclination to deal with it. In the 1920s, the high-technology sectors of the German economy that depended for their survival on systematic inventing and innovation—chemicals, electricals, and parts of mechanical engineering—had not yet developed the clout they would acquire later on. For many other industries, especially the politically dominant heavy iron and steel industry in western Germany, cultivating good labor relations with professional employees was a low priority and even frivolous at a time when rolling back the welfare state with its high labor costs was at the top of the agenda.[42]

The Berlin-centered RDI, in contrast, was something of a stronghold of the electromechanical and other engineering and advanced manufacturing industries, for which intellectual property matters were of necessity both more familiar and more important. The Patent Office and more than three-quarters of all patent attorneys were in Berlin, the nerve center of Germany's intellectual property establishment. Siemens, which had perhaps the greatest stake in the inventor agreement at this time, was also based in Berlin. Its chief, Carl Friedrich von Siemens, was a vice-chairman of the RDI and an active supporter of Köttgen's initiatives. The RDI's appreciation of patent and inventor issues was further strengthened by the existence of its intellectual property law committee, and the fact that between 1925 and 1931 Carl Duisberg was the federation's chairman.

Since his days as an aniline-dye and pharmaceutical inventor at Bayer in the 1880s, IG Farben's most senior and visible executive had been deeply involved in the politics of inventing, including the 1891 Patent Code reform and the inventor-rights struggles before 1914. It was "with grave concern" that Duisberg in November 1928 urged the RDI membership "to take to heart the 'Regulations concerning Inventions and Rights of Protection' and energetically work to implement them." The chief reason, wrote Duisberg, was the RDI's "fear that the continual neglect of the issue in industrial circles, the continual lack of a comprehensive understanding between the parties, and the ceaseless agitation against industry's alleged exploitation of salaried inventors will one day result in a solution by way of legislation ..." Inevitably such legislation, uninformed by "practical experience ... would be unable to do justice to practical conditions and far more undesirable to industry than contractual arrangements based on practical experiences could ever be."[43]

Initial responses to appeals such as this were positive. Köttgen and the BUDACI had agreed on a number of exemptions to avoid difficulties in sectors, including heavy industry, where a patent agreement might be irrelevant or especially cumbersome. The chemical industry would stick with its own labor contract. But interest in the electrical and mechanical engineering branches, for which the understanding had been tailored, was high. Upon receiving word of the agreement, the Federation of the German Electro-Technical Industry (Zentralverband der Deutschen Elektrotechnischen Industrie, ZENDEI) immediately printed its own supply of the regulations and disseminated them to all its members. In March 1929, the ZENDEI voted to join the RDI in "urgently recommending" adoption of the regulations by all its member firms.[44] The machinery and instrumentation industries, the automobile and motorcycle manufacturers, the aircraft industry, watchmakers, and railway and streetcar builders all requested additional information and copies of the "Regulations." Other branches of industry, such as mining, ceramics, glass, natural rubber, paper, textiles, upholstery, and food were also interested. In addition, numerous individual firms, government ministries, chambers of commerce, publishers, and professional associations such as the Association of German Engineers (which Köttgen chaired), as well as employer associations such as the VDA and the VBM (both of which collaborated with the RDI in promoting the agreement), asked for copies. By the end of January 1929 the RDI had received 13,500 such requests, and had already distributed more than 25,000 copies of the regulations.[45]

Precisely how many firms and which industries actually implemented the agreement is difficult to say on the basis of the available evidence. There is no doubt, however, that implementation, though extensive, remained well below expectations. At the Siemens concern itself, where introduction of the regulations represented major progress (relative, at any rate, to the draconian patent agreement dating from the turn of the century), they were an overwhelming success. Carefully monitoring progress from the date of introduction in mid-March, by April the company reported an acceptance rate of 98.5 percent, representing more than 16,000 salaried employees who had signed at the two main subsidiaries, Siemens & Halske and Siemens-Schuckert Werke.

At least one other Berlin firm controlled by Siemens, however, did not sign, as it would have broken a labor contract with the BUTAB.[46] A July 1929 ZENDEI report to Köttgen showed the status of the agreement at ninety-one of its member firms. While fifty-two companies (57 percent) had introduced the regulations, thirty-two (35 percent) had refused. Seven other firms were still considering introduction. Some of the refusing companies explained they already had more generous inventor agreements, others indicated their hands were tied because of preexisting labor agreements, while the remainder rejected it because it was too intrusive or not applicable. Among those declining were the German subsidiary of Sweden's Electrolux in Berlin, which gave no reason, Robert Bosch in Stuttgart, Brown, Boveri & Cie. in Mannheim, and Schott in Jena, all because of existing collective bargaining contracts. Early fence sitters in Berlin included Telefunken, Osram, and Zeiss Ikon.[47]

Confronted by BUDACI officials about these and other problem cases at a meeting in early June 1929, Köttgen redoubled his efforts to make the agreement succeed, pleading with some of the industrialists he knew personally to adopt the regulations. In a letter to Count von Arco, head of Telefunken, Köttgen pointed out how he "personally ha[d] striven hard to bring about the new patent agreement, which does greater justice to the employees' aspirations..." Considering that both Siemens and AEG had decided to introduce the new regulations, he wrote, "it would indeed be appropriate if Telefunken were to do the same thing as well." To lend extra weight to this request, Köttgen got AEG's chief executive, Dr. Elfes, to write a similar note to Arco. Telefunken soon came around and joined, as did several other electromechanical companies.[48]

In other cases, however, Köttgen got nowhere. The response he received from Heinrich Cuntz, a senior executive at Krupp, concerning refusal of the Germania shipyard in Kiel to implement the agreement, was short but unambiguous. Krupp had taken its stand with the other Ruhr industrialists, Cuntz wrote, and "therefore declined to introduce this agreement." Germania, being one of Krupp's subsidiaries, had been instructed "to do the same thing."[49] The decision of the Ruhr industrialists not to honor the understanding proved fatal. They controlled a substantial part of the country's mechanical engineering industry, participation of which was vital to success of the agreement. Abstention caused precisely the situation Knorr had been afraid of in 1926, when he predicted the agreement would likely remain a "scrap of paper." The BUDACI saw its worst fears confirmed by the agreement's incomplete acceptance in the engineering industries. After a year of waiting and hoping for things to improve it gave up, notifying the RDI in October 1929 that it was walking away from the failed agreement.

The breakdown of cooperation between management and labor was a major defeat for Köttgen and his allies in the RDI. After working for three years to find common ground, their inventor strategy now lay in shambles. The Siemens director, however, could hardly have been too surprised by the outcome, given what he knew about the position of the Ruhr barons. As early as September 1928, when Köttgen reported to the RDI's executive council on the progress of the negotiations with the BUDACI, some of the Ruhr industrialists claimed they knew nothing about it and all this was news to them. Köttgen pointed out that Paul Reusch, one of the iron and steel industry's top leaders, had gone on record supporting the agreement; other representatives from heavy mechanical engineering, including Krupp director Preussing, had been deeply involved in the negotiations as well. But this did not satisfy directors Oskar Sempell of Vestag and Wolfgang Reuter of Demag. They were skeptical, wanted more information, and made no commitments.

Köttgen and his allies in the RDI realized they were being pushed into a corner but remained hopeful there was a way out. That the iron and steel producers themselves were unwilling to embrace the agreement was one thing, considering they were gearing up for their great labor battle, the Ruhr Lockout of November 1928. Köttgen thought he could handle that problem, in as much as the BUDACI might accept that heavy iron and steel were so different from engineering proper that the agreement did not apply to them. The engineering works controlled by heavy industry, however, would have to participate. Köttgen pleaded with Sempell to persuade his colleagues that they should make this concession, or else his negotiations with the BUDACI would "probably fall through." If that were to happen, Köttgen warned, "it would thwart our main purpose—to settle the matter by voluntary negotiations, primarily with the employee association that really does organize inventors, so we can refer to the settlement in future parliamentary debates that will come no matter what."[50]

Sempell replied that he and Director Helmuth Poensgen would try to help but could make no promises, since everyone in the west was preoccupied with the "impending wage battle in the Ruhr area." A meeting between Köttgen's team and a delegation of Ruhr employers planned for 23 October, just three days before the scheduled signing of the RDI-BUDACI understanding, failed to materialize when the iron and steel industrialists did not bother to show up. Given these signals, it is evident that, in truth, the inventor agreement had only the most precarious foundations and its chances of success were slim from the outset.

Realizing his plans were in trouble, Köttgen made a desperate final plea with the Ruhr industrialists, just three days after the RDI and the BUDACI signed their agreement. He understood that the industrialists in the west needed freedom of action in their war with labor, and he was prepared to exempt iron and steel. But the machine-building firms they controlled would have to be included, because without them the pact would remain a dead letter. He explained in detail how important the issue was for the engineering industries, how they depended for their survival on a steady stream of inventions and technological progress, how the electrical and chemical industry had long been wrestling with the issue, how top-level industry federations such as the RDI and the VDA had become supporters, and how much of his own energy and life he had devoted to the problem over the past nine years.[51] Neither this, nor further prodding of Krupp's Cuntz in June 1929, produced any result. The Ruhr industrialists had unanimously decided to defer introduction of the agreement. "Existing agreements about inventor rights," they said, "ha[d] so far not yet given rise to difficulties."[52]

In truth, the difficulties proved insurmountable. On 29 October 1929, one day after the crash on Wall Street, BUDACI leaders Drs. Konrad Ilberg and Hugo Kretschmar informed the RDI that the union was walking away from the inventor agreement. The occasion was the VELA's quest for inclusion in the agreement. For months, Ilberg and Kretschmar explained, the BUDACI had resisted such inclusion because it did not want to see a rival "become the effortless beneficiary" of its labors. This view stemmed from a belief that the agreement represented an "important step forward in the field of salaried inventor rights outside the chemical industry." But after a whole year had passed there was no escaping the "regretful conclusion that the RDI had failed to demonstrate a majority of its members were willing to implement the agreement …" Ilberg and Kretschmar recognized that Köttgen and certain others had "striven most loyally" to make the understanding a success. But "disappointment over the negative attitude on the part of larger and leading firms" had already done its corrosive work, giving rise among the BUDACI's membership to a somber feeling of "dissociation from the agreement …" Having come to the "conclusion that this patent agreement in truth exists on paper only," the

BUDACI no longer wished to block a parallel understanding between the RDI and VELA.[53]

The RDI's business managers Herle and Döring pleaded with the BUDACI not to give up hope so soon. A year was much too short to judge whether the agreement was a success or a failure, considering how difficult it was to change ingrained habits and bring about gradual reform on a voluntary basis.[54] Perhaps the RDI officials were right, but it comes as no surprise that the industrial scientists ignored a request whose timing could hardly have been worse, and which must have struck the BUDACI leadership as more than a little ironic. Nor do the records show that anything else happened with respect to the inventor agreement during the next few years—years that brought with them a wholly new political constellation and different priorities once again. And so with this discouraging exchange of letters in the fall of 1929 did the episode of the Weimar Republic's would-be compromise over the rights of salaried inventors in the engineering industries come to an end.

Notes

1. See ch. 3. On the changing balance of power following currency stabilization, see Feldman, *Great Disorder*, 754–858; Peukert, *Weimar Republic*, 107–46; Childers, "Inflation, Stabilization, and Political Realignment"; Weisbrod, *Schwerindustrie*; Maier, *Recasting Bourgeois Europe*, 440–58, 481–94, 510–15, 580–94.

2. Wolf Weigand, "Emil Guggenheimer (1860–1925), Geheimer Justizrat," in *Geschichte und Kultur der Juden in Bayern. Lebensläufe* (Munich, 1988), 195–201; Guggenheimer's anti-inventor activities before 1914: Gispen, *New Profession*, 246–97.

3. For this context, see Peukert, *Weimar Republic*, 107–28, esp. 112–13; for "rationalization boom": Nolan, *Visions of Modernity*, passim, esp. 132; for relations between engineers and management in the U.S.: Noble, *America by Design*; Carl Köttgen, *Das wirtschaftliche Amerika* (Berlin, 1925).

4. On IG Farben: Gottfried Plumpe, *Die I.G. Farbenindustrie AG: Wirtschaft, Technik und Politik 1904–1945* (Berlin, 1990), esp. 47–77, 200–432, 591–613; Peter Hayes, *Industry and Ideology: IG Farben in the Nazi Era* (Cambridge, 1987); Raymond G. Stokes, *Divide and Prosper: The Heirs of I.G. Farben under Allied Authority 1945–1951* (Berkeley, 1988), esp. 3–33; "sick economy" and political climate: James: *German Slump*; Peukert, *Weimar Republic*, 112–17, 122; Nolan, *Visions of Modernity*, pt. 2; Borchardt, "Constraints"; idem, "Economic Causes of the Collapse of the Weimar Republic," in his *Perspectives on Modern German Economic*

History and Policy (Cambridge, 1991), 143–183; von Kruedener, *Economic Crisis*; Kershaw, *Why Did German Democracy Fail?*

5. See notes 1 and 4.

6. Peukert, *Weimar Republic*, 113.

7. On various uses and meaning of "rationalization": James: *German Slump*, 145–61, 420; Nolan, *Visions of Modernity*, 133–226, esp. 153; Peukert, *Weimar Republic*, 112–17; Shearer, "Talking about Efficiency"; Charles S. Maier, "Between Taylorism and Technocracy: European Ideologies and the Vision of Industrial Productivity in the 1920s," *JCH* 5, no. 2 (1970): 27–61.

8. Köttgen, *wirtschaftliches Amerika*, esp. 46–51, 70–71.

9. Radkau, *Technik*, 269–84, 299–312; Hughes, *American Genesis*, 53–54, 180–83. For discussions of sectoral technological change in the 1920s and 1930s in Germany: articles by Harm G. Schröter, Hans-Joachim Braun and David Edgerton, Gottfried Plumpe, Bernd Dornseifer, and Paul Erker in *Innovations in the European Economy Between the Wars*, ed. François Caron, Paul Erker, and Wolfram Fischer (Berlin, 1995).

10. Perhaps the most spectacular case of misguided technological choice was IG Farben's project to develop synthetic fuel from coal: Hayes, *Industry and Ideology*; Plumpe, *I.G. Farbenindustrie*; Stokes, *Divide and Prosper*; idem, *Opting for Oil: The Political Economy of Technological Change in the West German Chemical Industry, 1945–1961* (Cambridge, 1994).

11. Cf. Peukert, *Weimar Republic*, 115–17. On fascination with the United States: Nolan, *Visions of Modernity*, passim. Hopes of emulating American technological and economic innovations should not be confused with uncritically embracing American civilization as such. Ambivalences about America and anti-Americanism were too powerful for that; see Dan Diner, *America in the Eyes of the Germans: An Essay on Anti-Americanism* (Princeton, 1996), esp. 53–77.

12. Herbert Axster, "Die österreichische Patentgesetz-Novelle von 1925 und die Regelung der Angestellten-Erfindung," *GRUR* 30 (Oct. 1925): 270–72.

13. Quoted in "Das österreichische Erfinderrecht und seine Bedeutung für die deutschen Verhältnisse," *Der angestellte Akademiker* 7, no. 12 (15 Dec. 1925), 81.

14. Ibid.; Axster, "österreichische Patentgesetz-Novelle."

15. Minutes of IG Farben patent committee meeting, 3 Sep. 1925 in Nuremberg, Bayer, 23/2.3.

16. Fischer, "Bericht," 178–80.

17. Fischer to Engländer, 24 Oct. 1925, SAA, 11/Lg709.

18. "Österreichisches Erfinderrecht."

19. Süssmuth to Weidlich, 23 Jul. 1925, Hoechst, 12/255/1.

20. "Österreichisches Erfinderrecht"; Potthoff to Köttgen, 14 Jul. 1925, SAA, 11/Lf391; "Das Sonderheft 'Angestelltenerfindung' der Zeitschrift 'Arbeitsrecht' vom November 1925," *Der angestellte Akademiker* 7, no. 12 (15 Dec. 1925): 84–88.

21. Süssmuth to Weidlich, 23 Jul. 1925, Hoechst, 12/255/1.

22. Weidlich to Süssmuth, 4 Aug. 1925, Hoechst, 12/255/1.

23. Herle and Döring to VDA, 27 Oct. 1928, SAA, Lf11/237/238.

24. RDI to Köttgen, 20 Nov. 1925, SAA, 11/Lf391; Knorr to Köttgen, 8 Jan. 1926, SAA, 11/Lf237/238. "Green Society" was the informal name of the German Association for the Protection of Intellectual Property (Deutscher Verein für den Schutz des gewerblichen Eigentums, DVSGE), whose publications had green covers.

25. Herle and Döring to VDA, 27 Oct. 1928, SAA, Lf11/237/238.
26. Köttgen to Arbeitnordwest, 29 Oct. 1928, p. 7, SAA, Lf11/237/238.
27. Ibid., pp. 5–7. Curschmann to Weidlich, 10 Jun. 1926, Hoechst, 12/255/1; BUDACI answers to ILO questionnaire, pp. 17–19, in ACID to Akademikerausschuß, Hoechst, 12/255/1.
28. Herle and Döring to VDA, 27 Oct. 1928. Besides Köttgen, the RDI negotiating committee included Jakob Herle and Dr. Döring, Stephan Oppenheimer, Dipl.-Ing. Free, Dr. von Bonin, and Dr. von Wienskowski, SAA, Lf11/237/238.
29. Knorr to Köttgen, 19 Apr. 1926, in ACID to Akademikerausschuß, 8 Jun. 1926, Hoechst, 12/255/1. Knorr's contract proposal followed the outlines of the Chemists' Contract. Salaries and other forms of direct compensation were excluded. The union sought uniform regulations about hiring and firing, hours of work, vacation, penalties for breach of contract, publications, no-compete agreements, and the rights of inventors. It also wanted a self-governing corporation of industrial scientists (Kammer), which would take the place of state-controlled arbitration and the courts, to resolve disputes between employers and professional employees. It demanded a "professionals committee," which existed in the chemical industry and was responsible for management relations with academically trained employees.
30. Curschmann to Weidlich, 10 Jun. 1926, Weidlich to Akademikerausschuß, 19 Jun. 1926; Staubach to ACID, 28 Jun. 1926; Doermer to ACID and Akademikerausschuß, 29 Jun. 1926, Hoechst, 12/255/1.
31. Curschmann to Weidlich, 10 Jun. 1926; Weidlich to Akademikerausschuß, 19 Jun. 1926; Köttgen to Arbeitnordwest, 19 Oct. 1928, p. 4, SAA, 11/Lf237/238.
32. Die Neuregelung des Erfinderrechts der Dienstnehmer: Gegenentwurf des Bundes der technischen Angestellten und Beamten zu den §§ 121 bis 131 des vom Arbeitsrechtsausschuß aufgestellten Entwurfes eines Allgemeinen Arbeitsvertragsgesetzes (28. Sonderheft zum Reichsarbeitsblatt, 2. Stück) nebst Begründung (Berlin, 1926), 3–12. Minutes of meetings of the Green Society, 21 and 27 Apr. 1926, SAA, 11/Lf391.
33. Köttgen to Arbeitnordwest, 19 Oct. 1928, pp. 6–7, SAA, 11/Lf237/238.
34. Minutes of Green Society meeting, 27 Apr. 1926, SAA, 11/Lf391.
35. Decisions of 24 Jul. 1926, BAK, 131/162.
36. Herle and Döring to VDA, 27 Oct. 1928. RDI negotiators: note 28; BUDACI negotiators: Dr. Gerichten, Dr. Kurt Milde, attorney Dr. Georg Baum, Dr. Ilberg, Dipl.-Ing. Heilborn, and Dr. Gallus, RDI to BUDACI, 27 Oct. 1928, SAA, 11/Lf237/238; Fischer, "Bericht," 181–85.
37. Bestimmungen über Erfindungen und Schutzrechte, SAA, 11/Lf237/238; Fischer, "Bericht," 176–77.
38. Georg Baum, "Die rechtspolitische Bedeutung des Erfinderreverses," Der angestellte Akademiker technisch-naturwissenschaftlicher Berufe, Nov. 15, 1928; supportive letters by RDI: 30 Oct. and 14 Nov. 1928, SAA, 11/Lf237/238.
39. Fritz Pfirrmann, "Ein neuer Erfinderrevers," Deutsche Techniker-Zeitung 11, 1 (4 Jan. 1929): 1–5; also, anonymous, "Verrat am Erfinderrecht," Ibid., 11, 12 (22 Mar. 1929): 136.
40. Herle and Döring to Köttgen, 20 Dec. 1929, SAA, 11/Lf237/238; BUTAB and DWV to VBM, 11 Mar. 1929 (the BUTAB's demands for parallel negotiations on the basis of its 1926 bill "almost had the effect of an insult" and were "completely unacceptable," according to Fischer in May 1929), SAA, 11/Lf237/238.
41. Köttgen to Arbeitnordwest, 29 Oct. 1928, pp. 7–8, SAA, 11/Lf237/238.

42. Weisbrod, *Schwerindustrie*; Peukert, *Weimar Republic*, 112–14; James minimizes the significance of political differences between older and newer industries, *German Slump*, 162–69.
43. Duisberg to RDI membership, 14 Nov. 1928, RWWA, HK Duisburg, 20-874-5.
44. Frese to Köttgen, 23 Mar. 1929; ZENDEI circular to member firms, 6 Apr. 1929, signed by Hans von Raumer, SAA, 11/Lf237/238.
45. Döring to Köttgen, 26 Jan. 1929, SAA, 11/Lf237/238.
46. "Betr. Patentrevers Stand 5.4.1929 12 Uhr," Köttgen to Free, 9 Apr. 1929, memoranda of Siemens's Sociopolitical Department, 10 and 24 May 1929, all SAA, 11/Lf237/238.
47. Frese to Köttgen, 31 May and 11 Jul. 1929, SAA, 11/Lf237/238.
48. Köttgen to Arco, 8 Jun. 1929; Köttgen to Elfes, 8 Jun. 1929; Elfes to Arco, 11 Jun. 1929; Frese to Köttgen, 11 Jul. 1929, SAA, 11/Lf237/238.
49. Köttgen to Cuntz, 11 Jun. 1929; Cuntz to Köttgen, 15 Jun. 1929, SAA, 11/Lf237/238.
50. Köttgen to Sempell, 5 Oct. 1928; Köttgen to Döring, 17 Sep. 1928; Arbeitnordwest to Köttgen, 1 Oct. 1928; Herle and Döring to Köttgen, 3 Oct. 1928; Köttgen to Cuntz, 17 Jun. 1929, SAA, 11/Lf237/238.
51. Köttgen to Arbeitnordwest, 29 Oct. 1928, SAA, 11/Lf237/238.
52. Köttgen to Cuntz, 17 Jun. 1929; Cuntz to Köttgen, 19 Jun. 1929; Grauert to Köttgen, 21 Jun. 1929, SAA, 11/Lf237/238.
53. BUDACI to RDI, 29 Oct. 1929, SAA, 11/Lf237/238.
54. RDI to BUDACI, 9 Nov. 1929, ibid.

Chapter 5

RATIONALIZATION, NATIONAL SOCIALISM, AND INVENTORS AT IG FARBEN, 1925–1933

*E*ven as Köttgen was trying to build bridges to the industrial scientists, an opposite trend was underway in the chemical industry. In November 1925, shortly before IG Farben came into official existence, the nascent concern's powerful Technical Committee had already decided to explore the possibility of rationalizing the employment contracts of its scientists. The initiative came from senior managers at the dominant BASF unit in Ludwigshafen, especially Professor Kurt H. Meyer, who sought to extend BASF's lean remuneration system for professionals to the entire IG.[1]

The crux of BASF's payroll efficiency was that, unlike most other IG divisions, it did not calculate special inventor compensation. The Chemists' Contract called for special, individualized agreements between company and inventor that stipulated the amount of compensation for each patented invention (typically 5 per cent of the net profit generated). Firms such as Hoechst and Leverkusen scrupulously adhered to this method, which built on their own traditions of profit-sharing incentives dating from the nineteenth century. The royalties thus earned constituted one of three components that made up the basic compensation package for scientists at Hoechst and Leverkusen. The other two elements were the annual salary and a special bonus.[2]

BASF had never adopted the system of special inventor compensation, which required a great deal of extra paperwork and was expensive to administer. Instead it used a simpler, two-part remuneration scheme: the basic salary and the bonus. The company told its scientists that the bonus included any and all rewards for inventive activity stipulated in the Chemists' Contract, but it neither broke out nor calculated specific amounts for individual inventions. This was the system that formed the basis of a draft contract BASF began circulating among all of IG Farben's units in December 1925.

Initial reaction of managers at the concern's various units was mixed. Some were lukewarm, others ignored the BASF proposal altogether. In February 1926, the plan went for further study to a committee of four senior executives representing the concern's constituent blocks: BASF's Meyer, Hoechst's Weidlich, Bayer's Karl Krekeler, and Agfa's Fritz Curschmann. Curschmann and managers at some other IG units were favorably inclined, but Weidlich and Krekeler opposed the plan. Krekeler explained that Bayer's scientists typically had three- to five-year contracts that guaranteed them the special royalties. Changing this would cause the greatest difficulties, especially because the royalties ran for the duration of the patent and were inheritable. The BASF bonus, in contrast, was annually renewable and applied to the employee only. If it were at all possible, Bayer (or Leverkusen, in IG parlance) would prefer to keep its old system, which had done so much to spur inventiveness and which management and professional employees alike considered a most valuable institution.[3]

Weidlich and his associate, Dr. Landmann, were even more hostile. Explaining the problem to Director August von Knieriem in May 1927, Weidlich recounted the issue's long history and acknowledged the administrative burdens involved, but then repeated Krekeler's arguments and pointed out that if ever challenged in court, BASF's system would never stand up. If nonetheless it was decided to adopt the system, IG Farben should be prepared to cancel the Chemists' Contract and see its scientists support a much more radical, legislative solution. Landmann in an internal review of early February 1928 repeated many of the same points and concluded there was "no reason to change even a single iota in our employment contracts for professionals."[4]

Despite such high-level opposition, BASF's managers were unwilling to give up. The merger made the "standardization of the incomes of professionals at the different works desirable if not necessary." This was possible only if the concern operated with just one method of inventor compensation—the BASF system. As a concession the Ludwigshafen managers were willing to phase in the new system so that currently employed scientists would be exempt.

BASF had support from several other IG units. Griesheim Elektron in Frankfurt a. M. favored the plan, because it eliminated the complicated calculations needed to determine inventor compensation based on profit sharing. Curschmann, in charge of the Wolfen facilities in central Germany, took the same position. Bayer's Justizrat Doermer, who by summer 1927 had assumed responsibility for the issue from Krekeler, also was somewhat more receptive to BASF's arguments than his predecessor.[5]

The upshot of these debates was an agreement in August 1927 that complete standardization was not feasible and would be dropped. Instead, IG Farben's various units would be free to choose their own language and method for implementing a few core decisions, one of which was gradually to simplify and reduce the cost of inventor compensation. Leverkusen would take the lead. In the fall of 1927 its lawyers went to work, designing a new compensation scheme that approximated the BASF system. Inventors would no longer get their royalties but only the bonus, to be paid in monthly installments and based on their preceding year's productivity as determined by their laboratory supervisor. Agfa began phasing in a comparable system at the Wolfen and Berlin locations in December 1927.[6]

Protests by Leverkusen scientists resulted in a modification of their new compensation scheme: it now showed in writing which parts of the bonus were based on which inventions.[7] This did not alter the fact that company accounting had been simplified and the inventors suffered a setback. Starting in early 1928, professional employees at a number of IG locations, especially in central Germany, began noticing similar trimming of their compensation packages. Weidlich seized on the tensions generated by the difficult contract renegotiations at Bayer as the reason for refusing to implement the new scheme at Hoechst. Professor Paul Duden, Hoechst's top manager, fully supported Weidlich's decision. Likewise, Director Hagemann of the concern's Mainkur unit (the former Casella & Co.) counseled against moving forward, in part because his BUDACI local had refused to go along with the new contract, on the grounds that it contravened the Chemists' Contract. In fact, reported Hagemann, the BUDACI was in uproar not only at Mainkur, but also at Wolfen and Treptow in Berlin. All three locals were demanding that the concern's top executive, Carl Bosch, get involved to decide the matter. Until then their members would not sign anything.[8]

It is unclear how the incident was resolved. By July 1928, however, scientists at IG Farben's laboratories in Leverkusen, Frankfurt,

Offenbach, Griesheim, Bitterfeld, and Wolfen began migrating to one version or another of a simplified and reduced compensation system. The BUDACI local at Mainkur, too, had reluctantly accepted the inevitable and could not prevent its members from changing over to the new scheme as their old contracts expired.[9]

Inventor compensation was only one of several ways in which the employment conditions of salaried professionals at IG Farben were rationalized during the second half of the 1920s. Increasing use of standardized salary scales, an overhaul and tightening of no-compete agreements, longer working hours, shorter vacations, lower retirement benefits, revaluation of company-financed mortgages, and a variety of other measures and approaches already in use for nonacademic employees came to define the new policy of IG Farben's management toward professionals, which was facilitated by the oversupply of most specialties of chemists and engineers.[10] The onset of the Great Depression after 1929 only accelerated the trend. The Professionals Committee, which had traditionally handled employment matters for employees with higher education, was relieved of its responsibility in May 1930. Its tasks were transferred to the more powerful Social Committee, which was chaired by Carl Bosch's right-hand man Ernst Schwarz and already supervised all other personnel and social-policy matters. The massive increase this caused in the anxieties and resentments of industrial scientists was only compounded by IG Farben's decision to raise dividend payouts to its shareholders in the spring of 1930.[11]

Even though it was only one aspect of a larger rationalization of their employment conditions, the inventor-rights issue remained among the most sensitive barometers by which industrial scientists measured their status as professionals. It is therefore not surprising that in June 1930 the BUDACI filed a formal complaint against the BASF scheme of inventor compensation, contending it was a violation of the terms of the Chemists' Contract. The prospect of open revolt in the laboratories and a breakdown in the orderly progress of research and development guaranteed that the union's complaint received the immediate and undivided attention of the concern's top management.[12]

A special meeting of 9 September 1930, attended by directors Schwarz, Curschmann, Doermer, Hagemann, Wilhelm Gaus, Fritz ter Meer, Erwin Selck, and others, decided to retain the BASF inventor-reward system. To disarm the BUDACI and avoid future accusations of contract breach, however, the executives agreed to modify the BASF system in one point. The new Leverkusen practice of indicating in each contract which part of the employee's bonus was based on which inventions would become standard practice. The amount in question constituted an entitlement that remained for the duration of the patent regardless of death, dismissal, or resignation.[13]

When the new rules were announced, only the scientists at Wolfen protested, demanding a return to the original inventor-compensation scheme. But they soon retreated and agreed to go along with whatever their BUDACI representatives at the next higher level decided. So it went across the IG. By early November, in the midst of soaring bankruptcy and unemployment rates, the union had withdrawn its complaint and grudgingly accepted the new system, which received management's final sanction in a meeting of the concern's all-powerful Technical Committee in November 1930.[14]

Cutting back on royalties was not the only change IG Farben made in the inventor-compensation system. A common technique for reducing the employee's reward was to designate the laboratory supervisor or another, highly paid scientist higher up the chain of command as co-inventor, thereby increasing the company's share in the invention. One had to be careful with this method, though, for sometimes it backfired.[15] To solve the problem, Director Gustav Pistor of the concern's Bitterfeld works in central Germany began tinkering with the definitions of company inventions and service inventions. Pistor, a highly accomplished scientist who has been described as the "father of industrial electro-chemistry," began contesting the long-established practice of defining most patented inventions as service inventions.[16] He sought to expand the scope of company inventions. Though some other IG managers advised him against it, Pistor went ahead and on one occasion questioned the service-invention status of a certain improvement in the development of synthetic fibers by a Dr. Karl Brodersen of the nearby Wolfen works. Pistor thought Brodersen's invention was not much and therefore deserved no special reward and ought to be classified as a company invention.[17]

There were several other incidents like this, and in April 1931 BUDACI secretary general Dr. Kurt Milde wrote IG Farben's management board in Frankfurt about the problem.[18] Frankfurt dismissed Milde's complaint, but when the BUDACI persisted and a follow-up letter by Dr. Gerichten mentioned specific examples at Bitterfeld and Leuna, Schwarz looked into the matter more closely.[19] The allegations were true, Schwarz noted, but he added that the employees in question had willingly accepted management's decisions and did not support their union headquarters' meddling. In fact, wrote Schwarz to chief IG lawyer Selck, the affected employees in a follow-up interview had "strenuously protested against the Bund headquarters making an issue of the matter, since it had no mandate to that effect whatsoever."[20] Schwarz probably spoke the truth, for what options did the company's industrial scientists have in the midst of the Great Depression?[21]

A partial answer may be found in the subsequent career of Wolfen inventor Brodersen. Aided by one of the German Labor Front's most aggressive inventor counselors, Brodersen in the Third Reich sued his employer for RM 1.5 million, on the grounds that the company had deprived him of his inventor reward. It had refused to market his crucial invention, using the patent only to block technological progress. Though in the end he received only a very small settlement, in 1937 Brodersen was able to join the Nazi party and continue his vendetta. In 1941 he became a member of the Labor Front's Reich Working Group for Inventing, which advised Nazi party functionaries in drafting the regime's compulsory inventor-compensation regulations of 1942 and 1943.[22]

Such unforeseen—though not altogether unforeseeable—consequences of management's inventor policies during Weimar were the subject of a retrospective assessment in 1948 by Dr. Harald Mediger, himself a former Nazi party member and IG Farben manager.[23] Mediger, who had received his doctorate in chemistry from the Technische Hochschule Dresden and joined the Wolfen Film and Synthetic Silk Company in Bitterfeld in January 1923 at age twenty-five, noted that before 1914 industrial scientists could still get rich from their inventor rewards. This changed after World War I, when the "number of inventors, the sum total of patent awards and the taxes mounted steadily, but the number of employee inventors becoming affluent or even wealthy went down steeply." Starting in the 1920s, Mediger observed, there was a "growing tendency among the entrepreneurs to move away from profit sharing compensation and royalties based on profits or sales." Instead, management increasingly imposed profit-independent inventor compensation—a change "prompted mostly by the onset of the economic crisis and the cost-cutting it forced ..." In actual fact, the new policy had not originated at the time of the Great Depression, but rather in the mid-1920s context of rationalization. Even so, Mediger was quite right when he continued that management, at that time in the grip of an "unfortunate intellectual vogue, had applied much more firepower than was necessary and used [economic necessity] as a pretext for squeezing also the qualified employees—those who form the

backbone of every enterprise—into the drab army of an undifferentiated worker mass, supposedly to reduce costs." If this sounds as
though it might have come from the mouth of a BUDACI activist, it
is because Mediger was precisely that in the 1920s, serving as one
of his union's negotiators when Wolfen management started phasing in the new inventor rules in late 1927.[24]

Mediger's BUDACI affiliation apparently did not hurt him,
because subsequently he was rapidly promoted, becoming a department chairman in 1931, receiving another promotion the next year,
and acquiring full power of attorney by 1933. By 1938, Mediger,
having meanwhile earned an additional doctorate in patent law, was
chief of the patent section of the concern's entire product division
for cellulose, artificial fibers, explosives, and photographic products,
the so-called Sparte III. In that capacity he not only supervised all of
IG Farben's American patenting in the photo and rayon fields, but
also played a role in shaping and implementing the Nazi regime's
various inventor measures.

By 1948, Mediger was a displaced person without money or a
job, having been evacuated by the Americans to Munich when the
Wolfen installations in central Germany were turned over to the
Soviets. Although his postwar allegiances in patent affairs were still
firmly in the managerial camp, Mediger had enough experience and
historical insight to note that in the 1920s, "incomprehensibly, substantial assets of imponderables [goodwill] were squandered away."
Management, he wrote, had "in places induced long-term crises of
confidence in the relationship between companies and their professional employees and provoked a radicalization of these employees,
which in the following years and in the era of the German Labor
Front may have cost a multiple of what at the time one believed to
have saved."[25]

If anyone was in a position to know, surely it was Harald Mediger. His comments are a rare example of an insider's admission that,
indeed, the festering sore of Weimar's inventor relations had led
directly to the Nazi regime's pro-inventor cures. To make this point
is not to argue that corporate inventor policy was also responsible
for turning all engineers and industrial scientists into avid followers
of Hitler before 1933. Nor was that the case. On the contrary, the
available scholarship on the Nazi proclivities of industrial scientists
and engineers indicates that in the aggregate they were no more
prone to join the NSDAP than other professions—that is, in relatively small numbers.[26] Even so, the inventor issue clearly played a
role in amplifying the group's elective affinities and insidious paral-

lels with National Socialism, or even in constructing from the ground up a cultural climate and attitudes that weakened the inner resistance to Hitler of most, and made true believers of some.[27]

The seamless intertwining of industrial scientists' anxiety over corporate rationalization, their perception of technological and economic mismanagement, and the growth of National Socialist sentiment is manifest in records of the BUDACI's Hoechst chapter from the late 1920s and early 1930s. The exact number of industrial scientists at Hoechst at that time is not clear, but a 1922 BUDACI petition seeking reinstatement of the annual bonus had some 240 signatures, which likely included most of the company's industrial scientists who were not managers. Among the signatories there were some seventy-five names that appear in a union leadership capacity between 1922 and 1932. The documents say little or nothing about the professional careers and political affiliation of any of these individuals. In the spring of 1933, however, the BUDACI's chapter was "synchronized" and fell into the hands of the Nazis, among them a certain Dr. Nicodemus. Nicodemus was one of the signatories of the 1922 petition but otherwise does not show up in any BUDACI leadership function. In 1933, he suddenly appeared as Hoechst's "chemistry cell leader." In that capacity he gave an address in June of the same year, in which he not only raved about the new regime's wholesome spirit and the need for obedience, but also described some of his party's organizational measures in and around Hoechst—specifically the Labor Front's initial steps in the synchronization of organized labor.[28]

All thirteen, formerly separate, associations for salaried technical employees in Germany, Nicodemus announced, had been consolidated into a new Federation of German Technicians (Deutscher Technikerverband, or DTV). The DTV's *Gauleiter* for Frankfurt, Wiesbaden, Mainz, and Giessen was a certain Dr. L. Mack. Mack had been elected first secretary of Hoechst's BUDACI chapter in 1929, become chairman in 1930, and been reelected for another term in 1931. (In those functions, the records show, Mack devoted a great deal of ink and energy to protesting the stream of givebacks management extracted from the industrial scientists.) The DTV's Frankfurt district was subdivided into several local groups, of which the one in Hoechst came under the leadership of *Diplom-Ingenieur* Schörg. Like Nicodemus and Mack, Schörg was a long-time Hoechst employee, and he had been elected to the BUDACI's position of third chairman in 1932. The DTV local group, in turn, had been subdivided into three branch cells: one for mechanical engineers and technicians, under Schörg; one for chemistry, under Nicodemus himself; and one for civil engineers and architects.

Nicodemus had organized his chemistry cell into smaller factory blocks or departmental blocks (*Betriebszellen*), each of which would serve as a "germ cell" of the regime's new, organic-corporate socioeconomic structure. Nicodemus had established thirteen such departmental blocks, each led by a head block leader and an alternate. Of the twenty-three scientists and engineers he mentioned in this context, two held managerial positions. Seven of the remaining twenty-one head block leaders, however, were among the seventy-five individuals who had served in a BUDACI leadership capacity between 1922 and 1932, and most of the new Nazi leaders had already been employed in 1922. Lest his audience wonder why he had not appointed more "old members of the NSDAP," Nicodemus explained he had "intentionally not wanted to burden the old fighters of the movement, who have already done so much for us and are still overburdened with offices today." Altogether, Nicodemus's speech contains the names of nine "old fighters" and thirteen new Nazi officials who came from the ranks of the BUDACI.

Obviously, the Nicodemus report is too sketchy to determine Nazi membership among industrial scientists at Hoechst before 1933. Still, it shows the link between socioeconomic frustration and political behavior. Resentment ran deep, and radical sentiments had been gaining the upper hand since 1929 at the latest, when the salaried inventors elected their first Nazi to the union's executive committee. The focus of such resentment and radicalism was a perception that management was on the wrong track, that it should do something other than cut salaries, inventor rewards, benefits, and ultimately the jobs of the scientists themselves.

Although industrial scientists became more outspoken as the Great Depression deepened, they seem to have experienced the economic crisis of 1929–32 not so much as a distinct episode but primarily as an intensification of the rationalization that had started in the mid-1920s. The local's correspondence with management throughout the entire Weimar period is one long litany of complaints about the socioeconomic conditions of the academically trained chemists and engineers. This was not merely about management's insufficient appreciation of their crucial role in the process of generating new technology and economic opportunities—issues ranging from inventor rewards, bonuses, salaries, and the effects of inflation to the quality of seating at IG-organized concerts, dinner invitations to the company restaurant, and release time to attend professional conferences. The union also complained about the intolerable harassment and baiting of scientists by blue-collar workers, about anonymous threats in the worker-published company newsletter, about physical intimidation, fisticuffs, and verbal insults by workers, and about the need for discipline to maintain the status of the academic elite within the factory. The evidence leads to the inescapable conclusion of a group that felt squeezed between two sides, dragged down by one and betrayed by the other. There never was a letup of such sentiments, but superimposed on them from 1926 was a growing concern with the problematic effects of rationalization and standardization. Combined with the preexisting discontent, this expressed itself largely as a critique of the establishment's technological and economic policies and an obsession with societal renewal through the liberation of inventive genius.

In December 1929 local chairman Dr. Wilhelm Müller and first secretary Dr. Mack submitted to Hoechst management an employee resolution. With only two votes against, the local had decided that the "recent reductions in workers and employees at the IG have nothing to do anymore with questions of economic rationalization, but are conditioned by failed policies of the concern's leadership." The union strenuously protested an announcement that Hoechst would be among the first IG units to suffer a "rigorous reduction also in the ranks of professionals," and found it "incomprehensible" that such "purportedly necessary dismissals" were being justified by "inadequate achievements by the colleagues in question," even though they had done their "work for years without giving cause for complaint."[29]

A few months later, in March 1930, the BUDACI's annual meeting in Berlin featured two plenary addresses by national leaders Kurt Milde and Hugo Kretzschmar. The arguments of both men centered on the idea that the sound core of the rationalization movement had degenerated into mindless personnel reductions and professional layoffs destroying the foundation of Germany's technological creativity.[30] Milde, a chemist and the BUDACI's secretary general, attacked the rationalization he saw everywhere around him as mindless "scrapping of intellect." A veritable "rationalization avalanche," would soon "destroy everything that at one time had constituted the foundation for the world reputation of German work and German production"—for the simple reason that nobody took the time to think through how many "values essential to the national economy are destroyed in the path of the avalanche."[31]

In a companion address, Kretzschmar called for better and more creative industrial leadership. There was a desperate need for precisely that "high degree of mental flexibility ... that in today's era of rationalization [led] a pitiful existence" on the margins of industry. The industrialists' contempt for technological creativity was evident in the pages of the *Handbook of Rationalization* published by Köttgen's National Productivity Board. If one opened this bible of the industrialists, said Kretzschmar, it was "impossible not to be gravely concerned for the purely intellectual component of technological progress."[32] To overcome stagnation and rigidity, industrialists should "liberate the human potential for creating productive values from the confining shackles of the rational methods—for how can the technological imagination flourish without intellectual stimulation, agility and nimbleness?" The anticompetitive combination of numerous industrial branches in many cases had "severely curtailed and even blocked the effect of technological creativity," Kretzschmar maintained. Quite apart from the "mechanistic tendency to seek profitability at any price," he continued, "the consolidation of industrial operations condemned many technical workers to an administrative routine that was no longer creative, required no initiative, and meant responsibility without joy." Monopoly also carried the risk of "reduced striving for technological excellence (*Mehrleistung*)."[33]

In June 1931 the BUDACI's Hoechst chapter submitted to Director Duden an eight-page, single-spaced paper by its "economics committee." "The Current Economic Situation and the Professionals" attacked the company's mindless rationalization, which was shown to have only deepened the crisis. The Hoechst scientists proved their point with a discussion about economic development under capitalism and offered a prescription for recovery that anticipated key elements of Nazi economic and technology policy. Influenced by theorists such as Joseph Schumpeter and Arthur Spiethoff, the essay's intellectual building blocks were the business cycle, rationalization, wage policy, the role of the entrepreneur as technological and economic innovator, and the phenomenon of the complex high-technology firm—with special emphasis on the latter's vulnerability to bureaucratic ossification and the role of salaried scientists and researchers in rejuvenating it through invention.

Following more general observations about the importance of stimulating domestic demand, the authors of the Hoechst paper came to the point: layoffs and salary cuts for industrial scientists were a terrible mistake. The fundamental error of management's policy, driven by the unimaginative mindset of accountants and other financial types, was to cut across the board, to "overlook the fundamentally different function of a specific circle of individuals" within the firm. The role of this "small circle [stood] in polar opposite to that of the mass" and differed from it as much as did "cause and effect." Just as earlier business successes had been "forged by the past labors of this small group, [so] the future fate of the company and that of the mass of its employees depend[ed] on the current work of this same group." The group in question comprised top management and their technological experts, the chemists and inventors who made it all possible.

The industrial scientists should be kept on and continue to receive their customary salaries and special incentives, because as inventors they formed an essential part of the entrepreneur's innovative function, needed to transcend the depression. "Precisely the current economic situation demands that we give it our all, especially now, in order to ensure that, besides preserving local production, we conceptualize and lay the groundwork for new combinations ... which will maintain the Hoechst facility as an essential pillar of the IG. In this crucial economic question the interests of our profession are identical with those of the employers." Since the industrial scientists themselves lacked the power to make the requisite decisions, it was up "to those gentlemen among the company's management who are close to us to mobilize our potential for the future upswing." Since the economy was at its nadir, the time to act was now. As Spiethoff had written, "the recession forces the introduction of cost-saving procedures and measures. Production facilities must be brought up to the latest technology and are to be perfected to the smallest details. In the same way, great inventions are a product of depressions, and their introduction is indispensable for the upswing. Capitalism depends on conquests, regardless whether material or spatial."

But instead of making conquests by investing in new technologies, the country's industrial leaders had opted for a misguided policy of rationalization, followed by ever-deeper cuts and ever more firings. The result had been a loss of confidence in capitalism and an increase in radicalism. The consequences could be seen in the Reichstag elections of September 1930, which had "resulted in a majority for such parties as have adopted 'socialist ideas' in their program." Could there be any "other economic interpretation of this expression of opinion," the BUDACI authors asked rhetorically, "than that a majority of the German people are ready to eliminate the capitalist form of economy?"

The authors emphatically disavowed that they, too, might advocate a socialist solution to the economic crisis. They made a sharp distinction between the bad and the good dimensions of capitalism, between its petrifying and destructive side associated with the iron cage of industrial bureaucracy and accounting greed, on the one hand, and its liberating and creative side associated with invention and technological innovation, on the other. "The economically blind bureaucracy of industrial capitalism," they wrote, "is the gravedigger of the type of economy that brought us the massive upswing of the last century and that today once again has shown its power to liberate from the misery of the postwar period." As the self-appointed exponents of this beneficent, technologically creative side of capitalism, Hoechst's industrial scientists concluded their essay with a final warning about the folly of further undermining their morale and ethos. "It is very dangerous to make special profits for a while at the expense of this ethos; in truth these profits are only the reverse of losses among the most valuable intellectual capital that the economy possesses."[34]

Industrial scientists at other IG Farben units articulated similar ideas. In November 1931 some five hundred technical professionals at Ludwigshafen and Oppau submitted a resolution protesting the latest announcement of impending layoffs and made detailed "recommendations for increasing the efficiency and profitability of the Ludwigshafen-Oppau factory community." Their very first point was an attack on the technologically stifling effects of existing laboratory organization. Many promising avenues remained unexplored, they wrote, "because the younger gentlemen had no opportunity to assert themselves against their various supervisors." To address the problem, these young scientists should not be fired, but given an opportunity to work toward the solution of their own or other promising problems. Not only would this preserve their jobs, but also "open up new areas of research for the company." They also urged improvement and innovation in a variety of technological processes, such as "finding improved or new production processes" for colorfast dyes and other products. They called for a systematic attack on the wasteful consumption of energy, and "reorganization of the Ludwigshafen facility according to a uniform plan to be drafted by the design offices" that would optimize energy use and capacity in accordance with "the most recent technology." Other proposals called for more research into the most efficient production of copper; expanding the domestic market for nitrogen fertilizers; better, American-style market research; more aggressive sales by changing accounting methods; and better utilization of inventors and other technical professionals.[35]

Yet another employee memorandum focused on the relationship between Germany's economic situation in the early 1930s and the preceding phase of rationalization. Rationalization had been forced upon Germany in an effort to underbid foreign competitors for the exports necessary to pay reparations. This had driven up unemployment as machines replaced workers, which in turn had reduced the "purchasing power of the home market." The high cost of investments in rationalized plants, therefore, had not had the desired effect. On the contrary, the increased volume of goods had remained largely unsold and absolute production costs had gone up rather than down. The fundamental "calculation error in the planning of rationalization measures" was to ignore that as late as 1930 "eighty-five percent of production still sold on the home market." Regardless, business leaders had stuck with their export orientation and "continued on this disastrous road, which, when taken to its logical culmination, must lead to the complete paralysis of the domestic market and with that the total collapse of the German economy." The situation called for "quick national self-help. Germany's will to live, or more prosaically, the will to strengthen the domestic market—purchasing power at home—must be victorious." The solution was to put an immediate stop to any further dismissals and salary cuts. Companies could strengthen domestic consumption if they eliminated all remaining dividends, premiums, and excessive compensation of executives, and used the savings to keep up employment. "We may turn and twist as we want, but we cannot escape the necessity of strengthening purchasing power, nor the first measure from which everything else must follow, namely arresting all further layoffs, to first bring the disastrous course of the economy to a halt."[36]

What is striking about all these papers is how far IG Farben's industrial scientists, who were thinking mostly about their own immediate interests, had ventured into territory to which the Nazis laid claim on the grounds that it represented the welfare of the national community as a whole. Their talk of autarky and the domestic market, of entrepreneurial-technological heroism, of the distinction between creative technological capital and unproductive finance capital were crucial elements also in Hitler's ideology and the language of Nazism. The industrial scientists were manifestly procapitalist in terms of entrepreneurial innovation but at the same time highly critical of management's financially driven decision making. This left the door open, first for the antisemitism that was only a small step away from there, and, second for a technological

activism that centered on the encouragement of invention and innovation freed from the shackles of private ownership, the latter being
perceived as either technologically stifling industrial bureaucracy or
finance-capitalist greed incarnate.[37] Both these potentialities, in
addition to the autarky implicit in the employees' advocacy of
domestic demand and purchasing power, would become the actual-
ities of economic and technology policy in the Third Reich.

There is no evidence of antisemitism in the BUDACI files at Hoechst
before January 1933. But in June of that year, when cell *Führer* Dr.
Nicodemus addressed the company's assembled chemists, that piece,
too, fell into place. Haranguing his colleagues with the party's corpo-
ratist ideology and the evil of shareholders' profits as "exploitation"
and "interest serfdom," Nicodemus proudly heralded the arrival of the
"frightfully difficult" but absolutely necessary "struggle against the
egotism of 'me first' (*Kampf gegen das eigene Ich*), against the Jewish
spirit in profession, economy, and capital..."[38]

Notes

1. Excerpt, minutes of Technical Committee (TEA) meeting, 26 Nov. 1925,
   Hoechst, 12/11/2; Koettgen to Weidlich, 30 Nov. 1927, Hoechst, 12/255/1.
2. Summary of employment contracts, correspondence, and memoranda;
   Hoechst's comments on Ludwigshafen plan, 19 Feb. 1926 and 30 May 1927, on
   Leverkusen plan, 17 May and 16 Aug. 1926, in Hoechst, 12/11/2 and 12/255/1.
3. Krekeler and Heymann to Ludwigshafen, 17 May 1926, Hoechst, 12/11/2.
4. Weidlich to Knieriem, 30 May 1927, Hoechst, 12/255/1; Landmann to Weidlich
   and Judge Orth, 7 Feb. 1928, Hoechst, 12/11/1.
5. Meyer and Villiger to IG units, 12 Jun. 1926; Griesheim's Jacobi to IG units, 7 Jul.
   1926, Doermer to IG units, 16 Aug. 1927, Hoechst, 12/11/2.
6. Excerpt, minutes of Legal Committee meeting, 1 Sep. 1927; Leverkusen's
   Brüggemann to Weidlich, Selck, Buhl, Knieriem, 5 Oct. 1927; report on Wolfen
   negotiations of 8 Dec. 1927; Curschmann to IG units, 31 Jan. 1928; all Hoechst,
   12/11/2.
7. Management-BUDACI meeting, Leverkusen, 9 Jan. 1928, Hoechst, 12/11/2.
8. Hagemann to IG units, 21 Jan. and 28 Feb. 1928; excerpt, minutes of Legal Committee meeting, 9 Feb. 1928, Hoechst, 12/11/2.
9. Hagemann to IG units, 28 Feb. 1928; Director Buhl (Frankfurt) to IG managers,
   24 Jul. 1928, Hoechst, 12/11/2.
10. BUDACI report for ILO, in ACID to Professionals Committee, 18 and 19 Nov.
    1924, Hoechst, 12/255/1. IG Farben's top executives discussed what to do

about the "vast oversupply of organic chemists" and shortage of inorganic and physical chemists in Nov. 1925, excerpt, minutes of TEA meeting, 26 Nov. 1925, Hoechst, 12/11/2.

11. Schwarz to IG units, May 7, 1930, Hoechst, 12/11/2; other rationalization measures in Hoechst 12/11/2, 12/255/1, 12/35/4; BUDACI's IG Fachgruppe to managerial board, protesting layoffs, bonus cuts, and dividend increase, 31 May 1930, Hoechst, 12/35/4.

12. Excerpt, minutes of Soko meeting, 2–3 July and 1 Aug. 1930, Hoechst, 12/255/1, 12/11/2.

13. Excerpt, minutes of Legal Committee meeting in Frankfurt, 10 Sep. 1930; Dr. Brüggemann to IG units, 11 Sep. 1930, Hoechst, 12/255/1.

14. Curschmann to IG units, 10 Oct. 1930, Hoechst, 12/255/1; excerpts, minutes of Soko, 3 Nov. 1930, and TEA, 4 Nov. 1930, Hoechst, 12/11/2 and 12/255/1.

15. State Superior Court President Stadelmann to Kammergerichtsrat Herbert Kühnemann (RJM), 3 Feb. 1937, BAK, R22/629, Bl. 44.

16. Karl Winnacker, *Challenging Years: My Life in Chemistry* (London, 1972), 81.

17. Pistor to IG units, 20 Sep. and 13 Nov. 1930, replies by Knieriem, Doermer, Hagemann, and Landmann, 17 and 25 Nov. 1930, Hoechst, 12/255/1.

18. Milde to IG Farben managerial board, 13 Apr. 1931, Hoechst, 12/255/1.

19. Schwarz and Knieriem to BUDACI headquarters Berlin, 6 May 1931; Schwarz to IG managers, 7 May 1931, Hoechst, 12/255/1.

20. Schwarz to Selck, 23 May 1931, Hoechst, 12/255/1.

21. Kurt Milde, *Abbau, ein Schlagwort und seine tiefere Bedeutung* [Sozialpolitische Schriften des Bundes angestellter Akademiker technisch-naturwissenschaftlicher Berufe e.V., series 1, no. 14] (Berlin, 1930).

22. Brodersen file in BDC; Pistor to IG units, 13 Nov. 1930, Hoechst, 12/255/1; Harald Mediger, "Gedanken zur Gestaltung des Rechts des nicht-selbständigen Erfinders im Lichte der Betriebsverknüpftheit seiner Erfindung" (Munich: unpublished typescript, 1948), pp. 45–8, BA-Zwi, B141/2793.

23. Mediger to Ludwig Erhard, 25 Oct. 1948, and related correspondence, BAK, Z22/187, Bl. 68–77.

24. Wolfen meeting, 8 Dec. 1927, between BUDACI (Mediger, Wallis, Matzdorf, Kuhrmann, Eckler, and Keiner) and management (directors Curschmann, Hofmann, and Lisco), Hoechst, 12/11/2.

25. Mediger, "Gedanken zur Gestaltung," pp. 35–38.

26. Jarausch, *Unfree Professions*; Ludwig, *Technik und Ingenieure*; Charles McClelland, *The German Experience of Professionalization: Modern Learned Professions and Their Organizations from the Nineteenth Century to the Hitler Era* (Cambridge, 1991).

27. How states' inability to manage their affairs and govern effectively can result in disillusionment with established institutions and elites, trigger a crisis of legitimacy, undermine confidence in national identity, and cause "identity panic," leading in turn to nationalist backlash, racism, and the quest for radical solutions, is analyzed well in recent literature on citizenship and antiforeigner sentiment in contemporary Europe; see David Cesarani and Mary Fulbrook, eds., *Citizenship, Nationality and Migration in Europe* (London, 1996); Brian Jenkins and Spyros A. Sofos, eds., *Nation and Identity in Contemporary Europe* (London, 1996); also: Jürgen Habermas, *Legitimation Crisis* (Boston, 1973).

28. "Ansprache des Zellenleiters Dr. Nicodemus in der ersten Vollversammlung der Zelle Chemie am 23. Juni 1933," pp. 4, 10, Hoechst, 12/35/4.

29. BUDACI to Direktion der I.G. Farbenindustrie A.G. Werk Höchst, Entschliessung der Werksgruppe Höchst des Bundes angestellter Akademiker technisch-naturwissenschaftlicher Berufe, 6 Dec. 1929, Hoechst, 12/35/4.

30. For similar critiques by engineering leaders such as Krupp technical director Karl Wendt and VDI chairman Enno Heidebroek, see *Technik, Ingenieure und Gesellschaft*, ed. Karl-Heinz Manegold and Wolfgang König (Düsseldorf, 1981), 333–34.

31. Milde, *Abbau*, 6, 15.

32. Hugo Kretzschmar, *Die Technischen Akademiker und die Führerauslese* [Sozialpolitische Schriften des Bundes angestellter Akademiker technisch-naturwissenschaftlicher Berufe e.V., series 1, no. 13] (Berlin, 1930), 9–11.

33. Ibid.

34. Mack and Holz to Duden, 29 Jun. 1931, "Die gegenwärtige Wirtschaftslage und die Akademiker," Hoechst, 12/35/4.

35. "Vorschläge zur Erhöhung der Arbeitskapazität und Rentabilität der Werksgemeinschaft Ludwigshafen-Oppau," 5 Nov. 1931, Hoechst, 12/35/4.

36. Memorandum on "reduction question," 14 Nov. 1931, Hoechst, 12/35/4.

37. Cf. Barkai, *Nazi Economics*, 23; Herf, *Reactionary Modernism*; Peter Schwerber, *Nationalsozialismus und Technik* (Munich, 1930), esp. 46–63; James, *German Slump*, 345.

38. "Ansprache des Zellenleiters Dr. Nicodemus."

Chapter 6

THE GREAT DEPRESSION AND THE ORIGINS OF NAZI PATENT REFORM

1928–1932

When the Nazis turned to the inventor issue in 1933, they did not approach it as a problem of labor law but as one of patent law. The discontinuity spelled a return to the politics of inventing as practiced in Wilhelmine Germany, and it brought quick results. By May of 1936 the National Socialist government had promulgated a new Patent Code, which combined increased powers of state intervention with a number of social features designed to protect and enhance the rights of the inventor.

The centerpiece of the 1936 Patent Code reform was adoption of the inventor's-right principle, which replaced the first-to-file principle that had been the foundation of the country's patent system since 1877. Under the system adopted in 1936, the patent went to the first *inventor* who filed an application.[1] In terms of patenting procedure, the change of system in 1936 had few practical consequences, preserving the simplicity and efficiency as well as some of the inherent anti-inventor bias of the first-to-file principle.[2] But philosophically, it was a revolution. The new law unambiguously stated that "the inventor or his legal successor has the right to the patent."[3] This finally recognized the principle of technological authorship and represented a crucial moral victory for inventors, whose names were henceforth prominently displayed on the patent documents.

The 1936 Patent Code incorporated a number of other pro-inventor features as well, such as easier appeals procedures, litigation and filing-fee subsidies, and lower patent fees.[4] By far the most important of these more technical changes was a decision to outlaw the company invention. Elimination of the company invention gave teeth to the new inventor-right principle and was the single most important change in Germany's patent system since its inception.

Although the 1936 patent reforms had an unmistakable pro-inventor thrust, they did not go as far as inventors themselves had hoped. Patent duration, for instance, continued to be calculated from the time of application rather than the later time of publication.[5] A longstanding demand for "technical judges" was ignored, preserving the monopoly of the legal profession in patent litigation. Most significant, the 1936 reforms failed to address the issue of employee-inventor rewards, which had been the crux of the 1925 Austrian patent reforms and remained the biggest bone of contention in Germany as well.

Regardless, the Nazis celebrated the 1936 Patent Code as a huge triumph. According to Hans Frank, president of the Academy for German Law and the future governor of German-occupied Poland, patent reform was a shining example of National Socialism's social progressiveness. In a decisive break with the decadent past, Frank asserted, the new regime had crafted a Patent Code that overcame the twin evils of "inventor bolshevism" and liberal capitalist exploitation and restored the nurturing harmony of life in the *völkisch* community. On the one hand, he wrote, the law's pro-inventor features encouraged and protected the inventor, so as to "bring forth the natural talents and abilities lying dormant in the *Volk*." On the other hand, the government's new power to seize patents was the "purposeful integration into community service of the forces and achievements thus awakened."

The new Patent Code, Frank continued, rejected "Marxist mass delusion and majority thinking." It codified into law Hitler's dictum from *Mein Kampf*, that it was "not the mass [that] invents and not the majority [that] organizes or thinks, but in everything always only the individual human being, the person." At the same time, the creative personality singled out for honor and protection had an obligation "to put his achievement at the disposal of the community." The old Patent Code had been "dominated by liberalistic concepts," which paid attention only to the invention. It had been "created expressly to deliver the inventor defenseless into the hands of superior capitalist power." The new law was "fundamentally different! … It places the inventor at the center of the patent law." This change was no mere external formality, according to Frank. Rather it "corresponds to the National Socialist conception of the nature of work, which shall receive not merely material compensation in the form of wages, but also bestows a right to respect and honor." Emphasizing the importance of honor as a "better incentive to bring forth the highest achievement than the prospect of material success," Frank barely hinted at the unsolved problem of inventor rewards. Instead he trumpeted fee reductions and subsidies for poor independents as evidence of the law's "consideration for the social condition of the inventor" and the regime's unflagging pursuit of social justice.[6]

In truth, the reforms of which Frank boasted were the work of National Socialist reformers to only a limited extent. Adoption of the inventor's-right principle, for instance, went back to draft legislation prepared between 1928 and 1932 by bourgeois and Social Democratic governments. Other aspects of the new law, too, had roots in the pre-Nazi past. Even so, the regime made a number of crucial modifications in the bills inherited from Weimar, and in the end those changes resulted in a Patent Code with different accents that was considerably more favorable to inventors than anything that might have become law in the Weimar Republic.[7] In spite of doing less than it claimed, therefore, the Nazi regime's patent reform represented an undeniable measure of social revolution.[8]

Patent reform had first reappeared on the political agenda in February 1928, when the cabinet of Chancellor Wilhelm Marx announced it had prepared a bill to overhaul the 1891 Code. Like its predecessors, the Marx government was a reluctant patent reformer. A variety of stopgap measures during and after World War I had taken care of the most urgent problems. Collective bargaining and the various attempts at labor legislation removed the other main reason for adopting a new patent system. The fear of upsetting a precarious economy and concern about the government's financial difficulties in the 1920s also made it seem prudent to tinker as little as possible with one of the few institutions that consistently generated large surpluses for the Reich treasury.[9] When amendment of the Patent Code proved unavoidable in connection with a change mandated by Germany's membership in the Paris Convention, an international patent treaty, the government prepared legislation that made as few changes as possible.[10]

Even before it reached the Reichstag, however, the 1928 bill ran into trouble in the Preliminary Reich Economic Council (vorläufiger Reichswirtschaftsrat, RWR). An advisory body designed to promote social consensus, the RWR had been expected to give its routine approval. Instead it raised all sorts of objections and refused to let the bill go forward. The principal obstacle in the RWR was Berlin engineering consultant Carl Hartung, chairman of the Association of German Consulting Engineers (Bund deutscher Civil-Ingenieure) and an outspoken defender of the interests of independent inventors. Hartung also was the publisher of *Efficient Technology* (*Wirtschaftliche Technik*), a monthly focused on questions of rationalization, technology, law, and politics. His friends included Otto Kammerer, former rector of Berlin's Technische Hochschule and an inventor in his own right, and Hugo Junkers, the aeronautical inventor-entrepreneur and exponent of democratic causes.[11] One of Hartung's chief complaints about the patent system was its "self-cleansing" system of progressively higher annual fees for patent renewal. According to Hartung and many other independents, the self-cleansing principle stifled the inventive enthusiasm of independents and, therefore, the nation's technological progress. Before 1928 his protests had remained in vain, but now he had an opportunity to hold up the government's bill and get some attention.[12]

"Under absolutely no circumstances," Hartung wrote in late February, should the RWR "give the bill … a positive recommendation without thoroughly evaluating" the changes proposed by the government.[13] The bill failed to address all the "questions that have concerned everyone involved in intellectual property protection for decades." It did nothing about the "intolerably high" fee system, even though most experts had long recommended a change, as cuts made in 1926 were totally inadequate. It failed to change the starting time of patent protection from the date of application to that of publication, even though the "German intellectual worker is worse off than in all the rest of the world." Finally, there were a number of other problems that hurt the "smaller inventor" and the "inventor who lacks capital." To address all those issues, Hartung demanded formation of a special RWR committee to hold hearings and make its own recommendations for Patent Code reform.[14]

Supported by groups such as the Munich-based German Inventor-Protection Association, the BUDACI, patent attorneys working for independents, the Junkers aircraft works in Dessau, and the Committee for Inventor Interests (a group he and Kammerer had formed the year before), Hartung carried the day.[15] The Economic-Policy Committee of the RWR appointed a nine-member subcommittee to study the government's patent bill. With three representatives each from industry, employee-inventors, and independents, the group was clearly weighted in favor of the inventor side.[16] Between March and June 1928, it held hearings and canvassed experts with the aim of broadening the scope of patent reform. Specifically, it concentrated on fee reductions, longer patent duration by moving forward the time when protection took effect, and adoption of the first-to-invent principle.

Not surprisingly, industry opposed all three changes. Pro-business experts such as Professor Hermann Isay, Walter Duisberg (the son of Carl Duisberg and IG Farben's principal expert on the American patent system), and Richard Weidlich portrayed the American first-to-invent system as an unmitigated disaster: the cause of endless litigation, delay and retarded innovation.[17] On the other side, inventors Christians, Frank, and Kammerer contended that the American system encouraged patenting by independents and generated far greater technological dynamism and progress than that of Germany.[18] When pressed, they conceded that independents would fare no better under the inventor's-right system than the first-to-file system. The real reason for a system change, however, was not so much to help independents as employee inventors, who were "being gagged" and "rendered ... without any rights" under the first-to-file system.[19] In the words of inventor Christians,

> we should ... take into account that the inventor is in principle always somewhat the underdog. If he is an inventor by nature, he cannot help but invent and will end up the loser in other matters, being treated as a child by his supervisor, etc. In my opinion, we must approach the issue primarily as a psychological question. I am not entirely familiar with the finer legal points, but I would like to ask you whether it isn't possible to ... aim for protection of the valuable inventor. It isn't all that easy to achieve this in our businesses. True, an employee has lots of rights, and he can demand to get recognition for his patent afterward. But that creates a great deal of fuss, and so it doesn't happen. It's pointless. In my opinion, we would help society if we could introduce a greater degree of compulsion to really mention the inventor.[20]

Other pro-inventor experts and committee members agreed. According to one witness, the salaried inventor was "severely harmed" if the patent application went forward without his name. "If his name as the inventor of an important innovation does not become known, he can never expect any material benefit."[21] Kammerer asked, "Why doesn't anybody know the names of the engineers, for instance, who built the Munich bridge, or developed the U-boat?"[22] Engineer Frommholz mentioned how as a young engineer he had been forced to sign a contract assigning all his inventor rights to the company and "seen dozens of such contracts that others had been forced to sign." Patents, said Frommholz, existed to benefit society at large. "But I ask you, does it serve society or industry when such forcible renunciation induces the employee to make no inventions at all? Obviously that is harmful to society." The answer to this and other problems, the inventors argued, was introduction of the first-to-invent system.[23]

The employers maintained that the problem should be addressed in labor legislation and collective bargaining, or by listing the inventor's name on the patent, rather than by revamping the entire patent system.[24] This only succeeded in antagonizing the inventors. The inventor's honor, countered Dr. Fritz Pfirrmann of the BUTAB, could not be satisfied by merely listing his name and keeping everything else as it was. Inventions were creative achievements. Their authors were entitled to the appropriate honor and protection, just as were writers, composers, artists, and sculptors. But where was the protection of the inventor's honor in the Patent Code?

> We do not see it. We would like to have it in the Code. That is what the entire struggle is about. We desire that the person who brings to life a technical creation have his name protected, exactly as our copyright laws protect the person who is the first to express a musical idea, a literary idea, a painting idea, a sculpting idea, or any other idea. First and foremost, that is the core question for us.[25]

The sources contain a great deal of additional language such as this, but after two and a half months of wrangling—during which the nation's political constellation shifted decisively to the left—the patent committee was ready to announce its recommendations. It embraced the first-to-invent principle (albeit in a weak version, rejecting the American inventor oath and search for the true and first inventor), voted to move up the beginning of patent protection to the date of publication, and proposed radical cuts in the fee schedule.[26] The full RWR adopted the recommendations without change and submitted them to the government in June 1928.

The Great Coalition's new Minister of Justice, Democratic Party chairman Erich Koch-Weser, rejected the fee reductions as too costly but incorporated the other RWR proposals in a revised patent bill, which he sent to the Reichstag in April 1929.[27] Scheduled to take effect in the fall of 1929, the bill immediately became the focus of renewed lobbying, attacks by the big business community, and objections by the intellectual-property-law establishment.[28] As a consequence, the Reichstag had not yet brought the bill to the floor when the Great Coalition fell in March 1930.

Engineer Oskar Pöbing and the Nazi Party Interpellation of 1930

In the meantime, the crash on Wall Street and the onset of the Great Depression had already turned the patent issue into a battering ram for Nazi attacks on the Republic. The historical agent in this development was a long-time National Socialist engineer by the name of Oskar Pöbing. Pöbing was affiliated with the German Inventor Protection Association (Deutscher Erfinder-Schutzverband, DESV), a Munich-based lobbying group of independent inventors. The demands of the DESV included financial assistance from the government to defray the expenses associated with filing and defending patents in foreign countries. The government, according to the DESV, also had an obligation to end the frequent abuse independents suffered at the hands of fraudulent patent agents, who charged steep fees to assist in obtaining patents or bringing inventions to market but frequently absconded with their clients' money. To end such practices the government should revamp the Patent Office into an agency not merely concerned with legal questions but dedicated above all to helping inventors in practical and financial ways, such as developing and marketing their inventions.[29]

The idea of providing independents with government support in the economic aspects of their inventions—partly realized in the United States through bodies such as the Naval Consulting Board, the National Research Council (both 1915), and the semipublic Batelle Memorial Institute (1929)—would become a central tenet of inventor policy in the Third Reich. In the Weimar Republic it was completely ignored. Undaunted, the DESV kept up its demands. In a 1928 commentary on the government's patent bill, for instance, it called for some of the same reforms as did the RWR. The DESV also demanded waivers of the fees and legal costs associated with obtaining and defending patents, especially in cases where impecunious independents sought to protect their patents against attack by powerful third parties.[30] This demand, too, was disregarded.

By 1929, the chairmanship of the DESV had passed to Pöbing. A combat veteran of World War I, ex-officer Pöbing was an academically certified engineer who before the war had been a student and assistant of Rudolf Diesel's. Since 1911, he had held the positions of "conservator" at the Technische Hochschule in Munich and supervisor of its Hydraulics Institute and Hydroelectric Power Laboratory. Perhaps the first faculty member of any institution of higher education to join the Nazi party in 1920, Pöbing had marched with Hitler on the *Feldherrnhalle* in 1923. Married with four children and just scraping by, he was also an inventor who worked on improving turbines. In the course of his career Pöbing had obtained a number of patents in that field. One of them, granted in 1921, had been infringed upon since 1925 by the Middle Isar Corporation, a large, semipublic electric power utility in Bavaria.[31]

To defend his patent, Pöbing had petitioned the Munich superior court for financial support to help pay the costs of legal counsel under the poverty law statute. When the court granted his request, Middle Isar filed suit to nullify the patent, which put the case on track to go all the way to the *Reichsgericht*. Suffering a setback at the appeals level, Pöbing again petitioned for financial support under the poverty law statute, for expert witnesses and better counsel, this time to the *Reichsgericht*. In April 1930 his request was denied, on the grounds that the Regulations for Civil Procedure from 1877 (*Zivilprozeßordnung*, ZPO) expressly prohibited using the poverty law statute in patent cases. Pöbing thereupon filed a complaint with the Ministry of Justice and the *Reichsgericht*. He argued that the 1877 ZPO was in direct conflict with the Weimar Constitution, whose Article 158 stated that "intellectual work—the right of the author, the inventor and artist—enjoys the protection and care (*Fürsorge*) of the Reich." Since the Constitution was the supreme law of the land, Pöbing contended, the ZPO should be amended forthwith, to allow impecunious inventors access to the poverty laws in patent cases. The government had a duty to give priority to the "achievements of the 1918 Revolution and the 1919 Reich Constitution over the laws from Imperial times," he wrote, since failure to do so constituted a clear *"violation of the Reich Constitution and the idea of the Republic on the part of the authorities."*[32]

Even prior to the unfavorable *Reichsgericht* ruling, the hapless engineer had embarked on an effort to drum up wider support for his case. He published articles in the DESV's periodical, *Deutsche Erfinder-Zeitung*, about the abuse independents suffered under the capitalist system. In one of those contributions, Pöbing criticized the big-business argument—that inventing was a question of systematic research—as completely "erroneous." Inventing, he wrote, was "always the result of personal talent, and personal ability ... the source and precondition of every creative achievement." Only the law did not acknowledge it. On the contrary, "current protection under the law" was instead an "unfair instrument of state power against the inventor ... fraud against the invention and intellectual property." Unlike conditions in the Soviet Union, whose patent system he described as a "glorious exception," patent protection in capitalist societies, Germany in particular, was dependent on the "inventor's capacity to pay." This represented "a purely capitalist and egotistical point of view, which takes direct aim at the inventor. Every German inventor irrevocably loses this fictional legal protection when he gets in trouble, such as when he cannot pay the standard patent fees." The impecunious inventor was "treated as a pariah" and faced medieval conditions. For in the Middle Ages, too, as the saying went, "the soul can pass through heaven's gate / as soon as coins jingle the collection plate (*die Seele in den Himmel springt, wenn das Geld im Kasten klingt*)." Sadly, no one cared one way or another, and all criticism of the system was in vain. Not even the professional engineering societies, such as the Association of German Engineers (Verein deutscher Ingenieure, VDI), bothered to come to the independents' assistance. Everyone, Pöbing charged, conspired to "hush up this painful matter," especially the country's various legislative bodies and the professional associations on which they relied for advice, "among them, in first place, the VDI."[33]

Pöbing's anger toward the nation's leading engineering association was not without ground. For all practical purposes, the VDI was a mouthpiece of big business in the question of patent reform. Throughout the 1920s, the engineering association steadfastly demanded changes that were high on industry's agenda, such as the introduction of "technical judges," the expansion of patent attorneys' rights, and the formation of specialized patent courts. But it equivocated on, or rejected outright, the recommendations of the RWR and virtually all other measures close to independent inventors' hearts.[34] Pöbing's own case was a perfect illustration. In early 1929 he had approached the engineering association for help in his effort to obtain fee waivers for means-tested inventors under the poverty law statute. But the VDI had turned down his request, on the grounds that the poverty statute was intended to secure the legal and political dimensions of citizenship and not socioeconomic claims, especially not monopoly rights such as patents.[35]

Let down by the VDI, the Ministry of Justice, and the courts, Pöbing finally got somewhere with the Nazis. Shortly after the elections of September 1930, in which the National Socialists became the second-largest party in the Reichstag, Pöbing managed to bring his case to the attention of Wilhelm Frick, the Nazi fraction leader in parliament. Perhaps he received help from Gottfried Feder, at that time the party's second man in the Reichstag and a fellow engineer from Munich whom he had known since 1919. Whatever the exact connection, on November 21, 1930, the Nazi Reichstag delegation queried the government about the Pöbing case.

Did the Minister of Justice know, the Nazis asked, that in direct violation of the Weimar Constitution's Article 158 impecunious inventors whose patents were challenged were denied access to state support under the poverty laws? Did the minister know that the intellectual property of the inventor without means was delivered into the hands of German and foreign capitalist big business? Was the minister prepared to issue immediate regulations giving inventors without means the same access to government support as other citizens and the same rights to protect their intellectual property as patentees with money? Did the minister know that the DESV had submitted detailed patent reform proposals to the government as early as 1928, and had these proposals been properly studied? Was the minister prepared to explain why urgently needed reforms of the Patent Code had languished since 1928? Was the minister prepared to implement the recommendations of the DESV and thereby "protect the intellectual property of the inventor without means against the violent acts of big capital?" Did the minister know that when a patent was nullified because of errors by the patent examiner the original patentee did not receive a refund of the fees already paid, even though the government's patent business generated huge surpluses? Was the minister prepared to provide immediate relief for the above problems by a combination of ordinances and legislative proposals? Finally, was the minister prepared to facilitate these reforms by appointing a special expert, an academically certified engineer with at least five patents to his name, who was to be selected jointly by the NSDAP and the DESV?[36]

Although it did not mention him by name, the job description in the last question defined someone with the qualifications of engineer Pöbing. This was because the author of the Nazi party's parliamentary questions was, in fact, none other than Pöbing himself. The inventor's hope to propel himself into the center of the patent issue and use his Nazi connections to boost his career, however, did not become reality. On the contrary, while he did play an important role in making patent reform an ever more contentious issue, Pöbing's own fortunes, like those of so many "old fighters," soon took a turn for the worse. This is evident from a letter Pöbing wrote to Frick in 1934, in which he reiterated his hopes of being given an important position at the Reich Patent Office, preferably the presidency itself, and seeing patent reform finally become a reality.

Pöbing began by reintroducing himself and listing his credentials. He, more than anyone else, was the originator of National Socialist patent reform. He had prepared the party's 1930 parliamentary interrogatory and corresponded about it with Feder. By embracing the cause of Germany's independent inventors, Pöbing maintained, the "party members who at that time signed the parliamentary questions had, before the entire German people, assumed the obligation to implement those proposals on behalf of the German inventor." Pöbing reminded Frick of Hitler's words on inventing in *Mein Kampf* and asserted that "precisely what the *Führer* wants can only become a reality when the current, anti-inventor Patent Code is reformed along the lines that you yourself supported in your interrogatory of 21 November 1930."

But patent reform was about more than keeping a promise, Pöbing continued. It was a vital economic necessity as well, which would help overcome the Great Depression. "The German people, who need German technology and the German economy, need the German inventor today more than ever." Even though many "exponents of the old system, disguised by synchronization, deny the need for patent reform," the facts spoke for themselves. The inability of inventors to maintain their patents in the Depression years had caused a "horrendous mass death of German patents because of annual fees that are much too high." This had also paralyzed the will to invent, as demonstrated by the steep decline of patent filings between 1930 and 1933. Pöbing had pointed all this out earlier to the relevant technical associations of the Nazi party, the Fighting League of German Architects and Engineers (Kampfbund Deutscher Architekten und Ingenieure, KDAI, organized in August 1931), and the National Socialist League of German Technology (Nationalsozialistischer Bund Deutscher Technik, NSBDT, established in May 1934), both of which he had helped found. But to date he had heard nothing from them, nor anything about the fate of patent reform. If the delay were caused by the absence of a requisite tough-minded reformer in the Reich Patent Office, Pöbing wrote, he himself would be willing to step forward to do the job. Although not a lawyer, he obviously knew plenty about patent law, and a strictly legalistic spirit was antithetical to the new order anyway. Moreover, he also had the time to become president of the Patent Office, because since October 1, 1933 he had been "unemployed owing to intrigues of reactionary and Jewish circles."[37]

The records do not reveal what became of Pöbing. But he never received any kind of official position—neither in the Patent Office nor in any of the various Nazi agencies concerned with patent reform—and it seems safe to assume that his career continued on its downward spiral. This is not to say that Pöbing's complaints and the Nazi party's parliamentary questions of November 1930 also remained without effect. On the contrary, the Nazi clamor for access to the poverty laws in patent cases caused the government to withdraw the 1929 bill in February 1931 and to submit a slightly modified version, the 1931 patent bill, six months later. The revised bill went to the cabinet in August and the Reichsrat in September. It went to the Reichstag in late April 1932, six weeks before parliament was dissolved in connection with Chancellor Brüning's fall and General Schleicher's ill-fated scheme for taming the NSDAP.

Considering the massive political crisis of those months, it is not surprising that the 1931 patent bill failed to become law before January 1933. This was not for want of trying by the inventors, however. The last few years and months before the Nazi seizure of power saw a veritable flood of petitions, letters, and initiatives by independents and their organizations to get the bill through the Reichstag as quickly as possible, or better yet, adopted by way of Emergency Decree. The catalyst was the economic hardship of many small and medium-sized businesses in the Great Depression. Small entrepreneurial technology firms, manufacturers of specialized products and components, independent laboratories, and individual developers of new ideas depended on the maintenance of a few crucial patents much more than large companies or businesses where intellectual property played no role. If independents could not keep up their patents or experienced long delays in bringing them to market, they essentially lost everything and could be wiped out from one moment to the next.

If the records in the files of the Ministry of Justice are to be believed, this is precisely what happened when the Depression brought much of the economy to a standstill. Between January 1931 and June 1932, when the Papen government provided a small amount of relief by Emergency Decree, the ministry received thirty-two petitions requesting immediate help for inventors through patent reform. While some were from individuals, others came from professional associations of inventors, engineers, and architects, as well as from engineering unions such as the BUTAB, the BUDACI, the VELA, and others. There were letters from semiliterates, crackpots, people infatuated with Hitler, reformers, utopians, and desperate individuals, but also from serious independents, entrepreneurs, academics, politicians such as Reichstag Social Democratic deputy

Siegfried Aufhäuser, who was the chairman of the Afa-Bund, and famous inventors such as Georg Graf von Arco, the head of Telefunken. All were unanimous in demanding speedy adoption of pro-inventor measures—measures, they pointed out, that had already been approved by the RWR, several cabinets, and the Reichsrat, and which had been on the drawing boards since 1928. The Democratic journalist and historian Erich Eyck, too, wrote a favorable evaluation of the bill for the *Vossische Zeitung*, praising, among other things, the first-to-invent principle and the plan to move the beginning of patent duration to the date of publication.[38]

To coordinate this movement, Carl Hartung in November 1931 organized the German Working Community For Inventors (Deutsche Arbeitsgemeinschaft für Erfinder e.V., DAFER), a confederation of all the associations of employees and independents that at one time or another during the Weimar years had concerned themselves with the inventor issue.[39] The DAFER, which included groups such as the BUDACI and VELA, pressured the government to adopt several key provisions of the 1931 patent bill by Emergency Decree.[40] The arguments Hartung used to convince the government centered on the economic disaster that would befall not just the DAFER's members, but also the countless employees and workers who would become unemployed if the many small and medium-sized firms that lived or died by their patents were to lose them and go bankrupt. The government did not remain entirely deaf to these pleadings, which made sense and were inexpensive to implement. In June 1932 it passed an Emergency Decree on intellectual property.

The measures only partially met the DAFER's demands and generally were too little, too late.[41] Undoubtedly they would have gone further if big business and the intellectual property establishment had not opposed all patent reform to the bitter end. This included not just the minor reforms promulgated in 1932, but especially demands to pass the entire 1931 patent bill by Emergency Decree.[42] The Association of German Machinery Builders (Verein Deutscher Maschinenbauanstalten, VDMA), for instance, urged Justice Minister Joël to avoid doing so at all costs.[43] The RDI objected as well. So did the Green Association, the Association of German Patent Attorneys, the Reich League of German Technology (Reichsbund Deutscher Technik, RDT), the VDI, and the Reich Patent Office.[44] Intimidated by this phalanx, the Ministry of Justice in December 1931 decided against introducing the patent bill by Emergency Decree. The strength of the opposition also caused Joël to delay bringing the bill to the Reichstag. This destroyed whatever chances of passage by ordinary legislative procedure the legislation still might have had before the summer of 1932.[45] Passage of the scaled-down emergency decrees in June further reduced the pressure to do more. Pro forma gestures to reintroduce the bill in the last six months of the Republic's existence continued, but the ministerial records clearly indicate that the issue was no longer a priority.[46]

Weimar and the Origins of Nazi Inventor Trusteeship

What, if any, was the connection between the failure of patent reform and the rising popularity of National Socialism as the means of last resort in the early 1930s? The case of engineer Pöbing, a Nazi since 1920, was clearly an exception, even though it put patent reform on the Nazi party agenda. But there can be little doubt that a great many other engineers and inventors were profoundly alienated and increasingly desperate as well. The frantic organizational activities of Hartung and Kammerer and the support they received from countless independents and small businesses in the DAFER bear witness to this development. This chapter will conclude by reviewing some additional fragments of documentary evidence, to shed light on the question of why so many of Germany's inventors, engineers, industrial scientists, and others seeking patent reform— people who were not necessarily enamored of the Nazis—could nevertheless welcome Hitler as a savior when he became Chancellor in January 1933.[47]

Consider the case of Telefunken's Count Arco, a man who was anything but a Nazi yet still found himself in serious disagreement with his country's political and technological cultures. A well-known pacifist and member of the Society of Friends of the New Russia, Arco had more than 400 patents to his name by the time he died in 1940. In 1929, a time when many of his colleagues still had their American trips fresh in mind, Arco made it a point to celebrate his sixtieth birthday in Moscow, to tell the world where he stood. In December 1931, however, Arco found himself on Hartung's side, pleading in vain with Justice Minister Joël to pass the patent bill by Emergency Decree. His main reasons were the current system's unfairness to inventors, and his hope that stimulating invention might help overcome the Depression. Counting patent duration from the time of filing, Arco wrote, "abridges the rights of inventors in a way that does injustice to their interests." At a time when "Germany, more than ever before, depend[ed] on the consequences of intellectual activity of its citizens," he concluded, the country should not hesitate to adopt the new system.[48]

There is also the case of Manfred von Ardenne, the talented but opportunistic independent, who was only twenty-four in October 1930 when he drew the ire of the electrical-engineering establishment with his revolutionary proposal for a combination of cable and FM broadcasts to lower the cost and increase the availability of radio reception in Berlin. Ardenne, who in his 1972 autobiography recounts the event with undiminished resentment for the large electrical and broadcast companies that ridiculed and killed his plan, had enough good ideas and patents to weather this particular attack as well as many others. But he seems to have come away with an enduring hatred for the large capitalist corporations, such as Siemens and AEG, which were more interested in protecting their investment in current technology than promoting radical technological change. In part with the help of Wilhelm Ohnesorge, a fanatical Nazi and talented inventor who in 1936 became the regime's Minister of Telecommunications, Ardenne made a great name and career for himself in the Third Reich. And in 1945, although he had plenty of opportunity to flee west, he opted instead for the east, going on to do nuclear research for the Soviets and later for the GDR.[49]

In 1931, Ardenne, along with the Osnabrück rocketry pioneer Reinhold Tiling, lent his name to a scheme that would become an essential component of Nazi inventor policy: "Utilization of the Inventor Forces Present in the German People," the title of a pamphlet submitted to Chancellor Brüning in October 1931. The proposal it contained—to exploit the inventive forces of the German people more effectively than did society with its current institutions—was stock-in-trade among the circle in which it originated: the profession of consultants and facilitators who helped small independents realize and make money off their ideas.

This somewhat disreputable occupation was known by various high-sounding names, such as "patent utilizers" (*Patentverwerter*), "patent economists" (*Patentwirtschaftler*), or "consulting patent engineers" (*beratende Patentingenieure*). Unlike the prestigious patent attorneys (*Patentanwälte*), patent economists did not specialize in the legal aspects of securing or defending patents, but in assessing the economic viability, developing, licensing, or otherwise marketing inventions. Undoubtedly, the wreckage and the tragedies that patent economists witnessed during the implosion of capitalism in the Great Depression were far greater even than they already were in more ordinary times. It is understandable, therefore, that as a group they tried to publicize the problems—including their own—encountered on an almost daily basis. It is equally understandable that they began to suggest remedies "for the reconstruction and increase of national wealth" along rather unorthodox lines.[50]

Such was the background of the 1931 publication Ardenne endorsed. Written by Münster law professor Andreas Thomsen, the document was in reality the brainchild of Thomsen's friend, engineer Hans Meissner, chairman of the Association of Consulting Patent Engineers in Bremen and editor of the journal *Technology and Industrial Property*. Although Meissner himself did not go on to achieve any kind of power or influence in the Third Reich, others in his line of work would. Patent economists and patent engineers were the driving force behind the quasi-socialist system of inventor care (*Erfinderbetreuung*) that sprang from their professional experience and concerns in Weimar and became the core of National Socialist inventor policy.

The basic plan and concerns were outlined in Thomsen's 1931 pamphlet. Thomsen began by noting that inventions had the potential to create unlimited wealth, as evidence for which he pointed to the achievements of Justus von Liebig and the German miracle drug Baier 205, also known as Germanin, which cured sleeping sickness and made the tropics inhabitable. Inventions such as these, Thomsen asserted, were not just the product of systematic research by professional inventors, but could also be had from lay inventors, because, as Albert Einstein had said, "the fundamental source of all technological achievements is divine curiosity and the play instinct of the tinkering and brooding researcher." With those words Einstein had not meant the professional scientist or researcher, Thomsen contended, but rather the lay inventor. He quoted approvingly from an essay by the National Socialist publicist Peter Schwerber, who had also drawn attention to the "enormous treasures" of native and intuitive technological talent that slumbered in the German nation. Those powers, Thomsen said, should be mobilized, first, to get rid of the country's reparations burden, and second, to overcome the depression. The "German people," Thomsen maintained, "undoubtedly have one of the greatest inventive talents on earth."[51]

In contrast to all this unused potential, there was the reality that prevailed in Germany. Thomsen's chief examples were the Pöbing affair and affidavits from other patent economists that at most 4 or 5 percent of all patents became economically successful. The rest represented a huge waste of talent, energy, and money—not because the underlying ideas were no good, but because the German inventor never even had an opportunity to realize them, was treated with hostility, considered a fool, or simply "cheated of the fruits of his labor." This hostile culture so discouraged independents that often they did not even dare to divulge their ideas anymore, and, in the case of industrial scientists, simply stopped inventing.[52]

Contrast this to the situation in the Soviet Union, Thomsen continued. The inventor system of the Soviet Union was characterized by an "exceptionally friendly overall disposition of the Russian legal and social system toward Russian inventors." The Soviet Union was "a shining example for the German Reich," which had never even fulfilled the promises with respect to inventors made in Article 158 of the Constitution. What the Soviet Union had that was lacking in Germany were state-funded institutions to encourage inventing, support inventors, and develop their inventions. Thomsen mentioned Soviet institutions by the name of ZBRIS and PRIS, which took the place not only of Western-style corporate laboratories but also of the semipublic Kaiser-Wilhelm-Gesellschaft institutes. In addition, they performed many of the functions of a national patent office and helped inventors in a variety of practical and material ways. The new Soviet Patent Code of 1931, which explicitly mentioned the great contribution ordinary workers could make to technological progress and aimed to encourage the talented inventor wherever he might be, Thomsen wrote, was "a great idea."[53]

What should Germany do to obtain some of the same benefits? First, the government should start an educational campaign to promote a "culture of inventing." Then it should rescue all the patents that were lapsing because of unpaid fees, both at home and abroad, and make sure that all the important German inventions were securely protected by systematically filing for patent protection in the relevant foreign countries. Another measure would be to gather and publish all the important technological problems inventors might tackle, and to reward the most promising solutions. Yet other ideas included the establishment of a "state patent attorney office," which would function parallel to the Patent Office, examine the state's interest in inventions, support the patenting of promising ones, and note the ones that might later qualify for development at public expense. As a way to encourage creativity, inventors should be exempt from taxation on the income from their patents and inventions. Inventors should have the right to pay their patent fees in kind, by giving the government a share in the invention's profits. Patentees should be entitled to financial support under the poverty laws, as the Pöbing case had demonstrated beyond any reasonable doubt. The public interest demanded that salaried inventors receive special compensation as a way to encourage them to do their utmost and make as many inventions as possible. The current, "anti-inventor" Patent Code should be replaced by an "inventor-friendly" one, and existing efforts to that effect should be intensified many times. As in the Soviet Union, inventors should be publicly honored and supported, and receive titles, medals, and special orders.[54]

Thomsen's list of recommendations went on and on. His most noteworthy suggestion, because it would soon become reality in the Third Reich, was the idea of establishing a "Reich Invention Support Office" (*Reichs-Erfindungs-Pflege-Amt*, REPA). REPA, which Thomsen wanted set up immediately, would parallel and cooperate with the Patent Office, just as did Russia's ZBRIS, and help the inventor make his idea an economic reality. Thomsen did not mention it, but his REPA would have been a first step toward nationalizing the patent economists' occupation. The latter would be able trade their insecure existence in the private sector for secure jobs as government (or, as it would turn out, DAF and NSDAP) officials. While Thomsen was careful to avoid any reference to such selfish motives, the money question was never far in the background. Since the Reich Patent Office earned "no less than RM 7 million annually off the inventors," he maintained, "it would be no more than its 'damned obligation and duty' to return a small part of this profit ... as a benefit to the inventors, and use it to erect and maintain a REPA." Thomsen further proposed the establishment of additional "public societies for the encouragement of inventing." Though currently incapacitated by Germany's impoverishment, organizations such as the Association of Consulting Patent Engineers (headed by his friend Meissner) were an inspiring example for these future bodies.[55]

Detailed plans such as Thomsen's, which were a virtual blueprint for the agenda of Nazi patent reformers and inventor advocates after 1933, are a rare find in the archival records of the years before Hitler came to power. But there are any number of other documents—letters, recommendations, petitions—that express similar ideas, and the demands for help from small independents frequently had the same thrust as well.[56] All of this evidence directly prefigures the content of Nazi inventor policy. At the same time, and with the exception of Pöbing, very little of it identifies openly with the Nazi party. There are also few direct political references to the NSDAP. And there is as yet no mention of the blessings that Hitler would surely bring if only he were running things in Germany. Such explicit acclamation of the "Führer" would have to wait until after 30 January 1933.

Notes

1. This differed from the American system, which protects the rights of the first and true inventor regardless of time of filing.
2. Heymann, "Erfinder"; Georg Klauer and Philipp Möhring, eds., *Patentgesetz vom 5. Mai 1936, erläutert von Georg Klauer und Philipp Möhring* (Berlin, 1937), 116–27; Georg Benkard, *Patentgesetz, Gebrauchsmustergesetz; Kurzkommentar begr. von Dr. Georg Benkard; fortgeführt von Werner Ballhaus et al.*, 7th ed. (Munich, 1981), 43–49.
3. Patentgesetz vom 5.5.1936, par. 3, quoted in Klauer and Möhring, *Patentgesetz*, 116.
4. Annually increasing fees were cut by more than 13 percent, lowering the hypothetical cost of a patent lasting the full eighteen years from RM 7,120 in 1926 to RM 6,165 in 1936. Compared to the $40.00 fee for a U.S. patent, this was not much of a reduction, and German fees remained the highest in the world. They could also be postponed for the first six years in the case of impecunious inventors. Patentees willing to give a blanket license received a full rebate of patenting fees, plus a 50 percent reduction of maintenance fees. The cost of litigation also came down. The RPA and courts were authorized to reimburse inventors lacking means to cover court costs; there were subsidies to pay expert witnesses and attorneys' fees. It became possible to limit financial exposure to a fraction of the amount in dispute under Germany's loser-pays-all system; Heymann, "Erfinder," 112–13; fees in BAK, R131/155, Bl. 42 (Entwurf eines Gesetzes zur Abänderung der Gesetze über gewerblichen Rechtsschutz 4.2.1928), and BAK, R131/158 (Gesetz über die patentamtlichen Gebühren vom 5. Mai 1936).
5. This effectively reduced patent duration by up to two years. Detailed discussion of pros and cons of *Anmeldung* v. *Bekanntmachung*: BAK, R131/155 (RWR, Patentausschuß, minutes of Sitzungen des Arbeitsausschusses zur Beratung des Entwurfs eines Gesetzes über gewerblichen Rechtsschutz, 27–28 Mar., 2 Apr. 1928, esp. 1. Sitzung, 57–76).
6. Hans Frank, "Die Grundlagen des nationalsozialistischen Patentrechts," in *Das Recht des schöpferischen Menschen: Festschrift der Akademie für deutsches Recht anlässlich des Kongresses der Internationalen Vereinigung für gewerblichen Rechtsschutz in Berlin vom 1. bis 6. Juni 1936*, ed. Akademie für Deutsches Recht (Berlin, 1936), 7–8. For "honor of labor," see Alf Lüdtke, "'Ehre der Arbeit': Industriearbiet und Macht der Symbole. Zum Reichweite symbolischer Orientierungen im Nationalsozialismus," in *Industrielle Welt: Schriftenreihe des Arbeitskreises für moderne Sozialgeschichte*, ed. Reinhart Koselleck and M. Rainer Lepsius, vol. 51, *Arbeiter im 20. Jahrhundert*, ed. Klaus Tenfelde (Stuttgart, 1991), 343–92.
7. For comparable developments in biology education: Sheila Faith Weiss, "Pedagogy, Professionalism and Politics: Biology Instruction During the Third Reich," in *Science, Technology and National Socialism*, ed. Monika Renneberg and Mark Walker (Cambridge, 1994), 184–96; in medicine, health, and eugenics: Weindling, *Health*; Michael Burleigh, *Death and Deliverance: 'Euthanasia' in Germany 1900–1945* (Cambridge, 1994); Burleigh and Wippermann, *Racial State*, 1–73.
8. On "social revolution," see Introduction, note 16.
9. Profits of more than RM 8.6 million from revenues of just under RM 15m in 1924 and, following a fee reduction in 1926, RM 3.2m on annual revenues of RM 14.4m in 1927–29, BAP, 2336, Bl. 39c. Rationale for avoiding major changes: correspondence Dr. Curt Joël (RJM) and RPA president von Specht, 11 Jul. 1920,

The Great Depression

18 Jun. 1921, BAK, R131/153, Bl. 147ff., 160f.; circular letter by Reichstag deputy Dr. Bell, 18 Jan. 1927, BAK, R131/153. On Joël: Wolfgang Benz and Hermann Graml, eds., *Biographisches Lexikon zur Weimarer Republik* (Munich, 1988), 161–62.

10. Germany had joined the International Convention for the Protection of Industrial Property in 1903. Periodically amended, the 1883 Paris Convention to this day is the basis of international intellectual property agreements. One revision took place in the Hague in November 1925 and required ratification and revision of the Patent Code by the Reichstag. This prompted the RJM to consolidate various minor issues awaiting reform in an updated code; Benkard, *Patentgesetz*, 78.

11. Kammerer: *Reichshandbuch der Deutschen Gesellschaft*, 2 vols. (Berlin, 1930), vol. 1, 877. Olaf Groehler, "Hugo Junkers: Luftfahrtpionier, Industrieller und bürgerlicher Liberaler," in *Alternativen: Schicksale deutscher Bürger*, ed. Olaf Groehler (Berlin, 1987), 57–90; Olaf Groehler and Helmut Erfurth, *Hugo Junkers: Ein politisches Essay* (Berlin, 1989); Richard Blunk, *Hugo Junkers: Der Mensch und das Werk* (Berlin, 1943).

12. Carl Hartung, "Wie ist es mit dem Schutze geistiger technischer Arbeit bestellt," *Wirtschaftliche Technik* 5 (Jul. 1924): 35–37, RWW, 20-875-1. The Reichstag in March 1926 reduced patent fees by 29 percent, lowering the maximum from RM 10,055 to RM 7,120.

13. Reichsrat, no. 13, Tagung 1928, *Entwurf eines Gesetzes zur Abänderung der Gesetze über gewerblichen Rechtsschutz*, 4 Feb. 1928; Hartung to RWR, 28 Feb. 1928, BAK, R131/155.

14. Hartung to RWR, 28 Feb. 1928, BAK, R131/155.

15. Printed petition by Ausschuß für Erfinderinteressen, 21 Feb. 1928; various reprints from *Wirtschaftliche Technik* on burden of patent fees and other anti-inventor aspects; "Nochmals die Patentreform," *Der angestellte Akademiker* 10, no. 4 (15 Apr. 1928): 32–33; "Abänderungsanträge des 'Deutschen Erfinder-Schutzverbandes' e.V. München gegenüber dem Entwurf des Patentgesetzes 1928 nach den Vorschlägen des Reichsjustizministeriums," *Deutsche Erfinder-Zeitung* 12 (Jun. 1928): 1–9, BAK, R131/155.

16. Delegates from industry: Ludwig Kastl, RDI executive director; IG Farben's Carl Duisberg, RDI president; and Dr. Demuth, business manager of Berlin Chamber of Commerce and committee chair; employee delegates: engineer Breddemann, business manager of Verband deutscher Techniker; engineer Frommholz, executive board of Gewerkschaftsbund der Angestellten; and Fritz Pfirrmann, BUTAB business manager; independents: architect Kröger, chair, Association of German Architects; manufacturer Vögele; and Hartung. BAK, R131/155, RWR report, "Gutachten des Arbeitsausschusses zur Beratung des Entwurfs eines Gesetzes zur Abänderung der Gesetze über gewerblichen Rechtsschutz," 4 Jun. 1928, pp. 2, 11.

17. 1. Sitzung des Arbeitsausschusses zur Beratung des Entwurfs eines Gesetzes über gewerblichen Rechtsschutz, 27 Mar. 1928, 3–6, 7, 11, 14, 16–20, 43ff., BAK, R131/155. On the American patent system from a German perspective, and importance of the former for IG Farben's survival in the 1920s: Walter Duisberg, "Die Stellung der Erfinder der I.G. zum amerikanischen Patentrecht," 1926 (unpublished ms., Bayer, 22/12).

18. 1. Sitzung des Arbeitsausschusses, 13, 24.

19. Ibid., 16.

20. Ibid., 22.

21. Ibid., 25.
22. Ibid., 42.
23. Ibid., 30–31.
24. Ibid., 32, 18.
25. Ibid., 51–52.
26. Minutes of meeting of working committee, 24 May 1928; Gutachten des Ausschusses zur Beratung des Entwurfs eines Gesetzes zur Abänderung der Gesetze über gewerblichen Rechtsschutz, 4 Jun. 1928, pp. 55–60, BAK, R131/155.
27. Leipart to RJM, 18 Jun. 1928, BAK, 131/156; Koch-Weser to Reichsrat, 12 Feb. 1929, BAP, 2336 (RJM), Bl. 115; Hugo Cahn I, "Das zukünftige Patent- Muster- und Zeichenwesen," *Frankfurter Zeitung* no. 656 (3 Sep. 1929).
28. Evidence of pro-reform activity in RJM files includes letters and requests for speedy legislative action from BUTAB, Reichstag members Siegfried Aufhäuser (chair of AfA Bund) and Magdeburg representative Dr. Kulenkampff (trying to help his district's Junkers firm), individual inventors Carl Hartung, Hugo Junkers, Richard Weber, and Oskar Pöbing, patent attorney Georg Neumann, and the newly formed Association of Consulting Patent Engineers; BUTAB to RJM, 12 Jul. 1928 and 27 Feb. 1929, BAP, 2336 (RJM), Bl. 136, 146–47; Georg Neumann, "Patentgesetz Reform: Erleichterungen für unbemittelte Erfinder," *Vorwärts* 114 (4 Sep. 1928), ibid., Bl. 3; Richard Weber to RJM, 15 Nov. 1928, ibid., Bl. 165; Kulenkampff to RJM, 12 Dec. 1928, ibid., Bl. 75; Hartung and Junkers at Parlamentarischer Abend der freien geistigen Berufe und des Ausschusses für Erfinderinteressen, 31 Jan. 1929, ibid., Bl. 95–98; Aufhäuser to Koch-Weser, 18 Feb. 1929, ibid., Bl. 127; Pöbing's request in minutes of VDI patent committee meeting, 9 Mar. 1929, ibid., Bl. 153c. Anti-reform activity: minutes of meeting of DVSGE, Kleiner Patentausschuß, 10 Jul. 1928, BAK, R131/162; DVSGE to RJM, 27 Sep. 1928, BAK, R131/155; DVSGE to RJM, 18 Jun. 1929, BAP, 2336, Bl. 238; RDI to RJM 9 Apr. 1929, BAP, 2336 (RJM), Bl. 176; VDMA (Free) to RJM, 10 Dec. 1928, ibid., Bl. 57; Popitz to RJM, 2 Oct. 1929, BAP, 2337 (RJM), Bl. 72; RPA to RJM, 29 Aug. 1928, pp. 24–27, BAK R131/155; DVSGE to RJM, 27 Sep. 1928, BAK, R131/155.
29. Deutscher Erfinder-Schutzverband e.V. to RPA President, 25 Jun. 1927, BAK, R121/154.
30. *Deutsche Erfinder-Zeitung* 17 (Jun. 1928): 1–9.
31. Pöbing to Wilhelm Frick, 26 Apr. 1934, BAP, 2342; Pöbing to DAF, 17 Nov. 1934, BAK, R131/160.
32. Pöbing to RJM and *Reichsgericht*, 30 Nov. 1930, BAP, 2337, Bl. 137–42, italics in original. Posturing as the upholder of Republican institutions precisely to bring down the Republic was a favorite Nazi trick. That one did not have to be a Nazi to demand inventor relief under the poverty laws was proved by the BUTAB, which in August 1931 called upon Joël to do exactly what Pöbing had demanded the year before. The Regulations of Civil Procedure, the BUTAB wrote, contained a big "gap," which "for all practical purposes robs ... the inventor without means ... of patent protection." The BUTAB, "as a professional organization in the camp of the free trade unions," pleaded with Joël "to give the inventor without means, who is typically unemployed, the opportunity to invoke the poverty laws, so as to proceed in nullification complaints before the *Reichsgericht*." (The Reichsgericht on 12 Nov. 1930 decided to charge plaintiff Isar RM 1,500 for an expert witness report, which ended up siding with defendant Pöbing, who therefore in effect got the subsidized counsel he had demanded.) BUTAB to Joël,

6 Aug. 1931, BAP 2338, Bl. 155–57; correspondence Aufhäuser, Joël, and Justice Katluhn, 2 Nov. 1931 to 8 Dec. 1931, BAP, 2339, Bl. 62–80.

33. Oskar Pöbing, "Erfinder Heraus!" *Deutsche Erfinder-Zeitung* 19, no. 6 (Jun. 1930): 1–2, ibid., no. 7 (Jul. 1930): 1–2. On engineers' anticapitalism and atttitudes toward National Socialism in Weimar: Ludwig, *Technik und Ingenieure*, 15–102.

34. The VDI's failure to defend patent interests of its membership was the main reason for the formation by Hartung and Kammerer in Oct. 1928 of the Committee for the Interests of Inventors, which included some 800 technology firms of various sizes and numerous independents. Conflict between Hartung/Kammerer and VDI: VDI to RJM, 10 Jul., 15 and 20 Oct. 1928, 9 Mar. 1929, BAP 2336, Bl. 8, 42, 68a, 153c; Hartung to Brüning, 10 Nov. 1931, BAP, 2339, Bl. 4, pp. 10–14; "Zum Geleit," *Wirtschaftliche Technik* 18 (new series 8) (Oct. 1928): 1, BAP 2336, Bl. 155.

35. Minutes of VDI patent committee meeting, 9 Mar. 1929, p. 25, BAP, 2336, Bl. 153c.

36. *Reichstag*, fifth legislative period 1930, Printed Matter No. 270, *Interpellation Dr. Frick, Feder (Sachsen) und Genossen.*

37. Pöbing to Frick, 26 Apr. 1934, BAP 2342. Pöbing was forced into retirement by the Bavarian government over participation in the nudist movement. This ruling was rescinded in the Third Reich but Pöbing did not get his job back; Pöbing to Feder, 7 Mar. 1934, BAP 2342.

38. BAP 2339; Eyck's article, 19 May 1932, ibid., Bl. 2.

39. DAFER included (1) Hartung's and Kammerer's 1928 organization, the Committee for Inventor Interests, (2) BUDACI, (3) Bund Deutscher Architekten, (4) Bund Deutscher Civil-Ingenieure, (5) BUTAB, (6) Reichsvereinigung Deutscher Techniker im Gewerkschaftsbund der Angestellten, (7) Verband Deutscher Techniker Gewerkschaft der technischen Angestellten und Beamten, (8) Verband Deutscher Diplom-Ingenieure and (9) VELA.

40. DAFER demands were to (1) move up the starting date of patent protection to the date of publication, (2) allow deferment of annual fee payments for up to three years, (3) obtain waivers for court costs and charges for expert witnesses when defending nullification attacks (added Jan. 1932); DAFER (Hartung) to Brüning, 10 Nov. 1931, BAP, 2339, Bl. 12–14, and 7 Jan. 1931, ibid., Bl. 105–9.

41. The Emergency Decree temporarily reduced patent fees by amounts varying from RM 50 to RM 200. It suspended the penalty for late payments, in some cases allowing partial installments as evidence of good faith; and it exempted patent holders without means—though only in cases before the *Reichsgericht*—from court costs and expenses for expert witnesses. *RGBl* 1 no. 35 (1932): 295–96, Verordnung des Reichspräsidenten über Massnahmen auf dem Gebiet der Rechtspflege und Verwaltung. Vom 14. Juni 1932. Vierter Teil. Gewerblicher Rechtsschutz. Kapitel I–III, BAP 2339, Bl. 219–20.

42. BDCI (Hartung) to RJM, 13 Nov. 1931, BAP 2339, Bl. 22–23.

43. Free to Joël, 13 Nov. 1931, BAP 2339, 20–21.

44. The BUDACI also opposed passage of the 1931 bill, because it included an unacceptably broad definition of the company invention, which the government had quietly adopted at Ludwig Fischer's behest. The RDT had been established in 1918/19. Originally left of center, it was a loose coalition of engineers and engineering groups hoping to improve the social and political influence of their profession. Over the course of the Weimar Republic, it increasingly moved to the right. By the early 1930s the RDT had become an organization that advocated

neocorporatist social models as the way to improve the political power of technology and its exponents. Gottfried Feder became RDT chairman in April 1933, but shortly afterward it fell victim to synchronization; Ludwig, *Technik und Ingenieure*, 105–59.

45. RDI opposition noted in a ministerial comment on BDA (architects) to RJM, 23 Nov. 1931, BAP, 2339, Bl. 61; DVSGE minutes of meeting of 23 Feb. 1932, BAK, R131/163; DVSGE to RJM, 11 Oct. 1932, BAP, 2339, Bl. 299a–i, 300; VDPA to RJM, 19 Dec. 1931, BAP, 2339, Bl. 94; RDT to RJM, 19 Dec. 1931, ibid., Bl. 96; VDI (Director Hellmich) to RJM, 21 Dec. 1931, ibid., Bl. 98; RPA president Eylau to RJM, 3 Dec. 1931, ibid., Bl. 154–55; BUDACI's Dr. Kurt Milde to Brüning, 8 Dec. 1931, ibid., Bl. 86–87. RJM to DAFER, 30 Dec. 1931, ibid., Bl. 154–55; RJM, evaluation of objections and proposals, 30 Jul. 1932, BAK, R131/156, and idem, 1 Sep. 1932, BAP, 2339, Bl. 231–40.

46. Interior Minister von Geyl to Reichsrat, 15 Nov. 1932, BAP, 2339, Bl. 303.

47. Why and how Germans became "ripe" for Hitler, especially in connection with unrelenting delegitimization of Weimar and the rise of exclusivist nationalism, is discussed by Peter Fritzsche, *Germans into Nazis* (Cambridge, Mass., 1998); see also Peukert, *Weimar Republic*; Prinz, *Vom neuen Mittelstand*; Childers, *Nazi Voter*; Jenkins and Sofos, "Nation and Nationalism."

48. Gerhard Banse and Siegfried Wollgast, eds., *Biographien bedeutender Techniker, Ingenieure und Technikwissenschaftler: Eine Sammlung von Biographien* (Berlin, 1987), 297–305; Arco to RJM, 29 Dec. 1931, BAP, 2339, Bl. 103.

49. von Ardenne, *glückliches Leben*, 93–95, passim.

50. Andreas Thomsen, *Denkschrift an den Deutschen Reichstag betr. Nutzbarmachung der im deutschen Volke vorhandenen Erfinderkräfte zum Wiederaufbau und zur Mehrung des Volksvermögens* (Münster, 1931), BAP, 2338, Bl. 195ff.

51. Thomsen, *Nutzbarmachung*, 5–7. On Schwerber, see Ludwig, *Technik und Ingenieure*, 79–85, 159, in the context of engineers' anticapitalist ideology.

52. Thomsen, *Nutzbarmachung*, 8.

53. Ibid., 9–10.

54. Ibid., 11–18.

55. Ibid., 18–19.

56. Art-and-Craft-Metal-Workshop-Richard-Weber to Reichstag, 15 Nov. 1928, BAP 2336, Bl. 165; Otto Haentsch to RJM, 4 Aug. 1932, BAP, 2339, Bl. 223; Das Deutsche Erfinderhaus e.V., Hamburg (Heinrich Jebens) to RJM, 27 Jan. 1931, BAP, 2337, Bl. 196a–b; Fortschritts-Akademie e.V. Hamburg (Heinrich Jebens) to Chancellor Kurt von Schleicher, 29 Dec. 1932, BAP, 2340.

Part III

THE THIRD REICH

Chapter 7

HEINRICH JEBENS AND THE REICH INVENTOR OFFICE

When Hitler became Chancellor on 30 January 1933, the patent bill was waiting to be resubmitted to the new Reichstag, elections for which were scheduled to take place 5 March.[1] The Nazis conducted a frenzied electoral campaign, accompanied by systematic violence against the left and culminating in the Reichstag fire of 27 February. The next day, President Hindenburg signed the Emergency Decrees that suspended civil liberties and became the legal foundation of the Third Reich's police and SS terror. Hitler, meanwhile, gave constant campaign speeches and traveled all over Germany. Seeking to rally the German people with a final blitz of propaganda, Hitler did not forget to mention the technological miracles produced by human inventiveness and the genius of inventors. "A single inventor, a genius," he proclaimed, "can mean more for a nation than hundreds of millions, or billions, of apparent capital."[2]

Although the March elections failed to produce a majority for the Nazi party, Hitler's pro-inventor rhetoric appears to have made a big impression on those who felt his comments applied to them. The files of the Reich Justice Ministry contain numerous letters addressed to Hitler, mostly from small independents, thanking the new Chancellor for singling out the inventor and the individual's technological creativity as the ultimate sources of German national strength in the depths of the depression. Besides acclaiming Hitler personally and bursting with enthusiasm for the Nazi movement, the letters typically also include impatient offers to help fix Germany's unjust patent system and give the inventor his due. Similar writings, sometimes addressed to Hermann Göring, Joseph Goebbels, Gottfried Feder, or other high Nazi and government figures, continued to arrive throughout the first three and a half years of the Third Reich.

All of the communications in question welcome Hitler's accession to power and exhibit strong resentment of the existing legal situation. They all make specific recommendations for reform, such as lower patent and legal fees, introduction of the first-to-file principle, statutory rewards for salaried inventors, government funding to help undercapitalized independents develop and market their inventions, and other measures to help small business in its uneven patent contests with large corporations. Some of the letters and petitions leave it at that. Others blend their enthusiasm for the new regime and their reform proposals with a pronounced hatred of the Jews and the liberal-capitalist-Marxist free-for-all they equated with the Weimar Republic, forming an amalgam that is typical of the revolutionary energy of National Socialism.

It is not practical to review all this material in detail, but a few examples will serve to convey the flavor of the response elicited by Hitler's pro-inventor remarks in early 1933. One of the first to congratulate Hitler was Carl Hartung, chairman of the Bund deutscher Civil-Ingenieure and the driving force behind the RWR's patent reform proposals of 1928. Hartung made no antisemitic remarks in his letter. Nor did he salute Hitler with typical Nazi phrases such as "my Führer," "Heil," or "mit deutschem Gruss," using instead the traditional salutation of "Most respected Herr Reich Chancellor" and closing with the usual "Respectfully." Signaling his distance from the party with these formulations and other signs of restraint, Hartung was nevertheless most excited about Hitler's praise for the inventor. He was also optimistic that the new regime would finally succeed where the Empire and Weimar had failed. "It is our great pleasure to learn," he wrote, "that you are particularly interested in ... technical creation, and we are even happier about your views on the value of the individual and responsible initiative in politics and the economy." For many years, Hartung's organization and the "affiliated inventors from all sectors of industry and especially from small and mid-sized business" had been working toward the same end as did Hitler, fighting to improve the economic opportunities for the individual technological pioneer by "reforming the intellectual property laws." Making progress on that front was the only way to reach the "highest goals," on which "rest the highest achievements of our industry, which in turn are the only ones [immune to foreign] tariff barriers." Ever since 1913, patent reform proposals had gone to the

Reichstag, but for the past twenty years the "political parliament" had been incapable of resolving the issue. "We hope," Hartung concluded, "that you will take an interest in this question, whose satisfactory resolution will contribute like no other to the reduction of unemployment, and [we hope] it will then be an easy thing to quickly pass" the long-discussed patent bill.

In a separate enclosure Hartung informed Hitler about the organization he chaired. As the nation's leading association of independent engineers, the Bund Deutscher Civil-Ingenieure "greet[ed] Germany's national renewal with the strongest approval." Within the limits of its "modest powers and its professional ability and nature," Hartung wrote, the Bund "eagerly placed itself at the disposal of the national government." Doing so was only natural for an organization that since its founding in 1909 had been steadfastly patriotic and had remained patriotic even when patriotism went out of style. A second reason for the Bund's natural affinity with the new regime was its long-standing belief that the "basic principles of a free profession are incompatible with a Marxist orientation and collectivist forms of economic organization." The Bund had "always relentlessly and publicly rejected everything that tended in that direction, both from above and below." On behalf of the entire membership, therefore, Hartung could only conclude with an "expression of satisfaction that a government ha[d] finally been formed in Germany that corresponds to our ideal" in its views both of patriotism and Marxism.[3] Hartung's letter to Hitler was dated 31 March 1933, well after the first waves of state-sponsored terror and murder and SA actions against Jews had already taken place.

If someone like Hartung—successful, upper-middle-class, left-of-center, and a well-known public figure—could embrace the Nazi regime while turning a blind eye to its misdeeds, it is not surprising that others further right on the political spectrum and down the socioeconomic scale seemed almost delirious with excitement about the inventor measures the "Führer" would surely take. One of those enthusiasts was the marginal Stettin inventor Hans Koenig, a vocationally trained engineer and member of the Reich Inventor League, an organization of small independents. Koenig began his pathetic handwritten letter of 1 August 1933 with the messianic observation that the "German people's road to Calvary has come to an end." Hitler's "positive deed," wrote Koenig, had "freed the German people from its humiliating chains of slavery." Millions of people were "ready to give their best for the fatherland … among them thousands of German inventors, to whom you, most respected Herr

Reich Chancellor, paid tribute in your Berlin Sportpalast speech of 2 March 1933."

Koenig then explained that as a small and anonymous independent, he was writing to request that Hitler do everything he could in support of Germany's inventors—for instance, implement the pro-inventor program of Professor Andreas Thomsen mentioned in the previous chapter. "Surrounded by a world of enemies who envy us for the German inventor spirit—a spirit that has already created such infinitely great and admirable things—we German inventors are fighting a hard struggle for existence," wrote Koenig. To ameliorate their plight, Thomsen had submitted his study to the Reichstag as early as 1930, and Hitler ought to know that in this work "Thomsen developed very worthwhile proposals to support and promote Germany's inventor community." Other measures Koenig urged Hitler to take included "a fundamental reform of German patent law" and taking advantage of the wisdom and experience of men such as himself, who, though unemployed and beaten down by the Great Depression, had only one goal: "true to your lofty example, to place our God-given talents ... in the service of our suffering fellow countrymen and to create work, so that our beloved German fatherland may be resurrected." Koenig concluded by reminding Hitler of his "call for help from those who are of good will" and by reiterating that he was one of the ones prepared to "follow you, unshakable in our trust in you as our exalted Führer and steadfast in our faith in Germany's resurrection and in the help of the Almighty. So help us God!"[4]

Another memorandum was addressed not to Hitler personally, but to the Reich government as a whole. It hailed from a certain Konrad Pfreundner, to all appearances a failed small inventor in Munich who described himself as awaiting social-insurance payments and who made the "National Labor Day" of 1 May the occasion for his good advice. Accompanying a detailed twelve-page proposal for government-funded evaluation, development, and marketing of independents' inventions, Pfreundner's letter teemed with antisemitism, especially with resentment of the community of patent attorneys and lawyers specializing in patent matters. If only the government followed his suggestions for patent reform and state-supported inventor assistance, Pfreundner wrote, Germany would quickly emerge from the economic doldrums. But even more important, those measures would "especially strike hard at the Jew, who has developed very special roots and knows how to survive in this field, who cheaply buys up or copies the best inventions as soon as the usually impoverished inventor conceives of them, and who earns millions in the process while the true inventor goes hungry."

According to Pfreundner, "thousands of Jews here play their destructive game, aided by the inadequate facilities on the part of the Patent Office," which concerned itself only with evaluating the legal, as opposed to the economic, aspects of inventions. To counter the nefarious activities of what he called "patent hyenas," a comprehensive inventor-support institution should be established, which would take care of both the legal protection and the economic realization of "German intellectual matter, free from Jewish greed, borne by national pride and braced by the motto: 'German creations for the Germans first!'" If other countries wanted to buy German intellectual property, for instance, this should not happen "as it has until now, by way of Jewish shysters," but in an orderly manner and only after payment of its true value as determined by the state's comprehensive inventor agency.[5]

Then there was an otherwise unidentified Louis Bastian, who also praised Hitler for his "fantastic speech in the Berlin Sportpalast," which he called "unsurpassed in the happy resonance and the degree of enthusiasm" it generated among the German people, especially when the "Führer" had spoken up for "individualism and the creative energies of the individual to benefit the nation as a whole." The best way to unleash those energies, Bastian told Hitler, was to bring about a "fundamental revolution in the field of the patent system."

The current system, according to Bastian, amounted to "state-sanctioned corruption of the worst kind." Poor inventors—the vast majority nowadays—had no chance to obtain a patent on their own, because the Patent Office erected the "greatest obstacles" and made the "most impossible demands to justify a rejection" of their applications. The inventor, provided he had enough money, was therefore forced to engage a lawyer and "entrust himself to the hands of patent attorneys who were often completely unscrupulous, mostly Jews and masters in the art of scalping their victims."

To end the abuses, which the heartless and arrogant bureaucrats in the Patent Office could not care less about, Hitler ought to shake up that "ossified" agency and appoint new, dynamic leadership. Hitler should also establish a "strike force on National Socialist footing," whose chief mission it would be "to protect the less affluent inventor against exploitation by unscrupulous professional parasites." The same body, however, should also have an "advisory and controlling role in the Patent Office." Bastian concluded with a request that Hitler empower him to establish the strike force in question, which would present the Chancellor "with appropriate recommendations and guidelines."[6]

Similar proposals came from the small independents Th. Gosch and Heinz Klee. Gosch thought that one of the most important reforms, now that the "liberal-capitalist *Weltanschauung* had fortunately been transcended," was to establish an institution for the protection of the "impoverished inventor against unscrupulous finance capital." Inventors who turned their ideas over to the government, for instance, should be entitled to regular annuities. Gosch also wanted to endow the Reich Patent Office with a staff of dedicated National Socialist counselors, who would prepare patent applications for free and so "make superfluous today's patent attorneys (predominantly Jews)."[7] On a similar note, Klee, the son of a Hanau butcher and himself a devoted National Socialist, urged Göring in December 1933 to establish a "Center for German Inventors, to promote the German inventor spirit for the overall benefit of the German people." The 27-year-old, self-taught inventor, who had followed a commercial apprenticeship, reinforced his arguments by quoting Hitler's words about inventors from *Mein Kampf*. Hitler's book, he maintained, established beyond any doubt the value of the individual for the life of the nation, from which it followed that an inventor-support agency was the natural complement of any National Socialist inventor policy.[8]

Precisely how many people wrote letters such as this is difficult to say. There are some seventy of them in the Ministry of Justice's files alone.[9] In the aggregate, they present a revealing picture of the small independent's mentality in the late 1920s and early 1930s. While this contributes to the understanding of National Socialism's popularity, the same material also contains clues that help solve the puzzle of how exactly the Third Reich came to play such a decisive role in restructuring Germany's patent system. The documents in question do more than register sentiment; they also tell a story of action born of that sentiment—action that marked the beginning of National Socialist inventor policy.

Among the numerous outsiders hoping to become insiders who wrote to Hitler about inventing in early 1933 was a certain Heinrich Jebens, at the time a 37-year-old "patent economist" from Hamburg. The son of a Holstein farmer, Jebens had left school after obtaining the "middling maturity" at about age sixteen and was apprenticed as an electrician. When World War I broke out, he was a navy-engineering apprentice at a private shipyard in Kiel, and by

April 1916 he had become a Navy engineer-candidate. From that time until October 1918 he was on active duty, and then attended the Navy's Engineering School in Kiel until it closed in January 1919 because of the Revolution. In 1920 Jebens entered the police academy in Potsdam. After graduating, he worked as a police lieutenant in Prussian government service until 1924, consistently earning high marks from his superiors. In the meantime, however, Jebens had become interested in intellectual property law and the marketing of inventions, so he resigned his commission in 1924 to become a patent consultant.[10]

Having joined a mutual aid society for small inventors in Hamburg, Jebens by mid-1925 managed to get himself elected chairman of the organization in question, which he renamed the German Inventor House. As chairman of the Inventor House, or "Director," as he liked to describe himself, Jebens showed a great deal of entrepreneurial energy and talent. He rapidly increased the membership, wrote essays about the plight of inventors, and started a journal that marketed inventions. He attracted qualified collaborators and built up a staff of more than fifty employees, who assisted small independents by evaluating the economic potential of their ideas, making models, preparing patent applications, finding buyers, etc. His income derived primarily from the monthly membership contributions that inventors paid to belong to the Inventor House, but also from various other fees and sales commissions.[11]

Jebens' dynamism and success soon caused him a certain amount of trouble with the law. In early 1926, the Reich Patent Office turned down his request to represent clients before the Patent Office, because that was a monopoly of the legal profession and licensed patent attorneys and he lacked the proper academic credentials. When he continued to process and sign patent applications on behalf of others anyway, the Patent Office sent him a warning that his business would be closed down if he did not desist. Jebens temporarily solved the problem by contracting with various Berlin patent attorneys who would sign off on his applications for a fee. This earned him a certain notoriety with the Patent Office, but since the practice was difficult to prove, patent officials could do little besides monitor Jebens's activities and start a file on him.[12]

A second problem seemed initially more ominous. Other patent consultants in the Hamburg area began complaining about unfair competition, unethical tactics, false advertising, and fraud on the part of Jebens's German Inventor House. Patent attorneys, meanwhile, were angered that unscrupulous colleagues signed off on

Jebens's applications. When attacks in the local press remained ineffective, the Association of Consulting Patent Engineers, the professional organization of "patent economists," filed an official complaint against Jebens in early 1932. This caused the Hamburg prosecutor's office to launch a criminal investigation. Neither this nor another, earlier investigation by the Leipzig prosecutor found evidence of criminal wrongdoing, even though "in a number of instances the advertising activities of the accused come dangerously close to the limit of what is allowed."[13]

Another complaint about Jebens's subcontracting with patent attorneys, filed with the Reich Patent Office by the Association of Consulting Patent Engineers, also went nowhere. As Patent Office president Johannes Eylau concluded in November of 1930, it was not his responsibility to investigate allegations about unnamed patent attorneys laundering Jebens's applications. In the meantime, "real transgressions of the German Inventor House ... against its clients have not been brought to my attention—certainly not by the Association of Consulting Patent Engineers."[14] Still, filing patent papers over the signature of outside patent attorneys must have presented certain difficulties. Whether it became too expensive or because his sources dried up, sometime in 1931 Jebens once again began filing applications on his own and, expectedly, immediately got into trouble again. This time, the Patent Office would not even consider the applications that had already been filed. During a interview with President Eylau, Jebens begged for acceptance of the applications already submitted, promising in return to abandon the legal aspects of the patenting process once and for all and to limit his future activities to assessing the economic potential of inventions. It was to no avail, and the Patent Office turned down his request in August 1931.[15]

The upshot was that Jebens developed a profound hatred for the Reich Patent Office well before the Republic entered its final death throes. This, and the mounting difficulties of his own business and those of small inventors all over Germany during the Great Depression (which, among other things, raised the relative cost of patenting and caused a sharp contraction in filings and maintenance), led Jebens to become increasingly politicized. Initially, this took the form of a feverish writing and publishing campaign, which started about the time he received the Patent Office's final rejection.

Imagining himself to be some new prophet of technology who had found the answer to Germany's economic problems in its patent system, Jebens in August of 1931 published his first book, *The Phi-*

losophy of Progress: Against Spengler, Against Hitler. This work, as well
as other essays and long letters and his 1933 pamphlet, *The Road to
Total Power: Through Progress and National Socialism*, contained a
great deal of gibberish.[16] Jebens's *Philosophy of Progress*, for instance,
was a bizarre amalgam of Christian ethics and technology as objectified spirit; of a "true National Socialism" that was technocratic and
rational, above the parties and antidemocratic but also based on
internationalism and pacifism, as opposed to Hitler's rabid antisemitism and violent anticommunism, which would inevitably lead
to destruction; of an unshakable faith in eternal technological
progress from mindless matter to immaterial mind, based on man's
innate reason and inventive potential and refuting Spengler's pessimistic philosophy of standstill and decline; and of the scientific
proof of everything he said in terms of the principles of electricity.[17]

Amid all the nonsense, however, Jebens also preached the message that the nation could recover only if inventions, before being
patented, were tested as to their economic viability and potential.
Only the truly promising ones should be allowed to go on for a
patent application. What this relatively simple step would do,
according to Jebens, was two things. First, it would help independents by saving them untold misery as well as a great deal of
money—money and energy that were currently wasted on patents
and hopes that rarely panned out. Second, and more importantly, it
would turn the Patent Office from its current function as the
"gravedigger of the economy" into the fountainhead of economic
energy and recovery that it ought to be.[18]

Jebens's argument centered on the belief that the current stage of
the age of technology was characterized by such a massive flood of
inventions and innovations that entrepreneurs and small business
drowned in a sea of too much progress, which was too fragmented
and did not distinguish between good and bad ideas. As a consequence, according to Jebens, there was a kind of churning of inventions, an "inflation of inventing" inversely proportional to significance
and economic yield, so that no more than about one-half of one percent of all inventions contributed any longer to technological progress
and economic success. This minute yield, he claimed, was proof of a
"catastrophic lack of success in inventing." But no one was doing
anything about it, least of all the Patent Office, which continued to
"print new, worthless patent documents," very much like the printing
presses that at the height of the inflation had turned out ever more
bills that were worth ever less. Instead of strengthening industry,
patents had become "the grave of industry."

All in all, maintained Jebens, shifting metaphors as well as his argument, there was too much undisciplined technological progress, which by its "wild violence progressively destroys everything," as though one had "put half a dozen wild prairie horses in front of a light carriage." There was an "excess of power, and only the taming of these powers is still lacking." Fortunately, Jebens knew just how to do it, thereby solving in one fell swoop the economic problem of stagnation and the moral problem of why technological progress had so far failed to bring about mankind's happiness. Until now, he pointed out, patents were strictly legal documents, verifying only newness but saying nothing about value. The consequent devaluation of patents and slowdown of industrial progress could only be arrested and reversed if, prior to the legalistic test of novelty, the government introduced a "pre-examination of economic merit and achievement potential, so as to tie technology more closely to economic needs." This, according to Jebens, was "the core problem of our technological era," which he said he had gradually come to understand through his experiences with the German Inventor House in Hamburg. "Technology should no longer be allowed to grow like a weed and without supervision on the economy's terrain," he wrote, but needed to be "chained to the achievements of the economy." In other words, "the time had come to complement the current Patent Code with a law about economic pre-examination." This would eliminate 80 percent of all the poor and doomed inventions. The remaining patented inventions would not merely thrive, but would reconquer Germany's technological lead over the rest of the world and hence expand its exports, and so, in the end, also achieve its economic recovery. In sum, Jebens concluded, there was only one solution: "economic policy must be complemented with technology policy."[19]

Jebens was not satisfied with having discovered the answer to Germany's deepest problems; naturally he also wanted to see his solution implemented. To that end, he wrote letter after letter to the country's highest officials. In August 1931 he sent a copy of his *Philosophy of Progress* to Eylau, along with a sarcastic letter about the Patent Office president's "fabled energy and wisdom," which guaranteed that sooner or later the government would figure out how to "steer the right course."[20] In October 1932 he outlined his ideas in a long letter to Chancellor von Papen, and in December he sent a similar letter to Papen's successor Schleicher, suggesting they meet to discuss the problem and announcing that he and "the circles close to him" would not rest "until the Patent Office as one of the most important agencies of the Reich comes out of its quiet corner and has become that which it must be in our age of progress." Jebens enclosed an open letter he had sent earlier to President Hindenburg, all government ministers and Reichstag delegates, the various industry and engineering associations, and the press. In it, he again made the case for economic evaluation of inventions and accused the Patent Office of being a mindless "factory of certificates, the breeding ground of an invention-inflation that destroyed industry and economy." The Patent Office "wasted ninety percent" of the millions inventors spent on the "worthless paper" it issued as patents, meanwhile earning millions of marks in patent fees for itself. The abuse would end only if the president of the Patent Office mustered enough courage to "subordinate the principle of legalism to the principle of achievement."[21]

Having adopted the language of National Socialism well before 30 January 1933, Jebens followed his words with deeds a few months later, joining the NSDAP in May 1933.[22] From that moment he was a true believer, devoted to the cause like no one else and determined to realize his plans for patent reform and inventor help at any cost. Jebens's conversion, manifest in the sudden adoption of SA-style antisemitism and populism, revealed itself in his letter "to the Reich Chancellery" of 8 June 1933. "Only one question:" he began unceremoniously, "would our marvelous Hitler movement ever have succeeded if we had not had telephone, railroad, automobile and airplane—in short, modern technology?"

> No, never! We would have needed ten people's Chancellors to speak to the people everywhere. But there is only one of them, and he is called … Adolf Hitler. And the same Adolf Hitler has repeatedly placed great emphasis on the importance of the self-creative spirit in our nation of inventors, which in the same vein has freely given birth to the automobile, the airplane, etc. Even so, the "mass undertaker" of the economy that calls itself the Reich Patent Office is allowed to continue its frightful trade of destruction, to the benefit of the trusts. It is a grave situation, even a scandal, that none of the relevant Reich agencies bothers to deal with this Jew-infested breeder of destruction of the people's creative spirit.

The tragedy would end quickly and painlessly, Jebens continued, and the nation save "hundreds of millions each year," if he personally received the power to reorganize Germany's patent system and introduce the economic evaluation of inventions. Moreover, he was willing to take the job for free, on an honorary basis and as a special commissioner who grew "hair on his teeth" and knew how to "bite off the mandarin tail of the Patent Office's legalism." In the

Third Reich, this was probably the only way that the "Justice Ministry would begin to understand the concept of the achievement principle." As an agency of technology, Jebens added, the "Patent Office belongs to the engineer, not to the lawyer." In case the message was not yet clear enough, he repeated it once more "in plain German." It was "high time," he said, "for the Chancellor himself to rattle some brains with his walking stick," and, a few sentences later, "to sweep out the remnants of the old system." Only Hitler himself had the vision and the strength to see through the clever arguments of the jurists in the Patent Office, namely, that Jebens's ideas were nonsense "just as the Hitler movement itself had [supposedly] been 'nonsense.'"[23]

To increase his chances of success, Jebens in early June 1933 did not merely write to the Reich Chancellery, but to various other top figures of the new government as well, describing himself at one point as a "glowing National Socialist" and offering his services to anyone who would take them.[24] Basically, he mailed everyone the same text with minor variations in the cover letter, in which he amplified his proposals with attacks on "festering sores" such as the Patent Office and the big concerns or "trusts." But he also attacked "Marxist trade unions" and "lazy communists," who even in the depths of the depression continued to draw their contract wages while "unemployed S.A. men ... [were] working on behalf of the state without pay."[25] Fortunately, developments were moving his way in the Third Reich. One illustration was the "overnight explosion of the little patent-attorney-society's mostly Jewish executive committee, which had so closely been affiliated with you," he wrote to the president of the Patent Office.[26] He told Göring to "crack some nuts," a task for which "only one Minister has been chosen this century." Specifically, Göring should crack the "one nut no one has cracked so far," to wit, the Patent Office, under whose shell there "lie hidden values on the order of billions each year." In case the Minister of Air Transportation was too busy with other tasks, Jebens repeated his offer to do the job for him as unpaid commissioner.[27]

The documents do not reveal exactly what Jebens was up to in the second half of 1933, but at one point he met with Wilhelm Keppler, Hitler's plenipotentiary for economic questions. In December he resurfaced as the newly installed head of the grandiose-sounding Reich Inventor Office (Reichserfinderamt, or REA), a new inventor-support agency under the jurisdiction of the division for social affairs of the German Labor Front (DAF). Jebens's Reich

Inventor Office was located in Berlin, at Alexandrinenstrasse 137, just around the corner from the Patent Office in the Gitschinerstrasse. The new office opened its doors to the public on 2 January 1934, with an initial staff of ten. In a brief announcement of the Reich Inventor Office's birth, the *Deutsche Allgemeine Zeitung* of 15 December 1933 told its readers that "Henceforth, every people's comrade who has inventions and innovations can turn to this office prior to filing for a patent or a design patent—especially before filing for foreign patents—to get solid advice as to the chances of his idea's economic success."[28]

Astonished to learn about such basic developments in their own field from the newspapers, officials in the Patent Office invited Jebens for an interview the next day. At this meeting, Jebens informed his interlocutors that Robert Ley, the head of the German Labor Front, had ordered the establishment of the Reich Inventor Office two weeks earlier.[29] Ley, he said, had decided, "to utilize technology for the purpose of creating productive employment in accordance with the principles of the National Socialist state." Ley, an ex-chemist in IG Farben's Bayer division, had also approved the nomination of Jebens as head of the new agency and given him a mandate to establish the Reich Inventor Office.[30]

The REA's purpose, Jebens explained next, was not to infringe on the territory of the Patent Office or patent attorneys, but rather to practice something like the eugenics of technology. Without mentioning how it would do so or how much it would cost, Jebens said the REA was there to foster the economic development of good inventions and "to eliminate useless ones." Thus, on the one hand, "positive technology" would receive support in accordance with National Socialist principles. On the other hand, the elimination of "negative technology" would relieve the overburdened Reich Patent Office. Good inventions would be sent on to patent attorneys to prepare patent applications, while useless ones would be returned to the inventors. The most important thing, Jebens stressed, was to prevent the waste of precious national capital on the patenting (especially the expensive process of filing multiple applications abroad), development, and marketing of inventions whose economic prospects were clearly hopeless.[31] To place the REA in the best possible light, Jebens mentioned one final benefit of the new agency. It would lead to the elimination of Germany's infamous and fraudulent patent agency system. Independents, he said, would henceforth turn to the official REA when they needed help with developing and marketing their inventions.[32]

In truth, it was not Jebens who was about to eliminate the private patent-agency system and other abuses associated with capitalism, but his old enemies and established government bureaucrats who initiated the process of Jebens's own rapid demise. As early as 23 December a group of former competitors from Hamburg pointed out in a letter to the Reich Patent Office that Jebens was "at the very least a very controversial individual" who had "not exactly made a favorable impression." How could someone who as recently as 1931 had written a book "against Hitler" suddenly be a power broker and the head of an official agency in the new regime? Perhaps it was time for the Patent Office to contact those who might look into the situation a bit more deeply.[33] Likewise, in January 1934 a disgruntled inventor with bad memories of the German Inventor House wrote to the Reich Chancellery to warn the government that Jebens was a crook and a charlatan.[34] Also in January, the Reich Ministry of the Interior complained to the NSDAP about the name of Jebens's agency. Since "the designation 'Reich-Office' [could] create the impression of an official government agency," which in this case might also be mistaken for the Reich Patent Office, it was requested that the party see to it that the 'Reich Inventor Office' be renamed immediately. Within a few days Jebens's REA became the rather more modest Section for Inventor Protection in the DAF's division for social affairs.[35]

It is unclear whether these incidents by themselves would have sufficed to unseat Jebens. It soon became evident, however, that they were part of a broader pattern of self-promotion, self-aggrandizement, and empire building on the part of a freelance revolutionary who caused more trouble than his bosses thought he was worth. Among his various initiatives, Jebens issued guidelines to all patent attorneys concerning "voluntary" fee waivers for impecunious inventors. He also placed unauthorized announcements about his agency in the newspapers—press releases in which he shamelessly promoted his vision of the problems with Germany's technological culture, his theories of "invention inflation," his plans for state-funded invention-development institutes, and his own name.[36] He published interviews with himself as the principal subject, and he refused to slow down or adopt a lower profile when told to do so. Not surprisingly, these heady days did not last long. Jebens was fired on the spot, his career as the nation's technological savior abruptly terminated on 27 January 1934.[37]

After resigning from the DAF, Jebens made one last desperate attempt to interest the Reich Economics Ministry in resurrecting the "Reich Inventor Office" under its dominion. But by now the authorities were on to him, and the plan went nowhere.[38] Empty-handed, Jebens returned to Hamburg, where he tried to reestablish himself in the private invention-exploitation business. In 1935, when he surfaced one more time with an "Institute for the Promotion of Industrial Achievements" and advertising for "Inventor Competition," the NSDAP's Department of Technology (Amt für Technik) quickly moved to shut him down, as Jebens was now a "great danger for the public good, especially for the mostly poor inventors."[39] At this point, Jebens vanishes from the sources, although he seems to have survived the Third Reich unscathed and in 1948 resumed his old habit of publishing schemes for Germany's salvation.[40]

What, if any, is the historical significance of the rise and fall of Heinrich Jebens in the first year of Hitler's dictatorship? At first glance, the episode might seem emblematic for interpretations that emphasize the regime's rapid conversion from its revolutionary anti-capitalism at the outset to pragmatic collaboration in a partnership or "power cartel" with the country's existing business leaders and economic institutions. As Karl-Heinz Ludwig writes, for instance, "in 1933 the forces for reform inside the NSDAP did not disappear overnight, but increasingly found themselves in complete contradiction to the demands of pragmatic politics. To consolidate its position, the party needed a general economic upswing and visible reduction in unemployment. This meant that all uncertain reformist experiments in the technological-economic arena were by definition out of the question."[41] In other words, the quick demise of Jebens in this interpretation would likely be seen as evidence for the Nazi party's speedy dissociation from wild-eyed revolutionaries of the SA variety and its rapid reorientation toward the needs of big business and allied state institutions after 1933.

In truth, Jebens was thrown out more because of his own tactical mistakes and delusions of personal power than because of any real slackening in the revolutionary ardor that drove the politics of inventing in Nazi Germany. Unlike its one-time head, the agency that Jebens had founded with the blessing of Ley's Labor Front in December of 1933 did not disappear from the scene. On the contrary, it survived to become one of the crucial sources of dynamism in the Third Reich's pro-inventor policies from 1934 to 1945. To be sure, it now bore the less provocative name of Section for Inventor Protection, gave up the office space in the Alexandrinenstrasse, and supposedly limited its responsibilities to fighting fraud among private "patent economists" and helping independents assess the economic potential of their inventions. But in spite of this temporary retrenchment, the Section for Inventor Protection quickly reemerged as the core of National Socialism's system of "inventor trusteeship" (*Erfinderbetreuung*). Inspired by hostility to the inherited capitalist patent system, "inventor trustreeship" was an attempt, not merely to help inventors realize their ideas and protect them from exploitation, but also to gain control over all the nation's sources of invention, both among independents and in the corporate laboratories. In conjunction and partially in competition with other pro-inventor revolutionaries in the Nazi party's Department of Technology and Reich Legal Office (Reichsrechtsamt), Jebens's successors in the Section for Inventor Protection developed the initiatives that made for the most important change in Germany's patent system since its inception in 1877.[42] This historical role of the forces from below would be demonstrated in the events, to be discussed next, that led to the Patent Reform Act of 1936.

Notes

1. Von Gayl (RMI) to Reichsrat, 15 Nov. 1932, BAP, 2339; memo of RJM's Kammergerichtsrat Kühnemann, 31 Dec. 1932, BAP 2341.
2. Hitler, in his Sportpalast address of Thursday 2 Mar. 1933, proclaimed it was "not democracy [that] creates values, but personalities. Democracy has always destroyed and demolished the value of the personality. It is madness to think—and a crime to propagate—that suddenly a majority can replace the achievement of the individual genius … Each people must recognize that its most capable people are its greatest national asset, because that is the most eternal value there is. A single inventor, a genius, can mean more for a nation than hundreds of millions, or billions, of apparent capital." *Völkischer Beobachter*, no. 63 (4 Mar. 1933), p. 2. Cf. also Hitler's remarks at the opening of the International Automobile and Motorcycle Exhibition in Berlin on 11 Feb. 1933, quoted in Max Domarus, *Hitler: Reden und Proklamationen 1932–1945*, vol. 1, pt. 1 (1932–34) (Munich, 1965), 208–9; also Hitler's Königsberg speech of 4 Mar. 1933, *Völkischer Beobachter*, no. 67 (8 Mar. 1933).
3. Hartung to Hitler, 31 Mar. 1933, BAP, 2340.
4. Hans Koenig to Hitler, 1 Aug. 1933, BAP, 2340.
5. Konrad Pfreundner, "Versorgungsanwärter in München," to Reichsregierung, 1 May 1933, BAP, 2340.
6. Louis Bastian to Hitler, 24 Apr. 1933, BAP, 2340.
7. Th. Gosch to RMI, 27 May 1933, BAP, 2340.
8. Heinz Klee to Göring, 5 Dec. 1933, BAP, 2297.

9. Files examined in their entirety: BAP, 2297, 2340, 2341, 2342, 2343; BAK, R131/158. Partially analyzed: BAP, 10129, 10130, BAK, R131/160. Of the seventy-odd letters, sixteen are addressed to Hitler directly, with the remainder going to Göring, Goebbels, Feder, Frick, Gürtner, Schmidt-Ott, Keppler, Ley, Seldte, or RPA president Klauer.

10. *Curriculum vitae* in Jebens to Eylau, 22 Jan. 1926, BAP, 2340.

11. Harting to Gürtner, 28 Dec. 1933, BAP, 2340.

12. Ibid.

13. Ibid. (Staatsanwaltschaft Hamburg to VDP, Hamburg, 20 Sep. 1932, p. 1, BAP, 2340.)

14. Ibid. (Eylau to RJM, 11 Nov. 1930, p. 1, BAP, 2340.)

15. Ibid. (Note about Jebens meeting with Eylau, 4 Aug. 1931; Jebens to Eylau, 27 Aug. 1931, BAP, 2340.)

16. Heinrich Jebens, *Die Philosophie des Fortschritts: Contra Spengler, Contra Hitler* (Hamburg, 1931); idem, *Der Weg zur Allmacht: Durch Fortschritt und Nationalsozialismus* (Hamburg, 1933).

17. *Philosophie des Fortschritts*, passim.

18. Ibid., 52–55, 210–12; "Open Letter to the Reich Patent Office of 10 December 1932," in Jebens to Schleicher, 29 Dec. 1932, BAP, 2340.

19. Heinrich Jebens, *Die Bezwingung des Herrschers Technik und die dadurch ermöglichte Erschließung eines neuen Volks- und Welt-Wohlstandes* (privately printed, 1933), 6–15, in Jebens to President, RPA, 6 Jun. 1933, BAP, 2340.

20. Jebens to Eylau, 27 Aug. 1931, BAP, 2340.

21. Jebens to Schleicher, 29 Dec. 1932, BAP, 2340.

22. BDC, Jebens file.

23. Jebens to Reichskanzlei, 8 Jun. 1933, BAP, 2340. For parallel efforts by outsiders trying to become insiders, including similar letter writing, see Beyerchen's discussion of Aryan Physics in *Scientists under Hitler*, 79ff., esp. 100, 115–22.

24. Jebens to Hugenberg and State Secretary Fritz Reinhardt of RFM, 8 Jun. 1933; Jebens to interim RPA president Harting, Reinhardt, and Göring, 9 Jun. 1933; BAP, 2340.

25. Jebens to Reich Chancellery, 8 Jun. 1933, BAP, 2340.

26. Jebens to Harting, 9 Jun. 1933, BAP, 2340.

27. Jebens to Göring, 9 Jun. 1933, BAP, 2340.

28. *Deutsche Allgemeine Zeitung*, 15 Dec. 1933; *Berliner Börsen-Zeitung* 35 (21 Jan. 1934); Keppler to RWM, RJM, and RPA, 15 Dec. 1933, BAP, 2297.

29. On Ley, see Smelser, *Robert Ley*.

30. Minutes of Jebens interview at RPA, 16 Dec. 1933, BAP, 2297.

31. Ibid.

32. Ibid.

33. Reich Association of German Patent Realizers to RPA, 23 Dec. 1933, BAP, 2297.

34. Richard Richter, Leipzig, to Reich Chancellery, 17 Jan. 1934, BAP, 2297.

35. RMI to Verbindungsstab der NSDAP, 6 Jan. 1934; undated memorandum signed by Karl Peppler, head of DAF division for social affairs, clarifying the name "Section for Inventor Protection"; RMI to State Secretary Franz Schlegelberger of RJM, 15 Feb. 1934, BAP, 2297.

36. E.g., "Das perpetuum mobile beschäftigt noch immer viele 'Erfinder'. Die Aufgaben der neugegründeten Abteilung 'Erfinderschutz'," *Berliner Börsenzeitung* 35 (21 Jan. 1934), Jebens to RPA (copies of handouts, flyers, and information

sheets), 21 Jan. 1934; DAF, Sozialamt, Abt. Erfinderschutz flyer, 27 Jan. 1934; "Rundschreiben an die Patentanwälte," 10 Jan. 1934, BAP, 2297.

37. RWM to RPA, 19 Feb. 1934, report of visit of 9 Feb. 1934 by DAF functionary Schwemann about Jebens's termination, BAK, R131/160.

38. Ibid.

39. Thomas Schuster (Heinrich Seebauer's deputy in NSDAP AfT) to DAF Reichsorganisationsleitung, 7 Jun. 1935, BAK, NS 022/000851.

40. Heinrich Jebens, *Der Kleinsthofplan: Das Fundament zum Volksneubau* (Hamburg, 1948).

41. Ludwig, *Technik und Ingenieure*, 142, 149, 179; Martin Broszat, *The Hitler State: The Foundation and Development of the Internal Structure of the Third Reich* (London, 1981), 193–240, esp. 204–10.

42. Cf. remarks by Reichsamtsleiter Erich Priemer in NSDAP, AftW, ed., *Tagung der Reichsarbeitsgemeinschaft Erfindungswesen im Hauptamt für Technik der Reichsleitung der NSDAP, Amt für technische Wissenschaften am 11.1.1944 im Haus des Deutschen Rechts in München* (Munich, 1944), 1–9.

Chapter 8

NAZI REVOLUTION

The 1936 Patent Code

Despite the clamor for action from independents and invention developers, patent reform was not one of the priorities of the Reich Ministry of Justice in 1933. The officials in charge of the patent bill were, from the top down, Justice Minister Franz Gürtner, Secretary of State Franz Schlegelberger, and Ministerialrat Georg Klauer. Klauer, the ministry's chief patent expert, had been in charge of the patent desk since 1924. In February 1934 he was appointed president of the Reich Patent Office, succeeding Johannes Eylau, who had been dismissed in July 1933. Klauer never joined the Nazi party—not because he did not apply, but because he was considered an elitist without real enthusiasm for National Socialism. He nevertheless retained the patent-office presidency until he retired at age seventy, following Germany's surrender in 1945.[1]

Klauer's successor at the Justice Ministry was Ministerialrat Erich Volkmar, a 55-year-old conservative jurist who was never admitted to the Nazi party either and had little interest in patent matters. Volkmar's immediate assistant was the energetic Kammergerichtsrat Herbert Kühnemann. Only thirty-five in 1934, Kühnemann was a former judge who had specialized in patent litigation. Mentioned by Raul Hilberg as a typical example of the law-abiding desk perpetrator for his role in expropriating Jewish patent holders, Kühnemann remained at his post as the government's leading patent expert throughout the Third Reich. Joining the Nazi party only in 1940, he continued in the ministerial bureaucracy after the war and crowned his career by serving as president of the Federal Republic's Patent Office from 1957 to 1963.[2]

The only thing Klauer and Kühnemann did in the first few months of 1933 was edit the 1931 patent bill, deleting most references to republican institutions.[3] They anticipated no significant changes, least of all those demanded with growing vituperation by small inventors. Patent Office personnel, too, remained mostly passive. Privately, they agreed with criticisms of the 1931 bill submitted by industry in May 1933: moving up the beginning of patent protection to the date of publication was a bad idea.[4] They recommended—and the ministry agreed—that this proposal be scrapped. The bill was revised accordingly and one of its more contentious pro-inventor features eliminated without another thought. Patent Office bureaucrats also concurred that adoption of the first-to-invent system as proposed was an empty gesture. But they were astute enough to recognize that saying so in public would be a political mistake. As one official pointed out in June 1933, the Patent Office ought "to come out in favor of the inventor principle more strongly than before, since supporting the interests of precisely the so-called small inventors might suit current political tendencies." This recommendation, too, was accepted.[5]

Apart from approving a few other, minor adjustments, the ministry did nothing else on the patent front during the remainder of 1933.[6] As it continued to drag its feet, however, something unexpected happened. The patent bill would have to be reviewed by a special committee of the Academy for German Law (Akademie für Deutsches Recht, ADR). Presided over by Hans Frank, Hitler's former defense lawyer and now Reich Minister without portfolio, the ADR was a new advisory body, established in June 1933 to make sure all laws conformed to National Socialist principles.[7] This applied to the patent bill as well, which the ministry submitted to the ADR's Reich Committee for Intellectual Property Law in January 1934.[8]

The chairman of the ADR's intellectual property committee was none other than IG Farben's Carl Duisberg.[9] Besides the 72-year-old Duisberg and Frank himself, the committee consisted of regular members and a special "Führer Council." The latter included Frank's official deputy, civil-law professor Wilhelm Kisch, the ADR's vice-president and author of a textbook on patent law; Frank's crony Karl Lasch, who had been made a director of the ADR; Justice Minister Gürtner; and Secretary of State Schlegelberger.[10] Most of the actual work was done by the regular members: Klauer and experts from the private sector such as Paul Wiegand, the successor of Ludwig Fischer as Siemens patent director; IG Farben's Franz Redies; Krupp's *Dr. Ing.* Louis; Arthur Ullrich, chairman of the new Chamber of Patent Attorneys; and Hamburg patent attorney Utescher.[11] As its composition indicates, the ADR committee hardly reflected the discontent with the nation's patent system among independent and salaried inventors. Instead it represented the views of industry and the traditional intellectual property establishment.

Confident of easy victory, the committee limited its review to two issues about which it still had concerns: patent litigation and the first-to-invent principle. It wanted to make sure that the new law would reflect the interests of science-based industry in both areas. Initially the chances of success looked excellent.[12] Management's disappointment was all the greater, therefore, when for different reasons it lost out on both counts.

The question of patent litigation centered on a long-standing demand by science-based industry for "technical judges."[13] The patenting process itself was handled by the Patent Office, which had at its disposal plenty of technical expertise and therefore rarely gave rise to serious complaints of incompetence. Patent litigation, in contrast, had to do with problems such as infringement and came under the jurisdiction of the ordinary courts, where technical expertise was in short supply and the verdict often unsatisfactory—especially in the eyes of technically trained management. As Duisberg explained it to Gürtner, "… there isn't a jurist who understands the first thing about chemistry or even knows its basic concepts."[14] But breaking the lawyers' monopoly, which was what adopting the technical judge would have amounted to, proved impossible even for a company as influential as IG Farben in the Third Reich. The Nazi party and Labor Front decided the issue was "not fundamental" ideologically and stood aside.[15] In the end the decision came down to a simple matter of expediency. Duisberg claimed the technical judge was the "most important part of patent code reform," but the stronger battalions were on the side of the lawyers.[16] Schlegelberger's argument—that the legal profession had "been laid low" by unemployment and nothing should be done to jeopardize its economic opportunities—carried the day.[17] The issue dragged on for almost two years, but the 1936 Patent Code reform made no changes in the jurisdiction over patent litigation.

Industry's defeat in the matter of the inventor's right had very different causes. When Duisberg and his ADR committee first turned to this issue in January 1934, there seemed little cause for concern. Some industry spokesmen once again attacked the bill's first-to-

invent principle, even though it had already been watered down substantially. Siemens's Wiegand, for instance, argued that the "proposed inventor-right measures mean[t] a grave danger for labor
peace." But a majority on the committee realized that rolling back the
inventor's right would be ill-advised. Gürtner, Kisch, Klauer, and
Schlegelberger counseled against retreating from current formulations. The inventor principle was "a question of political tact,"
according to Gürtner, who also pointed out that the employers could
draft "suitable clauses in the labor contract" to neutralize any negative consequences in practice. Krupp's Dr. Louis agreed: so long as
the bill retained the concept of the "company invention," industry
had little to fear. Klauer wrote that "the desires of salaried employees
in this matter [had] been much greater" in the past, and industry
could accommodate the proposed system change without any real
consequences. IG Farben's Redies argued that abandoning the principle of the inventor's right might unleash forces that would produce
an outcome far worse than the current proposal. At Schlegelberger's
suggestion, it was decided to make no changes in the inventor-rights
paragraphs. Any problems, the State Secretary said, could be resolved
quietly and to the employer's advantage in the labor contract.[18]

In the months that followed, Duisberg's committee firmly adhered
to this approach. This was the case even though, on the one hand,
industrial associations such as the RDI continued to oppose the
first-to-invent principle and, on the other hand, certain groups from
within the party and the Labor Front began to agitate for much more
radical pro-inventor changes.[19] The committee ignored a reform proposal that Lasch had received from an unnamed source. This document attacked the central point of the government's compromise on
the inventor's right, which was to introduce the first-to-invent principle but keep on the books the company invention, which would
prevent practical consequences. Paraphrasing Hitler's observations
in *Mein Kampf*, the anonymous author wrote: "A legal person cannot
have made an invention, and for that reason a legal person also
cannot register an invention. Only a person of flesh and blood can
invent something and [is] therefore also the only one who can file for
a patent or a design patent." If a salaried employee or a blue-collar
worker made an invention, the anonymous proposal continued, "the
invention may only be registered in the name of the actual inventor,
and not in the name of the owner if the latter is not himself the
designer." Inventions assigned to the employer should automatically
get compensation. Duisberg dismissed the plan out of hand: there
was "no intention on our part even to discuss the above ideas."[20]

Duisberg's committee completed its review of the patent bill in April 1934. Except for recommending the introduction of technical judges, it made no changes. With a great deal of fanfare and dressed up with laudatory prefaces by Hindenburg, Hess, Göring, Darré, and Frank, the report was published in the second issue of the *Journal of the Academy for German Law* in July 1934, on the heels of the murders of Röhm and other SA leaders.[21] Precisely what happened next can only be imperfectly reconstructed. In August, Hess's secretary, Martin Bormann, wrote the Ministry of Justice and demanded to see a copy of the patent bill forthwith. In September, Heinrich Seebauer, the administrative head of the party's newly created Department of Technology, made a similar demand.

The party's sudden interest appears to have been triggered by the unrest Duisberg's report caused among rank-and-file members and inventor advocates—including those employed by the Labor Front. Considering the constant references to the importance of the creative personality and the unleashing of individual creativity in the Third Reich, it must have been a particularly bitter pill to learn that the new regime intended to do no more for the inventor than the despised Weimar Republic. The unhappiness and potential public-relations problem appear to have been communicated up the party hierarchy, which in turn seems to have approved the counterattacks that were now being launched at the lower levels. This scenario, at any rate, is suggested by the testimony of a central character in the patent revolution that was now imminent.

The individual in question was Dr. Kurt Waldmann, a Munich lawyer. Not much is known about Waldmann's activities during the Weimar Republic, though as a 19-year old student he was apparently briefly involved with Kurt Eisner and the Munich Council Republic. A "March casualty" who joined the party in the spring of 1933, Waldmann made a political career in the Nazi regime that repeated the meteoric rise and sudden demise of Heinrich Jebens—although Waldmann lasted longer and achieved more before he, too, was brought down.[22] Having joined the local chapter of the National Socialist League of German Jurists sometime in 1933, Waldmann soon became leader of the chapter's research department. It is unclear whether Waldmann had much prior experience with patent issues, but his involvement with the politics of inventing after 1933 seems to have begun as pure coincidence. "One day," he recounted in a 1937 report,

the news came that a new patent code had been prepared in the Academy under the chairmanship of the chief of IG Farben, Privy Councilor Duisberg, and that it did not sufficiently take the inventors into account.

Thereupon a new draft law was immediately worked up. The direct stimulus was an agency of the German Labor Front that counseled the inventors, which had learned about me from an old SA leader and party comrade and asked me for help. We based our work on guidelines from the book *Mein Kampf.*

The co-author at the Labor Front to whom Waldmann referred was engineer Karl August Riemschneider (1906–1985), Jebens's successor at the desk for inventor protection in the Division for Social Affairs. Although he always used the title "engineer," Riemschneider, who joined the Nazi party only in 1937, did not have an academic engineering degree. Little is known about his background, but he was probably trained at one of Germany's non-academic engineering schools, the predecessors of today's *Fachhochschulen.* Unlike Jebens and Waldmann, Riemschneider was a bureaucratic survivor— at least until 1945, by which time he had evolved into the regime's most determined and successful guardian of inventor interests.

Riemschneider's main purpose and central vision became the creation of an interventionist amalgam of social and technological policy. In his system, which was embedded in the regime's *völkisch-racist* categories of thought and action, centrally directed inventor assistance and strict supervision of employers were synonymous with the generation of optimal rates of invention, efficiency, and technological progress. These ideas, of course, were not very different from other inventor schemes, such as the proposals of Thomsen, Jebens, and others. What was different was that Riemschneider had a chance to implement them. He steadily expanded the grasp and size of the Section for Inventor Protection, which in 1938 migrated from Social Affairs to a new and separate Labor Front office, the Bureau for Technical Sciences (Amt für technische Wissenschaften). The latter, in turn, moved to the NSDAP's Department of Technology in early 1942. Riemschneider's office soon became a publishing center for patent and inventor law, as the enormously energetic, ideologically committed engineer wrote and collaborated on numerous essays, books, and legal commentaries. He organized conferences on technological progress, rationalization, and employee motivation; managed a growing apparatus of economic invention evaluation; negotiated the details of the regime's inventor compensation schemes with industry and government; and built up a ramshackle empire of inventor counseling, complete with its own newsletter and inventor counselors in all the districts of the Reich and in most large companies. Riemschneider remained at his post until the bitter end, sending out inventor-counseling directives and defending the royalty rights of employees as late as April 1945, when the regime and his life's work were already in ruins.[23]

In the summer and fall of 1934, all this lay in the future. The issue at hand was to draft a counter patent-bill, which Riemschneider and Waldmann, "working towards the Führer," did in short order. They sent their draft to Duisberg's committee in early October. Included was a five-page memorandum in which they attacked the spirit guiding the Justice Ministry's bill and the work of Duisberg's group, contrasting them with the principles of a truly National Socialist patent law.[24] Waldmann and Riemschneider began by noting that the text of the official rationale for patent reform was lifted verbatim from the commentaries to the 1928 and 1931 bills. This could only mean, they wrote, that the bill's framers assumed "'*the interests of the state and the economy*' and '*this time of struggle*' [were] *identical* in 1928, 1932 and 1934.*" Further evidence that the bill was nothing but warmed-over Weimar legislation was a reference in the text to "approval by the Reichsrat," which the ministry had not bothered to delete even though the Reichsrat had been dissolved in early 1934. Third, they attacked the grudging spirit with which the first-to-invent principle had been adopted, citing language to show that this crucial change in system was considered acceptable only because the employer could undo it again in the employment contract. "The bill, therefore, does not even begin to understand that in the Third Reich protection of the creative human being must be the starting point of a patent law." Fourth, the government's bill had retained language, originally proposed in 1927 by the "Jewish-led" Green Society, making it practically impossible for small independents to meet the deadline for filing patent applications abroad. Since large-scale industry always had the money to get around the rule in question, Riemschneider and Waldmann wrote, the bill in effect aimed at "'protecting' the German inventor with general expropriation, to the benefit of any foreigner."[25]

In contrast to all this, the authors wrote, a "National Socialist Patent Code must take into account the following demands." First, the "community of the *Volk* is the foundation that nourishes the cultural growth of the people." This meant that "all inventions and technical improvements ... as well as works of art and science ... are the common cultural property of the entire *Volk*. The interests of the *Volk* and the State come before those of the creator of the work." From this it followed that, "insofar as the interests of society as a whole demand it," the bill should have greater ability to compel inventions' reduction to practice, their working, and forcible licensing.

Having anchored patent rights in the National Socialist principle that community interests came before individual rights, Waldmann and Riemschneider could now proceed to the crucial next step, by which they arrived at their real purpose: to secure the rights of the individual inventor, especially the salaried employee, vis-à-vis capital. The second basic consideration for a National Socialist Patent Code, they wrote, was this: "The creative forces of technological progress serve to the honor of *Volk* and the State." This meant that while immoral or harmful inventions did not qualify for patent protection, "the creative human being has the protection of the State; the rights that this law guarantees to him are not negotiable. The creator has the duties and rights of a trustee vis-à-vis his work. He is exclusively entitled to its economic exploitation." From this it followed that (a) inventors could not keep their inventions secret but must divulge them, (b) "the inventor, especially *also the salaried inventor*, retains an interest in his invention," and (c) patent fees should be restructured according to social considerations and the legal expenses of poor independents borne by the state.

Riemschneider's and Waldmann's final points dealt with restructuring the Reich Patent Office, which they re-baptized as the Reich Office for Creative Achievements and equipped with expanded duties such as the encouragement and assistance of artists. More importantly—and echoing Jebens's ambitions for a Reich Inventor Office—the Reich Office for Creative Achievements should also include a Reich Trustee Authority and Reich Culture Attorney Office "for the encouragement and economic exploitation of creative achievements." The principal task of these new agencies and officials would be twofold. First, they would take over the functions of patent economists, giving the state rather than the private sector the responsibility for developing and marketing the inventions of independents. In addition, these same institutions would "suppress and prevent" all sorts of undesirable projects in the creative sphere.[26]

The bill drafted by Riemschneider and Waldmann reflected these principles. It foresaw a true, American-style first-to-invent system. It eliminated the company invention and mandated "adequate compensation" for all inventions that employees assigned to the employer. It recognized far-going expropriation and patent-voiding powers by the state. It had low fees and made fee waivers easy to obtain. It gave broad intervention rights and development possibilities to the envisaged Reich Culture Attorney Office, and it further restructured the Patent Office along the lines suggested in the cover memorandum.[27]

This frontal attack on the work of the Justice Ministry and the Academy for German Law had the desired effect. Duisberg and his colleagues were profoundly alarmed.[28] "The company as a whole would suffer grave harm," wrote Redies, whose analysis showed that the Labor Front's proposals would essentially destroy private property in inventions and bring capitalist technological progress to a halt.[29] To deal with the situation, it was decided to meet with the authors as soon as possible. The Justice Ministry had already sent out invitations for a final conference on the government bill to be held October 30, after which it would go to the Reich Chancellery for Hitler's signature. Duisberg, therefore, scheduled a prior meeting of his ADR committee for October 20, to resolve the problems raised by the Labor Front's demands.[30] Exploratory discussions with Waldmann and Riemschnieder took place in Berlin the afternoon of 11 October 1934. The delegation that met with the two self-appointed tribunes of the people consisted of Patent Office officials Engeroff and Altpeter, Frank substitutes Lasch and Dr. Gaeb, Siemens's Wiegand, and finally Redies, who wrote a confidential report of the encounter for Duisberg and other top IG Farben executives.[31]

According to Redies, "the views of the Labor Front and the League of National Socialist German Jurists were essentially presented by Dr. Waldman," who "used to be a lawyer in Munich, appears to be an old Party comrade, and has good connections to people higher up in the party." Waldmann "told us that toward the end of the week he would discuss the new rules with Rudolf Hess, who takes great personal interest in the fashioning of the inventor-right questions." Fortunately, Waldmann seemed very reasonable, a man with "practical experience in intellectual property law and not [one of] … the pigheaded theoreticians." The delegation learned that the Labor Front had "intentionally drafted a polemical and extreme statement," in order to gain everybody's attention and highlight "those shortcomings in the Patent Code that … are most in need of reform." In truth, however, Waldmann had said, the Labor Front would be "pleased to be taught that, on one point or another, a practical implementation of its current formulation was not possible."

Redies continued that "the fundamental views of the Labor Front" on employee inventions were as follows: "The National Socialist Program demands that the depersonalization of the economy be slowed as much as possible, or that it be undone." Accordingly,

> the creative human being as such must stand in the foreground of the patent system, and … anonymous forms of patent protection such as the company invention must be curbed. Besides, the Labor Front is of the

opinion that the German economy over the next decades will depend heavily on new inventive undertakings. For that reason it places great emphasis on a legal arrangement—instituted and supervised by the state—which encourages the initiatives of inventors as much as possible. Such encouragement can be given in two ways, according to the Labor Front: (a) by emphasizing the personal right of the inventor (mentioning the inventor in the patent document, honor of the inventor) and (b) by letting the inventor participate materially in the yield.

According to Redies, the Labor Front was willing to accept rather modest practical consequences from these radical-sounding principles. Waldmann, he wrote, recognized that the employer should ordinarily be entitled to use the employee's inventions. The bottom line was that in future the patent should be issued in the name of the inventor instead of the company. In return for a guarantee of special financial compensation all material rights would be assigned to the employer. The net effect of this would "not be too great," since companies such as IG Farben had operated with a very similar inventor compensation system since 1920. In short, "the situation ... will in practice remain the same as it has been."

On the remaining issues, too, according to Redies, Waldmann and Riemschneider had been reasonable. They raised points about the "interests of the small inventor," while the experts from industry conceded the theoretical perspective and made minor practical concessions, but mostly explained why the proposals were impractical or too expensive. Time and again, the "gentlemen of the Labor Front allowed themselves ... to be converted to our point of view," and in the end the whole group had gone out to dinner. The meeting had "been very useful, because the situation already looks a great deal calmer than in the definition[s] of the original draft." Finally, Redies suggested "that it might be a good idea to invite Dr. Waldmann, who is apparently the decisive figure at the Labor Front, to the meeting ... of October 20, so as to resolve the main difficulties there, and not at the last minute in the plenary conference of the Reich Justice Ministry on October 30."[32]

This recommendation was accepted, as was the upbeat mood of Redies's report. Duisberg and Redies prepared for the conference participants a detailed agenda in which they consistently suggested solutions that favored industry and rejected the Labor Front's bill. This is not to say that management made no concessions at all. In fact, the kind of arrangement—modeled on the Chemists' Contract of 1920—that the majority of Germany's salaried inventors had fought for in vain during Weimar was now freely offered to them. At the same time, Duisberg made very sure that management held on to the decisive cards. The IG Farben executive did not for a single moment consider surrendering the company invention, the long-time anchor of Germany's pro-employer patent system. Instead he proposed to settle most problems in conformity with the stipulations of the Chemists' Contract or the RDI and BUDACI's voluntary agreement of 1928.[33]

Duisberg's agenda and decision to invite the Labor Front's spokesmen, even as he dismissed most of their demands, inevitably raised the stakes for the meeting of October 20. The status of the ADR committee as proxy for the combined forces of industry and ministerial bureaucracy only intensified this configuration. The participants themselves sensed that the meeting might differ from the usual, tedious routine and showed up in strength.[34] Few of them, however, could have predicted what lay in store: that the Leverkusen discussions of 20 October 1934 would be a turning point in Germany's long history of the politics of inventing.

The first and central point of discussion was the company invention. Duisberg's recommendation was to leave things as they were: retain the concept and definition of the company invention in the Patent Code and settle everything else in the employment contract. The first to speak was Waldmann. "The Third Reich seeks the breakthrough of all creative ideas, wherever they occur," he began. "That is, if I may say so, the deeper meaning of our movement." It followed that "the creative human being must be placed in the center of the need for protection." Waldmann conceded there was a significant difference between the inventions of independents and those of salaried inventors working in an "inventor factory," the latter typically being the beneficiaries of the company's internal state of the art. "But in the end there is no way on earth to deny the fact that the final inventive achievement issues from an intuition-like revelation, from a creative thinking process by a creatively gifted human being." It was a pure fabrication of the mind to "maintain that a company can invent," he said. "If I combat this mental fabrication it is because I most strongly condemn the depersonalization of the company …"

Waldmann also conceded that inventing in large companies could be systematized and rationalized to the point that it differed only slightly from exercising routine professional skills. But no matter how slight, the difference remained. One might therefore reduce the "appropriate compensation of an inventor to *almost* zero, but no further [italics added]. Inventor protection today must be tuned to the [individual] man." Another, more pragmatic consideration was that whenever a German invention was patented in the United States, "an inventor must be named. It must be possible for us, too, therefore, to find language that enables every decent human being to say: to the best of my knowledge and on my word of honor I declare that so and so must be considered the inventors ... I therefore reject the concept of the company invention."[35]

Waldmann's remarks provoked heated discussion. His supporters were mostly professional technologists and ardent Nazis, his opponents managers and officials in the ministerial and judicial bureaucracy. The former included Arthur Ullrich, chairman of the new Chamber of Patent Attorneys. The year before, Ullrich had successfully pushed the Jewish members out of the profession and now he worked hard, again with success, to make it illegal for Jewish patent attorneys who had emigrated to continue representing clients from abroad.[36] With respect to the current item on the agenda, Ullrich "completely agreed" with Waldmann "that there is no such thing as a company invention." Another supporter was Privy Councilor Ferdinand Engeroff, a chemist who had worked in industry before joining the Patent Office in 1927. Engeroff had become a member of the NSDAP in 1931 and was the leader of the Party's section at the Patent Office. He emphatically wished to "second the interpretation of Waldmann and Ullrich."[37]

But the majority of those present were against Waldmann and restated the case for organized and collective inventing. Wiegand concentrated his counterattack on the Labor Front's trump card: its apparent agreement with Hitler's *Mein Kampf*. Hitler had not meant to cover each and every type of routine mental activity, argued Wiegand. On the contrary, "the Führer's book, *Mein Kampf*, refers to the human being who is in fact truly creative. Unfortunately, such truly creative achievements are rare in industry. Generally the employee works in purely constructional [design-like] fashion with the means, both material and intellectual, that the company puts at his disposal." From this it followed that "truly creative achievement cannot be insinuated by the legislator either in such cases."[38]

Duisberg illustrated Wiegand's point by recounting how the process of coal liquification had come about. The example was different, but his argument the same as what he first said in 1909 to dissuade the Imperial government from adopting the first-to-invent principle.[39] Based on the work of Bergius, Duisberg pointed out, "hundreds of chemists collaborated for years until finally it proved possible to produce liquid fuels from coal, based on a multitude of

small, partly patented and partly unpublished individual advances."
What today's big-science projects such as coal hydrogenation
demonstrated, Duisberg continued, was that "the essence is systematic collaboration of the people who work in the respective company,
without it being possible to determine who in the final analysis are
the 'inventors' of the process that succeeded in practice." Duisberg
presented several other examples, such as the development of pharmaceutical products, medicines and synthetic dyes, all of which
proved that "in the real world even inventions representing great
technological progress frequently do not spring from the intuition-
like revelation of a definable 'inventor,' but often are the result of
systematic collaboration of the 'company'—activity the entrepreneur is entitled to demand from every employee on the basis of the
regular employment relationship."[40]

Louis and Redies agreed, as did government experts such as
Reichsgericht Justice Pinzger, Judge Kühnemann, and Volkmar. It
was to no avail. At one point Hans Frank interrupted the debate to
announce that the Labor Front was right and IG Farben and
Siemens and the Reich Justice Ministry wrong. "The National
Socialist German Workers Party considers it very important," he
said, "that the basic point of view of National Socialism is anchored
in the upcoming laws, that the inventor as the carrier of the nation's
creative development is guaranteed full entitlement to his intellectual property. This principle must, in my opinion, form the basis of
this entire legislation."

His sarcasm palpable, Frank continued to say how "extraordinarily pleased" he was to have learned from the debate that everyone
seemed in agreement on a formula that was also "perfectly acceptable from the National Socialist point of view." Naturally, an
"expression of gratitude on behalf of the movement" was in order, as
he was "absolutely convinced that we have achieved a legislative
step forward, from which the subsequent progress of the Patent
Code will develop like a leaf unfolding." Frank then launched into a
short speech about National Socialism's patent and inventor philosophy. During the preceding discussion, he said, "the name of our
Führer was repeatedly mentioned in connection with the remarks he
made in his book, *Mein Kampf*. Gentlemen, I can assure you that the
Führer will be very happy about this solution, because I think that
with the passage in his book, *Mein Kampf*, he had exactly this question in mind. The movement's most dedicated fighters, after all, are
the ones who have pointed out that not just the masses of the workers are in the thrall of capitalist power, but even those who are

active in the field of intellectual creativity. There can be no doubt that the Führer has a great interest in these observations."

The committee, Frank recommended, should now widely publicize its breakthrough, and inform the various other agencies involved, such as the Reich Justice Ministry, the Academy's business office, the Labor Front, and the RDI. The concept of inventing had always been a difficult one, he continued, for "in the end every invention is, indeed, a company invention, to wit, a product of the workshops of the beloved Almighty, who has equipped every individual for his journey with His resources and His facilities. Thanks to God, however, our dear Lord is not as greedy as the capitalist world. On behalf of the inventor, He renounces His eternal, divine right, to which He is entitled." Quite apart from the religious factor, Frank also wanted his listeners to know that "this concept of the company invention, in any shape or form, including the most recent one, was somehow a kind of diminution of the salaried inventor's prestige." National Socialism's top lawyer was therefore "happy to see that ... we have raised the social status of the salaried inventor a little," and felt obliged to thank Duisberg once again for the "dignified manner" in which the latter had presided over the meeting.

The discussion was over. The company invention was gone forever. Duisberg still had the poise to thank Frank for his "observations on the basic questions concerning the reform of the patent system" without specifying what, in truth, he must have thought about them. But for once the 73-year-old industrialist—and with him German industry—was beaten. Having fought to the bitter end to retain the theoretical underpinning of management's prerogatives, Duisberg now conceded, five months before his death, that, "the concept 'company invention' will be deleted from the Patent Code." With only service inventions and free inventions remaining, the details of employee inventor rights such as compensation would be settled in labor law. In that arena, the IG Farben executive gamely predicted, it would prove possible to "achieve our common goal— the encouragement of the creatively active human being and with that the encouragement of our German technology."[41] In truth, industry continued a stubborn rear guard battle precisely over the details, and only in 1942 would the regime finally impose a solution.

Compared with the battle over the company invention, the remainder of the Leverkusen meeting was anticlimactic. The afternoon session worked through the remaining issues without Frank's intervention. Waldmann made various additional demands, but without Frank's direct support he achieved little, apart from some minor concessions.[42] In the end a number of unresolved issues remained. But they seemed minor, especially in light of the breakthrough in the inventor's-right dispute, which overshadowed everything else.

Consolidation

It would take months of bargaining and negotiating before a revised patent bill met with the final approval of all the parties involved. In the immediate aftermath of the Leverkusen discussions, however, the Justice Ministry was optimistic about imminent passage of the legislation. The conference it had called for 30 and 31 October in Berlin essentially rubberstamped the decisions made ten days earlier. The patent system would adopt the inventor's right, and the company invention would disappear. The rights of salaried inventors would be dealt with in a separate labor bill, a draft of which had already been prepared by a labor-law committee of the Academy for German Law. This bill was approved in principle. Patent fees would be further reduced.

Clearly, things were beginning to move in the direction of independent and salaried inventors in the fall of 1934. Their gain was the industrialists' loss. Industry's defeat might have been partially compensated if it could have won in the matter of the technical judge. But on this front, too, the tide was turning against the technology companies. Duisberg, supported by the RDI, the Association of German Engineers, and the Patent Office, did all he could to persuade the different groups represented in the Berlin conference of the need for reform. He ran into a storm of objections. Seeking to protect the monopoly of the legal profession, the German Chamber of Commerce, the BNSDJ, and the Reichsgericht all opposed the technical judge and the right of patent attorneys to represent clients in nullification procedures. The Labor Front and Nazi party remained on the fence, which for all practical purposes doomed the proposal.[43]

Following the Berlin conference, Volkmar and Kühnemann went to work revising the bill and by January 1935 began circulating their latest draft.[44] Contrary to what might have been expected, it retained the language and spirit of the 1931 bill as much as possible, reducing the changes demanded by National Socialism to an absolute minimum. It is not known what the Justice Ministry officials thought they were doing, but it is clear from the reactions to their work that they had made a huge miscalculation. Duisberg and the ADR committee were unhappy that the technical judge had been dropped.[45] The Labor Front protested against language that watered down the inventor's right by allowing up to a year's delay in recording the inventor's name; it was also unhappy about the absence of any reference to legislation concerning employee inventions; it complained about insufficient fee reductions, protested the absence of poor-law provisions for impecunious inventors, and criticized the failure to strengthen the government's position in the matter of forcible licensing. Göring's Air Ministry and the Army were angered by the bill's failure to make it easier for the state to obtain forcible licenses, and its failure to strengthen the law's "prior-use" provisions, both of which the military demanded as an additional way to strengthen the state's hand against patent holders. The Ministry of Public Transportation also insisted on stronger prior-use provisions. The Finance Ministry complained that fee reductions were too generous and would cause a deficit. Martin Bormann rejected the draft outright, informing Gürtner "that from the point of view of the party the patent bill [was] intolerable." He and Hans Frank would sink the bill if it went to the cabinet in its current form.[46]

A follow-up letter from the party's Legal Department, signed by Department of Technology chief Heinrich Seebauer and a certain Dr. Bechert, reminded justice officials of Hitler's wishes as articulated in *Mein Kampf*. The party insisted on true protection of the inventor, especially the employee. The employer had certain rights, to be sure, "but taking advantage of the inventor is not allowed" and "putting an inventor in fetters is forbidden." Appropriate compensation of the employee and just profit sharing were essential. Seebauer and Bechert concluded by demanding jail sentences for those convicted of patent infringement and large fines and forced expropriation for those unwilling to license their patents.[47]

Hans Frank and Kurt Waldmann did not remain quiet either. In April 1935 Frank's deputy, Dr. Bühler, sent the Ministry of Justice a critique that mainly repeated Seebauer's criticisms. Waldmann sent the same letter on behalf of the BNSDJ. The inventor's right in the government's draft had been "weakened to an extraordinary degree," even though this was "the main demand of the National Socialist movement." The language permitting a year's grace period before recording the inventor's name was an outrage. There would have to be some provision in the bill to make sure patent applicants spoke the truth about their right to file, backed by jail sentences and fines. Frank and Waldmann criticized the bill's failure to sufficiently reduce costs for poor inventors, who should have access to the poor laws and a variety of other subsidies and refunds. The labor legislation drafted by the Labor Front and the ADR's labor law committee should be promulgated at the same time as the patent bill. Moving from protection of the inventor to the rights of the state, Frank and Waldmann demanded peremptory curtailment of patent rights wherever they might impinge on national defense and "the well-being of the national community." The clauses dealing with forcible licenses and expropriation would therefore have to be strengthened. To be sure, the patent holder should be compensated, but a simple decree should suffice to void a patent or transfer the technology in question to whatever companies the government selected.[48]

If one pauses for a moment to reflect on these criticisms of Germany's patent system in 1934 and early 1935, two facts stand out. First, National Socialism was making a concerted effort to improve the social and economic circumstances of the inventor. It did so for social reasons—because the inherited situation was perceived as inherently unjust—and for technological-economic ones—because the Nazis thought that innovation and therefore national strength hinged on the performance of the small independent and the morale of the salaried inventor. These were considered more important sources of invention than the organizational prowess and institutionally anchored know-how of big business. Second, National Socialism increasingly focused on strengthening the power of the "national community," i.e., the state and the military, in the patent system. The Nazis in power, at least the vast majority of them, were determined to eliminate any legal barriers to the development and production of strategically relevant technologies already in the hands of big business. The principal instrument for achieving this objective was the forcible license and, to a lesser extent, the right of prior use. Some Nazi officials objected to forcible licenses and expanded powers of prior use for fear they might strike at independents just as much as the big corporations, which were the real target, but they were overruled.[49] As strengthening the individual inventor, too, came mostly at the expense of big business, it was German industry that—by design—would pay the price for the regime's patent measures.[50]

Under attack from all sides, the Ministry of Justice finally saw the light and made the necessary course correction. It quickly redrafted the offending paragraphs and collaborated closely with the decisive Nazi agencies to get everything just right. Over the course of the spring and summer of 1935 this removed most of the remaining issues.[51] At the same time, two of the ministry's senior officials went public to show their total commitment to the principles of Nazi inventor policy. In an article for the March 1935 edition of *German Justice* Volkmar wrote about the importance of "strengthening the creative forces that are crucial for the German people precisely in its present condition." The inventor's honor and recognition of the inventor's name would have to be anchored in the law better and "more visibly than before." The impecunious inventor would have to be helped with appropriate measures as well. And the spirit of National Socialism also called for measures to make sure the inventor could "not exploit his patent in the liberalistic sense of the predominantly selfish point of view."[52]

In June 1935 Volkmar's boss Schlegelberger lectured at the University of Königsberg in the same spirit on the same topic. Speaking on "Business Law in the New State," Schlegelberger pointed out that "in its current economic situation, Germany depend[ed] in the highest degree on the utilization of all the intellectual powers of the people's comrades." There was no doubt that "the scope and the yield of industrial production correlate most immediately with the quality of the legislation concerning intellectual property." Whether national prosperity could be increased depended to a large degree on the success or failure of the effort "to give the German people a Patent Code that accorded with current legal thought."[53]

In October, Schlegelberger lectured on Nazi inventor policy at Karlsruhe Institute of Technology, offering a preview of the new Patent Code.[54] He began by praising the old patent system as an institution that had contributed "mightily ... to the industrial development of our nation." The features that were good about "the old and trusted law" would therefore be retained as much as possible. On balance, however, the impending legislation was mostly "a new work." There were two reasons for this, according to Schlegelberger, who outdid himself as a disciple of National Socialist inventor philosophy. First, the aim was to create a new "vessel for a new spirit." Second, the German people should be made fully conscious that "the tools that shall arm them for the struggle of existence in such an important arena do not come from a law museum but from the armory of the present."

The new Patent Code, continued Schlegelberger, was much more than just an ordinary tool of economic policy. "In truth, it [was] a law to protect the creative power of the nation." That power, in turn, was a function of the "creativity of the individual citizens," who together made up the "people's community." For that reason, it was a moral obligation "to awaken the creative will and strengthen the creative powers of the people's comrades, giving wings to [those energies], so they may soar to the highest possible achievement in the interest of the public good." The new code achieved these lofty objectives in two ways. One was honor, the other money. With regard to honor, the new law centered on the inventor, replacing the registrant's right with the inventor's right. It also lifted the cover of anonymity from the inventor by outlawing the company invention. "In our fast-paced time," Schlegelberger said, "the spiritual authors of even the most pioneering inventions are forgotten all too easily and all too quickly." Every invention, he quoted almost verbatim from *Mein Kampf*, "is the achievement of an individual person, and never that of some vague, anonymous collectivity, even if for legal purposes the latter possesses the qualities of a legal person." True, a business corporation "can acquire inventions and market them, and it can also encourage their conception with its facilities and make them possible with its means of support, but it cannot itself invent, because its intellectual abilities are the abilities of the people who work there." The company invention was a "purely intellectual and abstract construct," for which there was no place in the new code. The individual's contribution might be small, "but the fact remains that in the end the person who has taken the last step necessary to complete the invention deserves recognition."

The second way the new Patent Code encouraged innovation was by providing financial incentives and measures on behalf of inventors. Separate legislation for employees was being drafted, but until it was complete there was little to say. Schlegelberger talked instead about the social measures for the small independent. There was a veritable cornucopia of them, on the principle that "little inventors" should not be allowed to "fail because the smallness of their means prevent[ed] them from showing the greatness of their mind." Applying for patents and maintaining them would become less expensive. There were various exemptions, waivers, and refunds, and additional reductions for those prepared to offer a license on demand. Judges were empowered to reduce the financial exposure of the weaker party, lowering the risks of not settling out of court.

The new code adhered to the party line also with regard to the "basic principle that the interests of the people and the state come before the special interests of the individual." This principle, Schlegelberger explained, gained added force from the fact that the state now made it a special point to protect the inventor against exploitation and abuse. In return, the inventor had a duty to grant the state special powers when that was in the national interest. The principal tools at the government's disposal in this regard were broadening the doctrine of prior use and expanding its power to obtain a forced license.[55]

At the time Schlegelberger gave his Karlsruhe address, the only question that remained unresolved was the reform of patent litigation. The technical judge was dead, but variations on the idea continued to the last minute. Eventually a weak and unsatisfactory compromise was adopted.[56] By December 1935 the last remaining details had been resolved, and Gürtner submitted the bill to the Reich Chancellery on 14 January 1936.[57] There were some last-minute skirmishes when Hess demanded an end to "highway-robbery patents."[58] The problem was cleared up by March, and the bill became law on 5 May. It went into effect 1 October 1936.[59]

The regime took great pride in its new Patent Code, which represented a great public-relations coup and capped reform efforts first begun in 1913. Hans Frank exemplified the celebratory mood in his preface to *The Right of the Creative Human Being*, a volume published on occasion of the congress of the International Society for Intellectual Property in Berlin in June 1936. "The Patent Code of 5 May 1936 is eloquent testimony," Frank wrote, "to the high value and the encouragement that the National Socialist State gives to the creative individual and achievement."

> It punishes as lies all the suspicious and malicious talk that the Third Reich lets the individual person go to waste and that it represses the creative forces slumbering in its people. The Reich government openly and freely presents the new Patent Code for evaluation, in the consciousness of having made a modern and progressive arrangement, which will not remain without influence on the patent laws of other civilized nations.[60]

Frank's claim was not a complete lie. The new Patent Code brought a number of gains for the inventor, even as it gave the state enormous powers to vacate patent rights on grounds of national security.[61] It should be pointed out, however, that Western nations adopted comparable rules for forced licensing in wartime; and preparation for war was of course never far in the background in Germany, even before 1936. The code's draconian and totalitarian elements, therefore, should not obscure the remarkable fact that it was the Nazi regime that finally got rid of the peculiar notion that an organization as such could invent, anchored in the concept of the company invention since 1877. Moreover, while it did not exactly adopt the first-to-invent principle that is the basis of American patent law, the new code still introduced the principle of the inventor's right.[62] This gave a central role to the inventor, who had not even appeared in the old system.

Even so, the 1936 reforms fell short of the aspirations of Germany's most determined inventor advocates. There was as yet no word on labor legislation and the rights of salaried inventors, which had done so much to turn technical employees into enemies of the Weimar Republic. Nor had anything been done to help independents bring their ideas to market, introduce public inventor counseling, or assess the economic viability of inventions. In the summer of 1936 all these plans were still awaiting action.

Notes

1. Klauer: *Wer Ist's* 1935, 820; BDC, Klauer file.
2. Gürtner: Ingo Müller, *Hitler's Justice: The Courts of the Third Reich*, trans. Deborah Lucas Schneider (Cambridge, Mass., 1991); Schlegelberger: Michael Förster, *Jurist im Dienst des Unrechts: Leben und Werk des ehemaligen Staatssekretärs im Reichsjustizministerium, Franz Schlegelberger (1876-1970)* (Baden-Baden, 1995); Hilberg's comments in *Victims, Perpetrators*, 30. Kühnemann as president of the DPA: Deutsches Patentamt, *Hundert Jahre Patentamt*, 429-32; BDC Kühnemann file.
3. RJM to RPA, edited copy of 1931 bill, 11 May 1933, BAK, R131/156 and BAP, 2340; virtually identical draft of 23 Dec. 1933, BAP, 2341.
4. RDI to RPA, 26 May 1933, BAK, R131/156.
5. Oberregierungsrat Kühnast (RPA) discussing RDI objections, 17 Jun. 1933; Eylau to RJM, 11 Jul. 1933, BAK, R131/156.
6. Gesetz betr. Massnahmen auf dem Gebiete des gewerblichen Rechtsschutzes v. 28.12.1933 (RGBl. II, p. 1075), BAP, 2341; Kühnemann to independent inventor Wilhelm Beilke, 26 Jun. 1933; Kühnemann to RWM, 5 Aug. 1933, BAP, 2340.
7. Werner Schubert, ed., *Akademie für Deutsches Recht 1933-1945, Protokolle der Ausschüsse*, vols. 1-4 (Berlin, 1986-96), vols. 5-9 (Frankfurt a. M., 1997-1999); esp. vol. 9: *Ausschüsse für den gewerblichen Rechtsschutz (Patent-, Warenzeichen-, Geschmacksmusterrecht, Wettbewerbsrecht), für Urheber- und Verlagsrecht sowie für Kartellrecht (1934-1943)*.
8. RJM to ADR, 2 Jan. 1934, BAP, 2341.
9. Frank to Duisberg, 29 Nov. 1933, AS Duisberg.
10. On Lasch, see Niklas Frank, *In the Shadow of the Third Reich* (New York, 1991), passim, esp. 190-91.
11. Membership in minutes of 1. Sitzung des Reichsausschusses für gewerblichen Rechtsschutz der ADR, Bayer, 28/6.2.
12. A proposal to introduce the technical judge made it into the ADR's official 1934 report and was supported by a host of Nazi luminaries: Todt to Gürtner, 21 Aug.

1935, BAP, 10130, Bl. 56; Todt to Klauer and Todt to Gürtner, 7 Nov. 1935, BAK, R131/158; Schacht to Gürtner, 3 Apr. and 11 Dec. 1935, BAP, 10130, Bl. 12, 102; Seldte to Gürtner, 21 Oct. 1935, BAP, 10130, Bl. 69; Pietzsch to Duisberg, 30 Nov. 1934, AS Duisberg; Hess to Gürtner, 22 Jan., 16 Mar. 1936, BAP, 10131, Bl. 22, 36.

13. E.g., the Green Society's 1903 *Vorschläge zur Reform des gewerblichen Rechtsschutzes; Einrichtung einer Sondergerichtsbarkeit in Patentsachen,* in RWWA, HK Duisburg 20-119-12; Hans Heimann, "Zur Frage der technischen Gerichtshöfe," *GRUR* 15 (Jul. 1910): 197–205; Berthold Wassermann, "Technische Richter für Patentprozesse," *VDDIZ* 11 (1920): 62–66, 75–79; VDI to RJM, 9 Mar. 1929, BAP, 2336, Bl. 153c; VDI to RJM, 7 Mar. 1932, BAK, R131/156; Reusch to DIHK, 7 Aug. 1924, RWWA, HK Duisburg, 20-477-9; RDI to RJM, 26 May 1933, BAP, 2340; Ludwig Fischer, "Erfinderprinzip oder Anmelderprinzip?" *GRUR* 33, no. 15 (1928): 1–8; idem, *Patentamt und Reichsgericht.*

14. Duisberg to Gürtner, 21 Oct. 1934, AS Duisberg.

15. Delegate of Rudolf Hess and Hans Frank at RJM patent conference, 30 Oct. 1934, BAP, 2343.

16. Duisberg to Gürtner, 21 Oct. 1934, AS Duisberg.

17. Schlegelberger at ADR intellectual property committee meeting, Leverkusen, 20 Oct. 1934, p. 19, BAK, 133/157. Situation of the legal profession: Jarausch, *Unfree Professions.*

18. Minutes of 1. Sitzung des Reichsausschusses für gewerblichen Rechtsschutz der ADR (Duisberg Committee), 13 Jan. 1934, Bayer, 28/6.2.

19. 2. Sitzung des Ausschusses für gewerblichen Rechtsschutz der ADR, 3 Feb. 1934, Bayer, 28/6.2.

20. The unidentified author was probably Karl August Riemschneider, Jebens's successor at the DAF. Demands (and the decision to ignore them): Redies to RPA's Müller, 4 Apr. 1934, BAK, R131/157.

21. *Zeitschrift der Akademie für Deutsches Recht* 1, no. 2 (Jul. 1934): 44–74 (Duisberg report: 65–73).

22. On Waldmann: BDC. Waldmann was thrown out of the party and turned over to the Gestapo in 1937 because of a variety of irregularities, including lying about when he joined the NSDAP.

23. For a comparable, ideologically driven project, albeit one without positive, institutional legacy, see Beyerchen, *Scientists under Hitler,* 141–67.

24. "Charakteristik des Entwurfs des Reichsjustizministeriums—Charakteristik eines nationalsozialistischen Patentrechts," n.d. (late Sep. or early Oct., 1934), "Arbeitsfront, Entwurf eines Gesetzes über den gewerblichen Rechtsschutz," Bayer, 28/6.2; Kershaw, *Hitler,* vol. 1, 527-60, and vol. 2, passim.

25. Roederer, "Die katastrophalen Folgen der geplanten vorzeitigen Drucklegung der Patentanmeldungen" and accompanying letter from the DAF's Social Office to Volkmar, 1 Nov. 1934, BAP, 10129, Bl. 2–3.

26. "Charakteristik des Entwurfs des Reichsjustizministeriums—Charakteristik eines nationalsozialistischen Patentrechts."

27. "Arbeitsfront. Entwurf eines Gesetzes über den gewerblichen Rechtsschutz" (20 typescript pages), n.d., Bayer, 28/6.2. In the chapter on patent reform he wrote for the second edition of Frank's *Nationalsozialistisches Handbuch für Recht und Gesetzgebung,* Waldmann began with a discussion of patent law in the United States, claiming that this nation's first-to-invent principle was the product of German concepts of justice introduced by German immigrants. Germany should

return to these original, healthy legal instincts, from which it had deviated in the course of the nineteenth century. Kurt Waldmann, "Auf dem Wege zu einem nationalsozialistischen Patentgesetz," in *Nationalsozialistisches Handbuch für Recht und Gesetzgebung*, 2d ed., ed. Hans Frank (Munich, 1935), 1032–49, esp. 1034.

28. Wiegand to Klauer, 2 Oct. 1934, BAK, R131/157.

29. Redies, "Bemerkungen zu dem Entwurf eines Gesetzes über den gewerblichen Rechtsschutz (Arbeitsfront bezw. B.N.S.D.J.)," 9 Oct. 1934, Bayer, 28/6.2.

30. RJM to Hess, Keppler, DAF, the NSDAP's AfT, RPA, all government ministries, and industrial and professional associations such as the RDI, DIHT, Patent Attorney Chamber, Reichsrechtanwaltskammer, and VDI, 2 Oct. 1934, "Einladung zu einer Besprechung des Entwurfs eines Gesetzes über den gewerblichen Rechtsschutz in Verbindung mit dem Ergebnis der Beratungen der Akademie für Deutsches Recht, BAP, 2343, Bl. 15; Duisberg to ADR patent law committee, 11 Oct. 1934, BAK, R131/157.

31. Redies, "Besprechung in Berlin am 10. u. 11.10.34," 12 Oct. 1934, Bayer, 28/6.2; all citations in following paragraphs are from this document.

32. Ibid.

33. Duisberg to ADR patent law committee, 13 Oct. 1934, Bayer, 28/6.2.

34. Besides Duisberg, the following attended: Hans Frank (Bavarian Minister of Justice, Reichsjustizkommissar, President of ADR), Schlegelberger (State Secretary, RJM), Volkmar (Ministerialdirektor, RJM), Kühnemann (Landgerichtsrat, RJM), Freiherr Du Prel (Reichspressechef der Deutschen Rechtsfront), Bühler (Reichsjustizkommissariat), Lasch (Direktor, ADR), Heuber (Reichsgeschäftsführer, BNSDJ), Ranz (BNSDJ), Waldmann (BNSDJ, DAF), Riemschneider (DAF), Engeroff (RPA), Jungblut (patent director, Alexanderwerk, Remscheid), Klauer (President, RPA), Louis (patent director, Krupp, Essen), Müller (Regierungsrat, RPA), Pinzger (judge, Reichsgericht), Ullrich (Patentanwalt, chairman, Patentanwaltskammer), Utescher (attorney patent specialist, Hamburg), Wiegand (patent director, Siemens), Redies (secretary of ADR patent committee, leading patent expert of IG Farben); Sitzung des Ausschusses für gewerblichen Rechtsschutz der ADR am Samstag, den 20. Oktober 1934, in Leverkusen, Bayer, 28/6.2.

35. Minutes, ADR patent committee meeting, 20 Oct. 1934, 3–5, Bayer, 28/6.2.

36. Ullrich's pet project of mandating resident patent attorneys for foreign applications (*Vertretungszwang für Ausländer*), which subsequently became law, in minutes of ADR patent committee meeting, 24–25. After the war, Ullrich was disbarred and sentenced to a one-year jail term for his aggressive "uncollegialism." By 1949 Baden-Wüttemberg had readmitted him to the practice of patent attorney, and he resumed a successful career; Ullrich case, BAK, Z22/165, Bl. 64.

37. BDC, Engeroff file; minutes of ADR patent committee meeting, 20 Oct. 1934, p. 8, Bayer, 28/6.2. Klauer also favored dropping the company invention. American colleagues had alerted him to the weakness of German patents in the U.S. if the underlying invention was classified at home as company invention. Klauer to RJM, 15 Oct. 1934, p. 16 of 55-page response to report of Duisberg committee of Jul. 1934, BAK, R131/157; minutes of ADR patent committee meeting, 20 Oct. 1934, p. 5, Bayer, 28/6.2.

38. Ibid, pp. 5–6.

39. See ch. 2.

40. Minutes, ADR patent committee meeting, 20 Oct. 1934, pp. 6–7, Bayer, 28/6.2.

41. Ibid., pp. 12–14.

42. Ibid., pp. 18–30.

43. Vermerk über die Besprechung des Entwurfs eines Gesetzes über den gewerblichen Rechtsschutz am 30. und 31. Oktober 1934, BAP, 2343. Industry's only consolation was that the military's draconian proposals for forcible licensing found no support at this time. Later, industry lost on this point as well.

44. RJM to RPA and RJM to distribution list of numerous agencies, 21 Jan. 1935, BAP, 10129, Bl. 50.

45. Peppler to RJM, 7 Feb. 1935, BAP, 10129, Bl. 58; Redies to RJM, 11 Feb. 1935, BAP, 10129, Bl. 74a–b.

46. RFM to RJM, 5 Feb. 1935, BAP, 10129, Bl. 86; Deutsche Reichsbahn-Gesellschaft to Reichsverkehrsministerium, 9 Feb. 1935, BAP, 10129, Bl. 73; RLM to RJM, 11 Feb. 1935, BAP, 10129, Bl. 85; Bormann to RJM, 12 Feb. 1935, BAP, 10130, Bl. 3; Reichswehrministerium to RJM, 9 Mar. 1935, BAP, 10129, Bl. 95.

47. Seebauer and Bechert to Gürtner, 11 and 13 Feb. 1935, BAP, 2343, Bl. 63; Seebauer to Gürtner 15 Dec. 1934, BAP, 2343, Bl. 40–41.

48. Dr. Bühler for Hans Frank to Gürtner, 17 Apr. 1935, BAP, 10130, Bl. 3–4.

49. The principal source of inside opposition to expanded government powers with regard to prior use and forcible licenses was Postal Minister Wilhelm Ohnesorge (1872–1962), a fanatical National Socialist who had first joined the party in 1925 and was readmitted in 1934 with party number 42. An accomplished engineer and inventor, Ohnesorge was considered a confidant of Hitler's. As head of the Post Office he was in charge of German telecommunications from 1936 to 1945; BDC, Ohnesorge file. Ohnesorge on expanded state powers in patent matters: BAP, 10130, Bl. 31–32a, 28, 55, 93, 94.

50. "If Germany's economy and position in the world are to reach the level that matches its wealth in creative powers, [then the various patent abuses discussed above], which are a heritage of the liberalistic era …, must be eliminated. The entanglement of capital groups, which amalgamate the interests of healthy firms with those that are parasites in the body of the nation, makes it impossible to eliminate those abuses with individual measures against the firms in question. The only economic policy that remains, therefore, is to reform the patent laws and educate the agencies entrusted with their enforcement. It hardly needs pointing out that Jewish "commentators" of the patent law are responsible for these abuses. Precisely on the question of the scope of patent protection, Jewish patent attorneys have made nitpicking reinterpretations of the relevant articles, thereby fashioning the Patent Code, not into an instrument of the state, but an instrument of some of the interests of big industry." "Patentgesetz und Wirtschaft," *Völkischer Beobachter* (Berlin edition) (26 Jul. 1935), p. 15, BAK, R131/158.

51. E.g., NSDAP Reichsleitung, Reichsrechtsamt, Amt für Rechtspolitik (Hauptstellenleiter Dr. Heinrich Barth) to Volkmar, 23 Apr. 1935; Volkmar to distribution list, copies of redrafted articles and updated version of patent bill, 8 May 1935; conference of 18 Jun. 1935 of RJM officials with Heinrich Barth of the party's Reichsrechtsamt and Waldmann as Hans Frank's deputy, BAP, 10130, Bl. 5, 14, 31–32a, and passim.

52. Erich Volkmar, "Rechtspflege und Rechtspolitik: Welche Umgestaltungen erfuhr das Privatrecht seit dem Siege der nationalsozialistischen Revolution?" *Deutsche Justiz* 97, No. 10 (1935), ed. A, 355–86.

53. Franz Schlegelberger, "Das Wirtschaftsrecht im neuen Staat," in *Patentwirtschaft: Mitteilungsblatt der deutschen Patentwirtschaftler* 1, no. 2 (1935), RWWA 20-874-5.
54. Franz Schlegelberger, "Die Grundlagen des neuen Patentrechts," *Berliner Börsen-Zeitung*, no. 509, 30 Oct. 1935 (morning edition), BAP, 10130, Bl. 72 (all Schlegelberger citations are from this source).
55. According to the doctrine of prior use, someone who was already using an improvement, invention, or technical idea before another person obtained a patent on it did not violate that patent and could continue using it. The 1936 Code redefined "use" to include mere mention or discussion in writing of an idea. The issue was especially relevant with respect to government employees, including all academics, engineers, and scientists in government agencies such as the ministries of transportation and communications. The government's new powers to issue a forcible license were extraordinarily broad as well. If a patent was considered relevant to national security it could be taken over and reassigned at will.
56. The RPA could deputize one of its experts as technical state's attorney to participate on a case-by-case basis in patent litigation.
57. Gürtner to Lammers, 14 Jan. 1936, BAP, 10131, Bl. 3, 8.
58. Hess to Gürtner, 22 Jan. 1936; Gürtner to Hess, 22 Feb. 1936; Hess's office to RJM, 16 Mar. 1936; NSDAP to Volkmar, 23 Mar. 1936; telephone call of Hess to Gürtner, 31 Mar. 1936; Gürtner to Lammers, 4 Apr. 1936, BAP, 10131, Bl. 22, 23, 36, 43, 49. "Highway-robbery patent" (*Wegelagererpatent*) had to do with the situation when the right of one patent holder depended on another patentee's similar but prior right. It came into play when patent holder A began a procedure against patent holder B on the grounds that B's patent depended on, or in some way infringed on A's, as a consequence of which B owed A money or would have to desist from using the technology. The bill had left the determination of dependency up to the courts. Hess wanted it turned over to the RPA. Eventually the ministry gave in.
59. *RGBl*, 1936, pt. 2, 117ff.
60. Frank, "Grundlagen," 7.
61. The Bavarian Polytechnical Society (BPV), which had a tradition of assisting inventors, reported that inventors on balance were extremely appreciative of the new Patent Code. BPV, "Bericht der Vorstandschaft über das Jahr 1937," DM, BPV/DAF III 79-2 (2).
62. Cf. ch. 1, note 4.

Chapter 9

INVENTOR TRUSTEESHIP IN THE MAKING, 1936–1940

*E*ven before the new Patent Code took effect, the Third Reich's inventor crusaders were already turning their attention to other matters. In fact, they had never entirely abandoned those other projects, but from 1934 to 1936 the patent-law struggle had overshadowed everything else. Now these activists refocused their energies on two major pieces of unfinished business: statutory compensation for employee inventions, and the various initiatives that came under the heading of *Erfinderbetreuung* or "inventor trusteeship."[1] Whether the new Patent Code would have any real consequences for inventors depended to a large extent on the success or failure of these two projects.

Though closely related, inventor trusteeship and employee-inventor legislation developed along parallel but separate tracks. They merged into a single, integrated inventor policy only in 1942. The first part to come into being was inventor trusteeship, the origins of which went back to Jebens's Reich Inventor Office. When the Labor Front shut down Jebens's operation in early 1934, it assigned the task of assisting inventors to a newly established Department for Inventor Protection (Abteilung für Erfinderschutz) in the Bureau for Social Affairs (Sozialamt). Engineer Riemschneider was put in charge of the new unit. For the first few years, Riemschneider occupied himself mainly with patent code reform, but then turned his attention to his predecessor's legacy, which centered on the problems of small independent inventors trying to develop and market their ideas.

Historically, the development and marketing of inventions by small independents was the work of invention-promotion firms and patent economists, a line of business rife with fraud, deceptive practices, and unscrupulous operators. In all of Germany there was just one, semipublic agency to which small independents seeking impartial assistance could turn. This was the Bavarian Polytechnical Society (BPV) in Munich. Founded in 1815, the BPV was Germany's oldest technological association. As a general engineering society with a primarily regional orientation, the BPV had gradually lost out to nationwide competitors such as the Association of German Engineers and the various societies for technical subspecialties. As a consequence, the BPV had redefined its mission. By the mid-1920s it had developed a public advocacy orientation, relying on its members to study environmental hazards and undertaking a variety of outreach services. It also did consulting work for municipalities and government agencies. The most important of the BPV's public-service functions was the operation of an information desk for intellectual property. Backed by a large engineering and patent library, BPV employees and members stood ready to assist, free of charge, any and all comers with questions about how to patent their inventions and about their prospects of success. The society did not, however, aid with development or marketing.[2]

In 1933, the BPV fell into the hands of the Nazis, and soon afterward the party's Munich-based Department of Technology began using its public-service capacity for its own purposes. Specifically, in the context of National Socialism's various inventor-assistance initiatives, high-ranking Nazi engineers such as Fritz Todt and Heinrich Seebauer decided that the BPV should be harnessed to the task of preparing technical and legal evaluations of inventions by impoverished independents. Clamoring for precisely this kind of aid, and already encouraged by Nazism's pro-inventor rhetoric, inventors were coming forward in large numbers.

The BPV was only marginally equipped to meet the challenge. Membership declined from 790 in 1934 to 661 in 1935 and to 603 in 1936. Its financial situation also deteriorated—precisely at a time when additional funds were needed to pay for all the evaluations the society was told to make. The kinds of reports the BPV was now preparing, moreover, were very different from its earlier work. Instead of doing feasibility studies for municipal or other, sizeable projects, members now had to write evaluations of invention proposals by private persons, many of whom, as one BPV official cautiously indicated in 1935, came up with "only insignificant inventions." Even so, the society embraced its new duties in a patriotic spirit, hoping to avoid being "synchronized" out of existence. Increasingly dependent on subsidies from the party, it produced as many invention-evaluation reports as it possibly could and answered thousands of intellectual-property questions.[3]

The party's pro-inventor activities in Munich did not stop Riemschneider from building up a parallel inventor-support network in Berlin. The duplication of effort had several causes. Initially, the BPV limited its activities to Bavaria. It did nothing about development and marketing, which represented an independent's most difficult hurdles. The party and the Labor Front also emphasized different aspects of inventor assistance. The Labor Front acted more from a concern with social policy and sought to address the injustices of a capitalist economy, while the party was primarily interested in fostering technological progress.

Despite the differences, there was enough overlap that jurisdictional rivalry was inevitable. Robert Ley and his Labor Front proved to be the weaker party. In 1934 Ley tried unsuccessfully to kill or take over the party's inventor program. In 1936, having lost a second battle with the party's engineering leaders, he reluctantly accepted a compromise. Fritz Todt, the head of the party's Department of Technology, also became chief of the Labor Front's Bureau for Technical Sciences. In theory, Todt in his new job would be subordinate to Ley, but in 1938 the latter was unable to prevent Todt from taking most of the Labor Front's other technology-related programs—monitoring the technical professions, occupational safety, factory organization and rationalization, and continuing education—and putting them under the Bureau for Technical Sciences. Riemschneider's Department of Inventor Protection also transferred to the bureau, which Todt moved to Munich and installed in the offices of the BPV, whose independent existence he terminated at the same time. Ley's final defeat came in February 1942, when Martin Bormann took the bureau away from the Labor Front and gave it to the Nazi party's Department of Technology under Speer.[4]

With Ley's power eroding only gradually, Riemschneider had time before 1938 to pursue his own ideas of inventor policy, specifically, to clean up the invention-promotion business. Under National Socialism, the notion of cleaning up was, above all, a code word for getting rid of Jews. This was true also for the Labor Front's program to help independents translate their ideas into functioning technology. When the presence of Jewish patent advisers or invention promoters came to light, Riemschneider pursued those cases aggressively and tenaciously. In June 1936 he learned that a Jewish ex-patent attorney had found new work as a consultant in Dresden. Riemschneider promptly launched an investigation, which resulted in the firing of the "Non-Aryan Brann" by the end of September 1936.[5] Another case concerned a small invention-promotion business in Stuttgart, the "Working Group of German Inventors," run by a Polish Jew named Chonon Lewin. When he learned about it, Riemschneider set in motion an inquiry that caused the police to initiate deportation procedures against Lewin. A delay prompted Riemschneider's assistant to follow up on the case. It was "important to the Labor Front that the creative German being, not in the last place also the German inventor," wrote engineer Luyken, "be protected from such evil creatures [as Lewin]."[6] In yet another instance, an invention promoter came forward with information that four members of his association were Jews, and what should he do about it? Riemschneider replied that, as a matter of course, good Germans could "not possibly work with Non-Aryans."[7]

Despite these and other instances of racial persecution, which were integral to National Socialism, anti-Jewish measures were only a sideshow for the regime's inventor activists. The larger project centered on building the institutions that would let inventors thrive, especially small-time independents. The basic idea—curtailing the power of big business and protecting the little man—was spelled out in an article in the *Völkischer Beobachter* of 25 August 1936. With the new Patent Code taking effect in just a few weeks, the anonymous author wrote, the next step was to implement "inventor trusteeship" as a condition for "realizing German socialism." The nation could not "tolerate that the labor power of the [inventor], for whatever materialistic or other reasons, is blocked from engaging in creativity because he is not justly rewarded for his achievement." What was needed was a level playing field for competition among inventors. If creating fairness hurt "traditional power positions" and broke down the "walls of so-called blocking patents or patents held in reserve" (*Schubladenpatente*) by big business, it was all for the best, since the public interest could only benefit. Investment in new technology and innovation would come more quickly, and Germany's competitive position vis-à-vis foreign nations would improve. An economy that had "put in practice a functioning German socialism," the article concluded, could not tolerate "such barriers against securing the fruits of German intellectual labor for *Volk* and Reich …"[8]

In this spirit Riemschneider had first worked on the new Patent Code, and in the same spirit he now developed measures to help small independents. He drew up a long list of fraudulent invention promotion firms (his predecessor Jebens's one of them) for distribution to organizations such as the Association of German Engineers.[9] He developed a close working relationship with a Berlin engineer and invention promoter by the name of Willy Salge. Salge headed an organization known as the Federation of German Patent Economists (Verband Deutscher Patentwirtschaftler). With the blessing of the Labor Front, Salge's mission was to unite all of Germany's patent economists in his federation. Excluding crooks and charlatans and following codified procedures for evaluating and developing inventions would professionalize the occupation. The next step would be to elevate Salge's federation into the Reich's exclusively authorized "chamber for patent economy."[10]

The actual inventor-assistance mechanism, still based on Jebens's idea of evaluating inventions before small independents wasted all their time and money on patent attorneys and patent fees, would function as follows. The Labor Front would collect invention proposals—excluding those related to weapons, aeronautics, railroad technology, and farm machinery, which had their own addresses—from small independents everywhere. Riemschneider's office would turn over the proposals to Salge's office, which was supposed to do a preliminary evaluation of their economic potential and send promising inventions on to the federation/chamber's members. A patent economist thus engaged would then contract with the inventor to further examine, develop, or market the invention. The terms of such a contract were laid out in an official fee schedule, which resembled a standard architectural agreement.[11]

Within a year these plans began to founder. Without reimbursement from the Labor Front, Salge could not long continue to process the flood of invention proposals that started coming his way shortly after the program was launched.[12] Riemschneider therefore soon began looking for alternative solutions. In September 1936 he established contact with the Association of German Engineers, one of the few professional associations not synchronized out of existence. The engineering society, in the spirit of the times, had recently established an invention-promotion committee of its own, and Riemschneider began exploring whether this might not be a more suitable candidate for the chamber of patent economists he hoped to establish. An understanding was soon reached, and by October the two parties signed an agreement to start collaborating.[13]

After making some initial headway, this plan also collapsed.[14] The Labor Front's rivals in Munich got wind of Riemschneider's arrangement with the engineering association, and in March 1937 Todt sent an angry letter to Riemschneider's boss, Claus Selzner, in which he claimed to have exclusive jurisdiction in the matter. He demanded the Labor Front cancel the deal with the engineering society.[15] Initially the Labor Front refused to comply. Karl Peppler, head of the Bureau for Social Affairs, informed Todt that he had personally signed off on the arrangement, trusting the party approved because Riemschneider had spoken about it with Todt's deputy. Peppler concluded with the hope that the incident would cause the Department of Technology to become just as interested in a "definitive resolution … of the current difficulties of the patent development and marketing business" as was the Labor Front's Bureau of Social Affairs.[16]

This is, in fact, what happened, though not on the terms the Labor Front's leaders imagined. Riemschneider himself had too many contacts with party inventor advocates and legal experts to get locked into an unproductive and dangerous power struggle. He quietly dropped the arrangement with the Association of German Engineers and began sending the invention proposals accumulating in his Berlin office to the BPV in Munich.[17] This put the BPV under greater pressure than ever and succeeded only in causing its backlog to grow.[18] In turn, this increased the pressure on Todt and the other party engineers to deal with the problem head-on. They did so in early 1938, when Ley was forced into beefing up the Bureau for Technical Sciences and looking on as it was moved to Munich a few months later.[19] Todt put the bureau under trade-school engineer Erich Priemer, and "old fighter" and trusted lieutenant.[20]

Riemschneider's Blueprint

Even as these jurisdictional struggles were going on, Riemschneider's efforts to get inventor trusteeship off the ground continued unabated. In the summer of 1937 he presented a paper on the subject at the Academy for German Law, copies of which he sent to the party's Office for Legal Policy and the Office for German Raw and Intermediate Materials in Göring's Four-Year Plan. There is no indication that Riemschneider's paper triggered any immediate action. But it appears to have played a role in contributing to the 1938 reorganization and the more ambitious system of inventor trusteeship launched at that time. The system that emerged in 1938 conformed in all its essentials to the design Riemschneider had mapped out the year before. This and the author's expanded responsibilities suggest that his 1937 paper did, in fact, have a profound influence on policymaking.[21]

Riemschneider's paper addressed two questions: "Is the Establishment of Trustee Agencies (*Treuhandstellen*) to Encourage Inventions and Patents Advisable? How Can Such Trustee Agencies Be Established and Expanded Effectively?" The author spent most of his energy answering the first question in the affirmative, saying rather less about how his plan, if adopted, might function in practice. He began by looking into the reasons that only a small fraction of all usable inventions were ever adopted by industry. First, the economy might be at full capacity, as a consequence of which there would be little inclination to contemplate the introduction of new technology, an expensive and risky proposition under any circumstances. Second, companies might have invested in a given new technology. Before embarking on even newer systems, they first had to recoup their costs. It was therefore perfectly conceivable for certain new inventions not to be adopted. In fact, there were circumstances where the only responsible behavior might be to "fend off and even oppose new technological progress." If introduced too quickly, some inventions could put entire branches of industry out of business and cause unemployment and bankruptcy overnight. This proved, according to Riemschneider, that sometimes there was a need for external, government intervention to plan and manage a gradual introduction of innovations.[22]

A third reason for not adopting new technology was more sinister than the preceding two. It had to do with purely private motivations, with wanting "to avoid abandoning a given level of technology that yields good profits, especially in cases where reaching that given level of technology required great means." Considerations of this kind explained the drive by companies to acquire monopolies protected by patent rights, which aimed at "fortifying their field of activity with protective rights in such a way that the outsider is completely prevented from entering that territory." The only purpose of patents held in reserve and blocking patents was to "secure a monopoly vis-à-vis the competitor." But sometimes it was impossible to create technological and thus business stability with patent policies alone. In those cases corporations sought to control the market by making agreements or combining themselves in trusts, concerns, and cartels, which allocated market share on the basis of a company's financial resources rather than its potential achievements.[23]

Controlling the market by the above practices, according to Riemschneider, meant that patents had lost their original purpose of stimulating technological progress and economic growth, because there was little or no chance now that patents would lead to financially rewarding innovation. The incentive that a patent traditionally represented having thus been undermined, the drive to invent also suffered, "so the great danger arises that, for this reason alone, a stagnation in technological progress occurs." Such stagnation might not be evident immediately, "but it leads to a nation gradually becoming dependent on the surrounding nations that willingly devote themselves to technological progress." It was, therefore, "the duty of Germany's political leadership to make sure that technological progress is promoted with all possible means." The only feasible way to do this, according to Riemschneider, was "to encourage and protect the creative people's comrade not merely in theory but also tangibly in practice."[24]

The regime had already taken some steps in this direction, by adopting a new patent code centered on the principle that the "creative personality to a very high degree requires the support of the state." But neither the new code nor the employee-inventor legislation being drafted in the Ministry of Labor did anything to increase the *rate of utilization* of inventions and patents. Technological and economic developments, Riemschneider said, had brought about the current levels of economic concentration, in which people were completely dependent on one another and in which individual initiative or technological ability had become largely irrelevant. Certain groups could "completely and forcibly block the kind of unhindered progress of productivity that depends on patents. Often those groups use economic pressure to force creative people to drop their patent rights. And then those same powerful groups go the very road indicated and make lots of money for themselves from the formerly unused inventions."[25]

A large proportion of all inventions, not counting the many worthless proposals by laymen, he said, stemmed from the economically weakest segments of society. Most of those inventors were at risk of being "exploited by asocial elements." Often somewhat naïve when it came to business dealings, the inventors lost their ideas without adequate compensation to unscrupulous scam artists, which embittered them and prevented further creative efforts, even as the asocial elements profited and gained strength, "which [was] directly contrary to the interests of the national community." The problem was made worse by the fact that many inventors lacked the financial resources to properly develop their ideas in laboratories or workshops and bring them to market. Germany also had "few private means of financial support available to develop as yet unripe invention proposals."[26]

It was commonly assumed that salaried inventors were better situated in this regard, but that was not necessarily true either. It was difficult to persuade supervisors of the merit of new ideas, while experimenting on one's own was typically too risky.[27] Trying out a new idea was frequently forgone also because a company was too busy with other things. In such cases, Riemschneider said, there "ought to be … an opportunity for the inventor to develop his idea in a scientific institute of the state, since he obviously cannot go to a competing firm in the same line of business." Then there were all the cases where the development of valuable inventions was slowed down because the company, though aware of their advantages, feared customers would be reluctant to change over to the new product. Overcoming such resistance required a major and expensive marketing campaign, which might be beyond the company's means.[28]

If one took into account the various permutations of innovation-retarding conditions he had discussed, Riemschneider continued, there was no escaping the conclusion that they represented "a weakening of the German economy vis-à-vis foreign countries." This was "all the more significant today," he added with a reference to the policy of autarky, "when the spirit of German ingenuity must be a substitute, as it were, for the natural resources we lack." The sums being lost to inadequate exploitation of the nation's inventions were incalculable. Germany's synthesis of indigo in the last part of the previous century, for instance, had not only saved the millions spent on imports but also earned further millions in export earnings. The German patent on vacuum tubes of 1906 had created an entirely new industry with thousands of jobs, yielding huge export revenues.[29]

These observations set the stage for the second part of the paper, in which Riemschneider addressed the question of what should be done to promote innovation and invention. The best solution was to establish a central agency that would make sure valuable inventions ignored by industry would not be lost to the economy. Opponents contended that such an agency would have to be a "mammoth apparatus," for which qualified personnel and money were not available, and which, in light of the uncertain outcome of its endeavors, would not be worth the money and effort expended on it. But the example of France, where such an institute had been in effect for about fifteen years, showed that it was perfectly feasible. Likewise, England, the United States, and Italy had "taken measures to steer and plan invention." In Germany, too, the voices demanding action were becoming ever louder: invention was of "such exceeding importance for the state that it could not longer be left to private initiative."[30]

Opponents also claimed that an inventor agency was undesirable—or at best superfluous—because systematic research and development already guaranteed a steady and uninterrupted stream of innovations. But systematic development was often nothing more than a hollow slogan, since no one could look very far into the future and it was not known what the next invention might be. To be sure, systematic research and development generated a great deal of progress. But it also tended to ignore different possibilities that emerged along the way, and to be unduly rigid in its vision of the future. It had no use for the "opposing forces [that] inevitably also emerged," no use for the new impulses that refused to "trot along" with everyone else because they did not suffer from trained incapacity and "wanted to venture into new territory to realize better, new, simpler, and more economical ideas."[31]

For all these reasons, argued Riemschneider, it was desirable to "create a special Reich agency" that would (1) "secure the utilization of such inventions as concerned the general well being of the national community" and (2) "inspect and monitor those persons whose business it is to help bring inventions to market." Some progress had already been made in this direction. But what was lacking so far was an agency that would help develop inventions by "encouraging and supporting" them, especially with money for models or experiments, or the filing of foreign patents. To be sure, the Four-Year Plan and the War Ministry had funds to support and develop inventions of immediate military significance. But they, too, could only develop those inventions they knew about. If inventors or entrepreneurs held back with their ideas for the reasons he had outlined above, then these agencies were powerless also.[32] What was needed therefore was an agency to help and support all those inventors who, for whatever reason, were unable to help themselves but whose ideas might be valuable for the economy. Assuming that the steel industry had a trustee agency capable of monitoring all the patent filings in its field, Riemschneider hypothesized, it would be possible to help develop the deserving ones and speed up innovation. Germany had "Reich Trustees of Labor nowadays, why should not also Reich Trustees for Patent Utilization be possible!"[33]

Riemschneider concluded by reiterating that "establishing a trustee agency ... [could] be of decisive importance," that France had such an agency and spent big money on it because it appeared to work, and that Germany should do everything possible to avoid falling behind foreign countries with regard to technological progress. "Let us therefore take care of the social protection of the creative personality and the encouragement of inventions and patents in a practically feasible way. The German inventor will then happily place himself with total dedication in the service of the business at hand."[34]

Like many proposals, that of Riemschneider promised much. Whether it would deliver, and if so how much, remained to be seen. But his ideas proved sufficiently persuasive and dovetailed well enough with other economic preparations for war to be adopted— though it was only the Nazi party-Labor Front combination, and not the state bureaucracy, that took up the plan, and then only in principle.[35] The program that was actually established in the course of 1938 was considerably more modest than its architect had envisioned. Still, the 1938 reorganization was a major step on the road to a fullfledged system of inventor trusteeship.

Regulation of the invention-promotion trade, a key aspect of Riemschneider's pre-1938 activities that still figured importantly in his 1937 paper, fell by the wayside after the move to Munich. The perceived excesses of the business had mostly disappeared by 1938. Jews had already been weeded out, while the economy was controlled to the point that free-market entrepreneurship for the civilian market was becoming nearly impossible.[36] Instead of the earlier plans for a chamber of patent economists, distribution of the occasional list of reliable addresses sufficed.[37] This left the inventor activists at the Bureau for Technical Sciences free to devote more time to their other tasks, which were daunting enough.

There was an enormous amount of invention evaluation to be done. Independents kept submitting large numbers of proposals for feasibility assessment. Riemschneider became much more closely involved in this aspect of inventor trusteeship than before, now that he supervised the former BPV staff coordinating this work. The bureau was also responsible for a huge volume of legal advice and institutional support in disputes over the rights to inventions. Inventors, exporters, Nazi organizations, associations, and businesses of all kinds flooded the bureau with questions about how to proceed with their patent problems, and about who was entitled to what.[38] Many of the questions were taken care of with a quick letter, but many others took prolonged correspondence. Some dragged on for years and consumed incalculable amounts of effort before being resolved.[39]

A great deal of energy also went into reorganizing and building up the inventor-assistance bureaucracy across the country, and into publicizing the existence of the bureau and promoting the regime's culture of invention.[40] Riemschneider had a great interest in the propagandistic and informational aspects of inventor trusteeship. Getting out the word that the Patent Code had changed; that inventors now had specific rights such as being entitled to subsidies in the patenting process; that the Labor Front was actively involved in reaching out to independents; in short, that a new age had dawned and that the small inventor no longer need be discouraged—all this took a great deal of advertising and publicizing energy.

Riemschneider threw himself at the task with abandon. Already in 1936, he had published his first book, *The Economic Utilization of Inventions*, a hundred-page manual for small-time inventors on how to get their ideas patented and marketed. Although concrete technical advice made up the bulk of this work, Riemschneider did not ignore the ideological dimensions of the Third Reich's pro-inventor climate. "We possess, in the German *Volk*'s exceedingly large talent for inventing," he wrote, "a gigantic treasure of unborn creative achievements of the person, which are lying fallow, completely unutilized. To compensate for our lack of natural resources, this valuable asset of the German economy needs intense cultivation." Besides fraudulent patent promoters, the greatest obstacle to tapping all this talent was the "common but erroneous view that the small inventor (the lay person and the do-it-yourselfer) is not in a position to achieve anything worthwhile." The professional researcher in industry might have certain advantages over the amateur or the small independent, but frequently the expert could not see the forest for the trees. "The occasional inventor usually arrives at his proposals from a different vantage point. When treated with respect and properly developed, however, those ideas often turn out to be 'just the thing' ('*Ei des Kolumbus*') or, at the very least, most valuable."[41] Inventors should always remember, Riemschneider continued, that making the initial conceptual breakthrough was the most crucial step of all. It was, in "a word, the 'act of inventing,' the creative achievement depicted by the poet Max Eyth as that certain stirring of the human soul that must, indeed, be called truly spiritual." The act of inventing could take place "completely independently of mere knowledge and know-how, because it represents an inspiration, a conceptual and exhilarated sparking of an individual human being."[42]

"Worthless Inventions ... and ... Nonsensical Proposals"

How realistic were the premises on which Riemschneider based his advice to small-time inventors in *The Economic Utilization of Inven-*

tions? How good were the prospects of speeding up innovation by making Germany's technological culture more hospitable to independent inventors in the manner he envisaged? If he had known about the efficacy of comparable schemes in the United States during World War I, the Third Reich's chief inventor crusader might have tempered his optimism.[43] But Riemschneider soon got plenty of opportunity to see for himself. Beginning in 1939, he could read digests of the invention assessments made by the bureau's internal evaluators and outside consultants. In 1938, these experts had evaluated a total of 935 invention proposals by independents. Most were judged to be in a category of worthless, utopian, or otherwise completely unrealistic ideas. There were, for instance, lots of suggestions for perpetual-motion machines, automatic turn signals for automobiles and bicycles, automatic brakes of one kind or another, impossible or ridiculous safety nets, turbines of the most rudimentary and primitive type, energy conservation systems that had been around for a long time, and all sorts of little gadgets. Two fields where small-time inventors proved to be particularly "clueless," according to the report's author, were the refrigerator and railroad safety installations. "Honor the inventor, indeed!" he concluded, "but the person who does not research anything and then comes up with some chance proposal is not an inventor."[44]

The activity report for 1939 was even more discouraging. Riemschneider's staff had gone through some 1,237 proposals that year. "What should be said about the invention proposals," wrote engineer Wagner, "is that good and usable proposals, as in the past, were very much in the minority." The few good ideas, however, received an unusual amount of special attention and, "if at all possible, are supported by the Department of Inventor Protection, especially when there is the slightest chance they may be useful for the economy in wartime." But in this respect "the claims and expectations of the inventors are usually too great, or even exaggerated, or completely illusory." Especially "frightening," according to Wagner, was the "great number of proposals for perpetual-motion machines, which have recently been lining up as the proven means to solve all economic difficulties associated with the war." Among the most recent crop of proposals, his 1939 report continued, the only two that looked promising were for a typewriter and an electric dryer. There were some other ideas in specialized fields that might be usable, at least in principle. Then there were several without any prospects, because they were no better than comparable but established products. Another group had "only very minor significance, promising a modest yield at best and then only after a great deal of advertising." There was no way these inventions could be said to have "general economic significance." Finally, there was a "large group of proposals for completely worthless inventions, partly design proposals that have not mastered the difficulties of the relevant field, partly completely pointless things, and partly nonsensical proposals, such as the perpetual-motion machine."[45]

An interim report for the first half of 1940 that Riemschneider prepared for Priemer tried to put a more positive spin on such results. Riemschneider made much of the fact that there were at least some worthy inventions, whose development his department was supporting to the tune of RM 15,000. He also put a different light on the continuing stream of worthless proposals. They proved the value of the bureau's preliminary evaluation mechanism, in as much as it had saved an estimated RM 200,000 or more in unnecessary legal fees and other expenses in the first six months of 1940 alone. A follow-up report by Wagner in the second week of July 1940 mentioned that the department had processed 219 invention proposals so far that year, 194 of which had yielded a "negative result," among them "twenty-six ideas for perpetual-motion machines in different variations." The remaining twenty-five proposals had received a positive evaluation.[46]

There is no information on the bureau's invention-evaluation service for the years 1941–1944, owing to destruction of the relevant files in the last years of the war. A surviving, brief report for January of 1945, however, suggests that little had changed during the intervening years, other than the continued shriveling of this particular dimension of inventor trusteeship. Engineer Klee in a report for Riemschneider mentioned only that the routine evaluation of inventive proposals by small-time independents had yielded no valuable ideas, and that his Munich office had only had four visitors that month.[47]

What conclusion did Riemschneider and other inventor crusaders draw from these disappointing findings? Did they abandon as myth the notion of a special German genius for technological creativity present in the mind of even the humblest "people's comrade?" Their initial reaction was, indeed, an almost complete loss of faith in this most cherished principle of National Socialist inventor ideology. Disillusionment and realism briefly gained the upper hand.

In July 1939 bureau chief Priemer sent a memorandum to the *Gau* Officers for Technical Sciences about the inflated significance many inventors attached to their proposals, issuing a warning against being taken in too easily. Riemschneider became so quick to dismiss invention proposals from independents that he had to be reprimanded. In September 1939, in connection with an appeal from Göring to inventors, the bureau urged people to stick to problems they really knew something about, and not to try to reinvent the wheel. "Good ideas usually come only from a thorough knowledge of the relevant technical specialty," the bureau's clarification read. "No one denies that laymen have done exceptional work in many technical inventions. But usually these were men who possessed an intrinsic and special technical talent and who worked on a problem for many years, always pursuing everything that others contributed to it."[48]

Given the demoralizing experience of sifting through the countless numbers of useless ideas of would-be inventors and small-time independents, it was only natural that the regime's inventor activists would try to refocus their energies. De-emphasizing trusteeship of the independent, Priemer and Riemschneider increasingly shifted their attention to inventors in the factories and laboratories of organized production. Concern with salaried inventors, of course, had never been completely absent from their horizon. For much of the 1930s, however, it had been overshadowed by the effort to improve the fortunes of small-time independents, while the Ministry of Labor drafted employee-inventor legislation. Still, Riemschneider from the beginning had handled his share of questions about disputes between employees and employers and published about it as well.[49] In his grand design of 1937, too, he had already addressed himself to trusteeship in the factories—spurring inventiveness, protecting employees against employer abuse, monitoring industrial patent applications, and intervening when promising ideas were not pursued.

Beginning in the fall of 1939, Riemschneider and his cohorts set about translating such notions into reality. Unfortunately, they did not yet quite know how. In mid-September, the bureau sent preliminary instructions to all *Gau* Offices of Technology on how to deal with patent conflicts between employers and employees. But the bureau changed its mind in February 1940 and presented yet another, different plan to the Patent Office in October.[50] Another illustration of bungling at this time concerned an attempt to steer the nation's course of technological development by leaflet. In 1940, Riemschneider and Priemer conceived of the idea of printing and disseminating leaflets to inform inventors across the country of areas and problems where their efforts would be wasted, where innovation was still needed, and where an adequate level had been reached already. These so-called *Merkblätter*, which reflected in part the results of the bureau's own technology-evaluation efforts, did not earn high marks from people who were in a position to judge their usefulness. As one high-level examiner in the Patent Office reported to Klauer, most of the Labor Front's leaflets were "highly problematical." When Klauer politely explained to Priemer that some of his leaflets were inadequate, the embarrassed Labor Front official felt obliged to acknowledge the point and promised his agency would try to do better in the future.[51]

An even more humiliating episode of failed technology steering centered on Riemschneider's and Priemer's project to monitor corporate technological progress by going over the heads of management and reaching out directly to salaried inventors. Gaps in the sources prevent complete reconstruction of this scheme, but in April 1940 industry sent a rare protest letter to the ministers of Justice and Labor. Copies went to Wilhelm Canaris, chief of military counter-intelligence, Albert Pietzsch, head of the Reich Economic Chamber, and the Supreme Command of the Armed Forces. The Bureau for Technical Sciences, according to the complaint, was sending out questionnaires directly to industrial employees, requesting them to report back to the Labor Front on their research and inventions. This scheme was causing "considerable anxiety in business circles" and would lead to serious friction between management and employees. Since it was crucial to keep labor peace during the war, and because employees were in effect being urged to breach their fiduciary responsibilities by revealing company secrets, the practice should be stopped immediately. The letter concluded with the urgent plea that the government take "immediate steps to eliminate the danger caused by the circular, force the Labor Front to take back its questionnaire and prohibit further propaganda of this kind."[52]

The bureau was soon forced to retreat. Hess intervened to stop the mailing of any further questionnaires, and by June the crisis was over. Even before then, however, Riemschneider had already apologized to Siemens patent chief Wiegand, explaining that the program had been halted because the "expectations for utilizing such reports were too great for the initial framework," and that individual inventors appeared not quite to grasp the full extent of their own activities anyway. He gave his assurances that there had been no intent to spy on Siemens or any other company and that, of course, employees had to keep quiet about research in progress.[53]

The sources contain no further documentation about the questionnaire project. But it is obvious that the outcome was another humbling setback for inventor trusteeship and its ambitions to manage Germany's technological progress. In the spring and summer of 1940 the bureau's fortunes were at low ebb.

Notes

1. *Betreuen* combines the notions of "taking care of" and "looking after by means of supervising and monitoring." I translate *Erfinderbetreuung* and *Erfinderbetreuer* as "inventor trusteeship" and "inventor trustee."

2. BPV annual report 1926, DM, BPV/DAF II 36, 1 (2).

3. It answered 4,859 questions in 1934 and 4,565 in 1935; untitled BPV report, 2 Oct. 1935, DM, BPV/DAF III 79-2 (4); BPV reports 1935 and 1936, DM, BPV/DAF III, 79 (1).

4. Ley to Seebauer, 30 Oct. 1934, BAK NS 022/000851; Ludwig, *Technik und Ingenieure*, 132–33; Ley's Instruction No. 34/38 of 3 Mar. 1938, DM, BPV/DAF VIII 282 (1); file "Personalfragen der DAF 1938–39, 1941, 1945," DM BPV/DAF VIII 287 (2); Priemer to BPV, 15 Jun. 1938, and BPV to Oberbürgermeister Munich, 18 Jul. 1938, BPV/DAF 36, 2; NSDAP, Gauleitung Berlin, Amt für Technik to President RPA, 16 Apr. 1942, BAK R131/22.

5. Correspondence of patent attorney Heilmann of NSBDJ, Riemschneider, and other DAF agencies, Jun.–Sep. 1936, DM, BPV VIII 291 (3).

6. Correspondence of Bremen patent attorney Hans Meissner, Riemschneider's office, and other DAF offices, 22 Aug. 1936 to 12 Dec. 1937, DM, BPV VIII 291 (3).

7. Correspondence of Verband Deutscher Patentwirtschaftler e.V. (VDPW) and Riemschneider, 8, 14, 19 Apr. 1937, DM, BPV/DAF III 290 (20).

8. "Erfinderbetreuung und Angestelltenerfinderrecht als politisches Problem," *Völkischer Beobachter*, ed. A, Berlin ed., (25 Aug. 1936), unpaginated clipping in BAK, R22/629, Bl. 33.

9. Riemschneider to VDI, 23 Dec. 1936, DM, BPV/DAF VIII 290 (2).

10. VDPW flyer, 1 Apr. 1935, DM, BPV/DAF 290 (2); 1936 correspondence VDPW and Berlin Chamber of Commerce, DM, BPV/DAF VIII 292 (5); Riemschneider to VDI, 19 Oct. 1936, DM BPV/DAF VIII 290 (1, pt.2); Salge to DAF Office for German Raw and Intermediate Materials, 19 Oct. 1937, DM, BPV/DAF VIII, 290 (2).

11. Riemschneider to RPA president, 26 Jan. 1937, esp. copy of DAF "Merkblatt P" (Beurteilung von Erfindungsideen), DM, BPV/DAF 290 (2).

12. Summary of material in DM, BPV/DAF 290 (1) and (2). One inventor, having heard nothing for three and a half months, complained about having "fallen into the hands of a swindler firm ... It could really make one's blood boil, to have one's patience tested to this extent" (Weisbrod to VDPW, 14 Aug. 1936); correspondence of Robert Kahlert of DAF Reich Cooperative Iron and Steel and Riemschneider, 31 Aug. and 18 Sep. 1936, DM, BPV/DAF VIII 290 (1).

13. Correspondence of Riemschneider, VDI, and Salge, 24 and 28 Sep., 3, 15, and 19 Oct. 1936, DM BPV/DAF VIII 290 (2) and 290 (1, pt.2).

14. Correspondence of Salge and Riemschneider, 8 and 17 Dec. 1936, Salge to Riemschneider, 3 Nov. 1937, DM, BPV/DAF VIII 290 (2).

15. Todt to Selzner, 15 Mar. 1937, DM, BPV/DAF VIII 290 (2).

16. Peppler to Todt, 3 Apr. 1937, DM, BPV/DAF VIII 290 (2).

17. Correspondence of Riemschneider and Reichswirtschaftskammer, 9 and 22 Jun. 1937, DM BPV, 290 (20); Riemschneider to Amt für Deutsche Roh- und Werkstoffe, 29 Oct. 1937, DM, BPV/DAF VIII 290 (2); Riemschneider to RPA, 15 Nov. 1937, DM, BPV/DAF VIII 290 (2).

18. BPV 1937 report, DM, BPV/DAF III 79–2 (2).

19. Ley's Instruction No. 34/38; drafts in the files of the AftW, 2 and 3 Mar. 1938, DM, BPV/DAF, 282 (1).

20. Ley to Seebauer, 30 Oct. 1934, BAK NS 022/000851.

21. "Empfiehlt sich die Einrichtung von Treuhandstellen zur Förderung von Erfindungen und Patenten? Wie können solche Treuhandstellen wirksam eingerichtet und ausgebaut werden?" 23 Aug. 1937, DM, BPV/DAF VIII 290 (1).

22. "Empfiehlt sich die Einrichtung," 1–2.

23. Ibid., 2–3. On the use of patents and patent-pooling agreements as "weapons of monopoly" and methods of regulating or retarding costly innovation in the United States—including contemporary American critiques similar to Riemscheider's—cf. Noble, *America by Design*, 84–109.

24. "Empfiehlt sich die Einrichtung," 3.

25. Ibid., 3–4.

26. Ibid., 4–5.

27. Ibid., 6.

28. Ibid., 6–8.

29. Ibid., 8–9.

30. Ibid., 9–10.

31. Ibid., 10–11.

32. Ibid., 11–13.

33. Ibid., 13–14.

34. Ibid., 15.

35. On DAF jurisdictional grasp: Smelser, *Robert Ley*, 149–217; Mason, *Social Policy*, 151–77, 209–24.

36. Inventor promoter complaints: Riemschneider to Deutsche Arbeitsgemeinschaft für gewerblichen Rechtsschutz und Urheberrecht, Apr. 6, 1937; patent economist Reinhold Claren to VDPW, 4 Jan. 1937; both in DM, BPV/DAF VIII 290 (1).

37. VDI to Zweigstelle Berlin of Deutsche Zentralstelle zur Bekämpfung der Schwindelfirmen, 22 Oct. 1937, DM, BPV/DAF VIII 290 (1).

38. Summary in DM, BPV/DAF VIII 290 (1) and (2), and VIII 293 (1).

39. Numerous examples of the bureau's mediation efforts in DM, BPV/DAF VIII 293 (1) (A–N), 293 (2) (P–Z).

40. DM, BPV/DAF VIII 283 (2); Reichskommissar für die Wiedervereinigung Österreichs mit dem Deutschen Reich to DAF, Zentralbüro, AftW, 28 Nov. 1938, DM BPV/DAF VIII 282 (2); correspondence of Gauamtsleiter Ostertag in Munich and Priemer, 5 and 14 Dec. 1938, DM, BPV/DAF VIII 282 (3); DAF, Zentralbüro, Sozialamt, to AftW, 2 Feb. 1939, DM, BPV/DAF VIII 290 (2); Priemer circular to all party district administrations, Dec. 1938, DM, BPV/DAF VIII 282 (3).

41. Karl August Riemschneider, *Die wirtschaftliche Ausnutzung von Erfindungen* (Berlin, 1936), 5–6.

42. Riemschneider, *Ausnutzung*, 9.

43. For the disappointing results of the Naval Consulting Board, established in 1915 to collect and evaluate invention proposals from the public, see Noble, *America by Design*, 148–49; Daniel J. Kevles, *The Physicists: The History of a Scientific Community in Modern America* (Cambridge, Mass., 1995), 117–38, esp. 138. Of more than 100,000 proposals, just a single one was adopted.

44. 1938 report of Abteilung technische Arbeitsgemeinschaften by engineer Wagner, DM, BPV/DAF III 79-2 (3).

45. 1939 report of Abteilung technische Arbeitsgemeinschaften. Wagner, who appears not to have been a member of the Nazi party, was BPV former business manager, DM, BPV/DAF III 79-2 (3).

46. Report 1939/1940 of Abt. Erfinderschutz, 29 Jun. 1940, and report of Abt. technische Arbeitsgemeinschaften, 8 Jul. 1940, DM, BPV/DAF III 79-2 (3).

47. Klee reported that an air raid of 7 Jan.1945 had destroyed Munich's Institute of Higher Technical Education. Housing the 20,000-volume library that the AftW had taken over from the BPV in 1938, the building was about to collapse. The library on the Erhardtstrasse was in better shape, but its greatest risk was women stealing books and journals for heating fuel. Report by engineer Klee, 29 Jan. 1945, DM, BPV/DAF III 79-2 (3).

48. Priemer to Gauämter für technische Wissenschaften, 15 Jul. 1939, DM, BPV/DAF VIII 282 (3); Priemer to Riemschneider, 9 Aug. 1939, DM, BPV/DAF VIII 290 (2); AftW clarification in *Rundschau Deutscher Technik*, 28 Sep. 1939 DM, BPV/DAF VIII 288 (6).

49. Karl August Riemschneider, "Die begriffliche Abgrenzung der Angestelltenerfindungen," *Zeitschrift der Akademie für Deutsches Recht* 1936: 1004–5; idem, "Zum Recht der Angestelltenerfinder," *GRUR* 42 (1937): 393–95; Hugo Wilcken and K. A. Riemschneider, *Das Patentgesetz von 5. Mai 1936* (Berlin, 1937).

50. Correspondence of Priemer and Klauer, 16 Oct., 4 and 15 Nov. 1940, BAK, R131/19.

51. Chairman of Patent Division II to Klauer, 7 Oct. 1940, Klauer to AftW, 9 Dec. 1940, Priemer to Klauer, 30 Jan. 1941, BAK, R131/22.

52. Letter dated 26 Apr. 1940, BAK, R22/629, Bl. 175–175a. Other examples of employer resistance to DAF encroachment and worries about industrial espionage: Gregor, *Daimler-Benz*, 164–75; Mason, *Social Policy*, 151–78, 209–24.

53. RAM's Werner Mansfeld to AftW, 3 May 1940, Riemschneider to Wiegand, 6 Jun. 1940, Mansfeld to RJM, 22 Jun. 1940, BAK, R22/629, Bl. 176, 180, 182.

Chapter 10

Inventor Trusteeship and the "Production Miracle," 1941–1944

*E*ven as they were reeling from one fiasco to the next in 1940 and the first half of 1941, the inventor crusaders at the Bureau for Technical Sciences were making adjustments, going back to the basics, and trying to learn the ropes of the patent and invention business in industry. The most important step they took was to set up two study groups of outside consultants, one for "Questions of the Protection of Industrial Property" and the other for "Workforce Inventor Law." Exactly when these bodies were first formed is unclear, but it appears they dated from the fall of 1939, when the bureau was first beginning to reorient its energies from independents to employed inventors.[1]

Riemschneider chaired both groups, whose overlapping membership consisted of established experts on invention and patent law in industry and private practice. Also included were Nazi party and Labor Front engineers with broad experience in industry.[2] The two committees merged in 1942 to form the Reich Working Group for Inventing (Reichsarbeitsgemeinschaft Erfindungswesen). There is no record of the groups' early activities, but their labors bore first fruit in August 1941 with the publication of three technical papers. The first of these examined the results of participation by the Patent Office in patent litigation, a reform introduced in lieu of technical judges in the 1936 patent reform. A second paper dealt with the definition of the untranslatable concept *Betrieb* (business, works, concern, enterprise, factory, plant, place of work) insofar as it related to the employer's right to employee inventions. The third report, written by Viennese patent attorney Robert Hans Walter, was a detailed description of an inventor support system the author had created at the Semperit Concern, a large Austrian manufacturer of asbestos and industrial rubber.[3]

Robert Hans Walter's *Semperit-System*

While the first two papers contributed to closing the gap between Nazi inventor ideology and established patent discourse, Walter's study established a link between the inventor movement and the preexisting discourse and praxis of industrial psychology, labor rationalization, and social policy associated with the work of engineer Karl Arnhold of the German Institute for Technical Labor Training (Deutsches Institut für technische Arbeitsschulung, DINTA), which had been absorbed into the Labor Front after Hitler came to power.[4] It was Walter's study more than anything else that pulled the bureau out of the doldrums and pointed the way to promising new shores. As early as February 1938 and largely on his own initiative, Walter had instituted at Semperit the prototype of all later inventor-trusteeship programs in large companies. His description of what he called the *Semperit-System* began with a discussion of the reasons employee-inventors were often reluctant to come forward with their ideas. Many of them had meek personalities or self-doubts, as a consequence of which they tended to keep their ideas to themselves. Employee reticence of this type, Walter pointed out, was not his discovery alone. It had been the subject of a 1940 psychological study published in the journal *Deutsche Technik* by a Dr.-Ing. Willy Kovats of the Bureau for Technical Sciences. According to Kovats, the typical organization of industrial bureaucracy militated against employees reporting inventive ideas to their supervisors, especially if they were somewhat timid. Supervisors hardly ever took the time or had the patience to make an inventor's ideas their own. As a consequence, supervisors were rarely in a position to present an inventor's case with the necessary conviction to their own superiors. The result was that many inventions were "'throttled …'' and failed to become reality." Frequently, supervisors also added their own suggestions or amendments, which were more likely to water down the invention than improve it, thereby hurting the inventor's enthusiasm and undermining his desire to make further inventions. The solution, according to Kovats, was to provide inventors and employees with an alternative avenue for reporting their ideas.[5]

Circumventing the regular corporate hierarchy and opening up a confidential line of communication to an impartial evaluator-advocate of inventions were precisely the innovations Walter had introduced at Semperit starting in 1938. His measures had created the "precondition for truly positive, inventive collaboration of the entire workforce" and made it possible to "bring to the surface" … "Incredibly rich treasures of uncompleted inventions." While ordinary work could be organized in a regimen of discipline and compulsion, creative work required above all "inner joy, eager enthusiasm, soaring imagination, and faith in having chosen the right direction." Great inventive contributions, therefore, could "only be expected when the problem of inventor trusteeship is attacked from the psychological and social angle." Without such an approach one might still get a rich harvest of improvements from systematic research and development. But one would "have to forego the great, creative pioneer inventions of fundamental significance."[6]

The distinction between systematic research and development and truly creative inventing played a crucial role in Walter's argument. One could not insist "strongly enough," he wrote, "on making a sharp distinction between creative inventions and the results of development." While the former depended on the conditions under which all the different kinds of creative work came into being, the latter were the product of methodical deliberations, "which reach maturity at a certain stage and a certain point in time without being tied to a special, creative achievement." All scientific-technological development was systematic, logical progression, dependent on gradually accumulated knowledge. It could advance in a direction that was correct in theory, and yet reach a dead end in practice. Turn-of-the-century experts, for instance, had been convinced human flight would never succeed because the problems of flight control were impossible to solve; what would they say now? All systematic technological development traveled along a fixed direction, Walter contended, and a change of direction "can be effected only by a large, external force, by the force of a creative, unprejudiced inventive idea, which took shape—and that is the point—independently of such development."[7]

Many people were convinced that "inventing today is nothing more than systematically developing the foundations of technical knowledge, … assiduous searching on the basis of what is already known, methodical working of a field, and industrious studying thereof." One could "only smile" at all those who believed that inventing might be reduced to "an infinity of material permutations."

To be sure, "innovations do come about this way, even the majority of them, innovations of great usefulness, innovations that bespeak the hard work of the person who made them." To look for and to find such innovations by way of systematic research did indeed require technical training and technical specialization. "But none of those industrious technical workers," Walter wrote, "can ever be counted among the truly creative inventors, among the giants of the spirit and the heart, who shape their poems in steel and matter with the same passion as do their comrades in the world of art. Truly, a technically creative spirit, a real inventor of stature, is not to be thought of as any less than an important poet or composer." And just as no one ever asked a great poet or a brilliant composer about his qualifications, so the technical creator should not have to answer questions about his education and certification. What mattered was someone's "creative power, unhampered by questions about preparatory training and prior experience."[8]

Walter emphasized creative genius and disparaged formal training not because he thought that Germany should embark on an unfocused quest by amateurs for new, radical inventions that might launch as yet unborn or unimagined technological systems. Rather, he wanted to mobilize all workers at all levels in all companies on behalf of technological progress in a much more immediate sense. Thus, he bent the Romantic impulses driving Nazi inventor ideology to the very practical purposes of productivity enhancements, design modifications, and other innovations that might help the war effort. Technological progress, he argued, was too important to leave to the trained experts in the laboratories and the design offices. "We should abandon all traditional social views, those old errors, according to which only the man of education, only he who has sat long enough on a school bench, is entitled to act as a leader in intellectual matters, a leader in the realm of technological creation." Instead, "we should take the man in blue working clothes just as seriously as the technical expert, when both come to us as a creative inventors." Each and every inventive suggestion should be evaluated as to the creativity, on the one hand, and the developmental effort, on the other hand, that had gone into it. If that was done, one saw very quickly that creative inventions frequently hailed from creative minds whose occupational responsibilities could hardly have predicted those results. The results of systematic development, in contrast, stemmed almost invariably from people whose job it was to produce them, which did not preclude, of course, that "occasionally we also encounter development engineers in the role of creative inventor."[9]

Effective inventor trusteeship, Walter pointed out next, required not merely making the distinction between creative and developmental ideas, nor simply ranking the technological value of an idea objectively, but also weighing it in light of the person who was the author and where that person stood in the company hierarchy. Something might be "a creative achievement for a blue-collar worker that, for a professional development engineer, is merely the result of his normal work." Calculation of the financial award in those two cases would therefore come out very differently. In addition to the regular inventor reward, an initial small bonus should be paid right away, to reinforce the "state of happiness that accompanied the act of creation, when the inventor is still emotionally involved with his invention." Another crucial aspect of making inventor trusteeship work was the speedy and accurate identification of an invention's true author (or authors). Without this, the company invention, though outlawed in the 1936 Patent Code, was likely to slip back in through the back door, fatally undermining employee morale. Even if the company invention could be kept out, however, employees would never develop any confidence in inventor trusteeship unless correct and immediate identification of the inventor(s) was a top priority. Workers and employees responded positively if they saw that the company had a just and impartial approach to the facts of authorship. Then they readily volunteered their ideas, allowing the company to harvest, from a hundred suggestions, the one gem that was the program's ultimate rationale. "For only the great creative idea, after all, propels technological development forward with a gigantic, accelerating leap."[10]

Some people, said Walter, might object that lavishing all this attention on the inventor was counterproductive and could only result in an unmanageable stream of useless ideas, including any number of the usual crackpot suggestions for perpetual-motion machines. But his experience at Semperit had shown that such fears were groundless. During the three years that his system had been in effect, "not a single one" of more than 1,200 inventive proposals was for a perpetual-motion machine, which proved that "the foundations of the system [were] sound" and that its principles were being "strictly enforced." The key to this success, Walter emphasized again, was the "psychological, social, and spiritual" dimension of his approach. "Inventor trusteeship," he said, was "at least fifty percent a psychological and social problem, and only the second fifty percent [was] a technological problem."[11]

Creative people such as inventors, Walter continued, often were like "big children," who had a hard time figuring out the objective merit of their ideas. It was therefore essential to give them access to a person outside the normal company hierarchy, someone who would act as their friend and advocate, sharing their enthusiasms and frustrations and gently "guiding them toward the more important and promising problems, all the while strengthening their creative enthusiasm—that is and remains one of the most important tasks of inventor trusteeship ..." Given this mission, it was "clear that a great deal of trusteeship must be devoted to the little man." The latter had "more inhibitions than a managing engineer, not merely because of his subordinate position. He also has good reason to be shy about presenting his ideas because he is very afraid of embarrassment and, ultimately, he worries a great deal about being taken off the job." The further removed the inventor trustee was from a man's daily chain of command, therefore, the fewer were the inhibitions to reveal his ideas. Creating such a system at Semperit was precisely what Walter had done. He had managed to orchestrate all the necessary support from the company's central patent division as well as from its satellite patent offices in the individual works, which had gradually been staffed with the right kind of people. In fact, claimed Walter, he had made Semperit's inventor trusteeship so proactive and worker-friendly that its patent experts even helped employees with the patenting of their free inventions. And the system secured inventor rewards that were as high as 5 to 10 percent of the amount the company saved in one year as a consequence of the pertinent service invention. It also allowed for the testing of ideas under conditions of strict confidentiality, "completely removing the development of reported innovations, if need be, from the jurisdiction of the employee's supervisor." While success was rewarded, therefore, failure was kept quiet, avoiding any negative consequences an inventor might suffer from the wrath of his supervisor.[12]

Walter finally came to the bottom line. What had his elaborate system bought Semperit? Three years were only a beginning, he said, "for the trust of every individual workman has to be earned." Also, the program had gotten off to a slow start, because in early 1938 he had sensed "noticeable resistance from the company's then still Jewish directors."[13] Progress remained to be made, therefore. Even so, in the fall of 1941 Walter was "very happy to be able to report that the concern's top management has completely accepted the system's basic idea as its own." Others, too, had become interested. The city of Vienna had studied his system and was now in the process of introducing a similar system for "workforce inventors."

Likewise, the German Labor Front in the district of Neunkirchen had become acquainted with his system and begun to adopt it for other firms in the region. The reason for all this interest was the remarkable success he had achieved: 1,245 reported ideas, of which no fewer than 533, or 42 per cent, had resulted in usable inventions. Among the latter, nineteen inventions, or almost 2 percent, had resulted in "fundamental changes in individual branches of production." Semperit had fully or partially introduced 250 inventions into the manufacturing process. The development of another twenty-eight inventions had been completed but not yet introduced, and the final 236 inventions were either still under development or could be used only as peacetime innovations. Of the total, some sixty-one inventions were directly relevant for the armaments industry. Fully 44 percent of all the inventions came from the blue-collar workforce (with ordinary manual workers accounting for three-quarters of that fraction, and crew leaders and foremen coming up with the rest), which Walter described as a "very high incidence." He gave no further statistics to put the numbers in perspective, but did point out that Semperit's economic activities belonged to a patent class that ordinarily did not see many patent filings and that the above results were "very favorable" indeed.[14]

Walter concluded by casting his reforms once more as the one policy that would keep Germany on the cutting edge of technological progress. "The time of great inventions has not passed," he wrote, "even though technological development has reached the highest stage and technological progress is working in top gear. True, very great ideas are rare; but precisely for that reason the preconditions for new paths must be kept open—paths that revolutionize and overturn established traditions and secure the technological advantage of the German *Volk* in the future. Those paths can only be opened when the right atmosphere for creative inventing is established and the desire to do so is kept alive, for which the *Semperit-System* has proved to be the right foundation."[15]

Expediency and Ideology

Walter's paper was significant, not because it was the first to sketch a vision of inventor trusteeship in industry, but because it was the first to show how the ideas of Nazi inventor policy could be made to work in practice and produce results. There existed, of course, a powerful tradition of labor rationalization and productivism, dating back to Weimar and easily adapted to the needs of the Nazi regime after 1933. Even before the Third Reich, this German science of work had reflected an authoritarian bias and subsequently formed an essential component of the Labor Front's vision for the new Germany. Initially focused on problems of human fatigue and on shop-floor efficiency in the Taylorist tradition, the science of work had come to include a concern with social policy as the road to higher productivity as well as with questions of aptitude, motivation, and psychology.[16] From here it might have seemed only a small step to figuring out how psychological and motivational techniques could be used also to increase the yield from the creative process itself. But progress in this direction had been very limited at best. By 1940, Nazified industrial-psychology experts such as Karl Arnhold, under the pressure of growing labor shortages and the need for higher productivity, had begun to think about how to stimulate the inventive creativity of employees. Building on his older program of motivating labor by sports-like skills competition and indoctrination, Arnhold proposed to make inventing a sport, too.[17] However, Arnhold's idea remained untested and in any case was far less developed than Walter's sophisticated *Semperit-System*, which had proved its worth since 1938. It was Walter, therefore, who bridged the gap between the science of work and inventor policy. Walter integrated theory and practice in a system of human-resources management that fulfilled Nazi ideological objectives, increased productivity, spurred innovation, and promoted rationalization all at the same time. This was precisely what the regime had begun practicing in many other areas of social policy and in 1941 and 1942 was looking for in the realm of inventing. It is therefore not surprising that Walter's *Semperit-System* became the foundation of not only the bureau's continued existence, but also its quick revival and massive expansion in the years 1942–44.

When Armaments Minister Fritz Todt died in an airplane crash in January 1942, Albert Speer took over all his offices—including command of Priemer's and Riemschneider's Bureau for Technical Sciences, the responsibilities of which he rapidly expanded. The principal focus of this expansion was inventor trusteeship in the factories. Assistance for independents did not disappear but was increasingly overshadowed by a massive effort to boost technological creativity and labor productivity through intervention between employer and employee on the model of Walter's *Semperit-System*. To this end, special, party-sanctioned inventor trustees were introduced in the country's factories, employees were given statutory rewards for their service inventions, and employers were forced to file patent applications for all employee inventions to which they laid claim. This was the thrust of the 1942/43 Göring-Speer inventor ordinances, which introduced the inventor-trustee system as law.[18]

At the center of it all was Riemschneider's Department for Inventor Protection in the Bureau for Technical Sciences on Munich's Ehrhardtstrasse. From the spring of 1942 to the fall of 1944, when inventor trusteeship was at its height as an integral part of the German war economy, hundreds of people were directly or indirectly involved in its operation. The bureau produced handbooks for inventors and inventor trustees. It gave advice on the psychology of inventor counseling and motivating employees. It published legal commentaries on the 1936 Patent Code and the wartime inventor ordinances. It developed formulas for calculating the size of employee rewards. It intervened with employers on behalf of inventors who thought they had been treated unfairly. It negotiated tax advantages for the recipients of employee-inventor rewards. It managed the complicated apparatus of inventor trustees in the country's large industrial firms and enforced the inventor ordinances. It issued its own newsletter, *News of Inventor Trusteeship* (*Nachrichten über die Erfinderbetreuung*). It organized conferences on inventing and was in constant communication with industry, the Patent Office, and anyone and everyone who had anything to do with inventing.

There is no question that pragmatic considerations played a large role in explaining the bureau's reversal of fortunes, which began in late 1941. Inventor trusteeship in the factories dovetailed perfectly with Todt's and Speer's efforts to increase the output and efficiency of German industry for the war effort. The "production miracle" over which Speer presided between 1942 and 1944 resulted from a two-pronged attack on inefficiencies in the armaments industry. First, Todt and Speer made administrative changes at the macro level, such as creating a system of committees and rings, which was largely run by industry itself and eliminated interference by the military in the production process. Other organizational changes of this type included the establishment of unified control by the Ministry of Armaments over the allocation of raw materials, and a decision to replace the practice of paying industry on a cost-plus basis with a fixed-price system. Speer also achieved a reduction from the many different types and models of weapons systems to a smaller number of standardized designs. He shifted human and capital resources from the building of new production capacity to making armaments, and he benefited from prior investment in new facilities coming on line in the second half of the war.[19]

Second, there was a great deal of rationalization at the micro level. Firms made numerous internal reorganizations such as transferring and concentrating production; embracing shop-floor efficiencies and material-saving practices; adopting special-purpose machine tools, flow production, and longer production runs; stripping and simplifying designs; and converting production from less to more essential systems. In addition, there were crucial changes with respect to the use of labor. These included speeding up production lines; lengthening hours of work; substituting where possible less skilled workers, women, and foreign labor for skilled, male, German workers; reducing piece-rates; and segmenting pay-scales and differentiating the labor force with selective pay increases and bonuses in order to increase incentive and enhance performance.[20]

There is no doubt that all these measures in the aggregate accounted for the "production miracle." Their relative importance, however, is difficult to assess with any degree of precision and undoubtedly varied from industry to industry. On balance, the macro-level changes seem to have produced greater results than reforms on the plant level, which were often characterized by crisis management and organizational improvisation. An exception should be made for the labor-related measures, which appear to have played a disproportionately large role in explaining the "production miracle."[21] Designed with the active participation of Labor Front and Nazi party agencies, the labor measures functioned as a double-edged sword. While incentives to work harder were heavily laced with threats, fear, and terror at the lowest levels of production, for the more critical tasks those same labor reforms presented a more appealing face. The dimension of external compulsion was minimized, while that of enticements, special rewards, and the concept of the "just wage" took center stage. Those positive measures were calculated to maximize output by dissolving the solidarity of labor, individualizing strategies of socioeconomic mobility, inspiring trust, creating loyalty to the company, and bringing about the internalization of discipline and identification with the goals of the regime.[22]

It was at this juncture that Walter's system of inventor trusteeship came in and merged with the larger strategy of increasing labor productivity. By seeking to reduce distrust, establishing specific legal rights, and providing special financial compensation that included tax benefits for patented improvements, the inventor measures aimed at overcoming employee alienation, boosting morale, and improving attitudes. The result would be increased inventive energy for the war effort. Speer readily acknowledged the pragmatic motives that led him to embrace inventor trusteeship. As he put it in the foreword to a 1943 legal commentary by Riemschneider and Barth on the regulations concerning inventor trustees and inventor rewards, "Inventions by members of the workforce must be forcefully promoted, evaluated and protected, because they serve to increase the achievements of the German economy and especially armaments." The new rules, Speer continued, spoke to "all those creative German human beings who, as comrades-in-arms in a sworn community of achievement, want to help encourage technological progress and with that the victory of the German people and the construction of the greater German empire."[23]

But expediency and the Nazi science of work were not the whole story. A specific, inventor-ideological dimension survived independently of broader productivist strategies and remained central in all the writings, pronouncements, and activities concerning the inventor issue. Faith in the genius of the inventor and suspicion of the capitalist employer were the twin cornerstones of that belief. In the passage quoted above, for instance, Speer did not talk about inventor policy as just another means to strengthen the armaments industry. He praised it as the fulfillment of a moral command by Hitler, quoting at length from *Mein Kampf* to the effect that "all inventions are the result of an individual's work," and that the just society is the one that helps the inventor do his creative work.[24] Was this merely lip service? Or was the Nazi regime's high priest of industrial rationality and efficiency also a true believer? References to setting free the inventor and the engineer's technical genius also figured importantly in Speer's address to *Gau* leaders and party bosses when he assumed his new duties as Minister of Armaments in February 1942. On that occasion, he talked about expanding the system of production rings started by his predecessor as the best way to curb bureaucracy and give "the engineer and his energy, for the first time, an opportunity to be freely creative." Industry, said Speer, had at its disposal a vast pool of untapped talent for armaments production, "a gigantic mass of inventive genius and talent that so far has insufficiently been exploited for military purposes."[25] Given the continued existence of private industrial ownership and bureaucracy, it was not a contradiction to set free the engineer with totalitarian measures.

The autonomous, ideological foundations of inventor trusteeship in industry are evident as well in the comments by other influential Nazi figures. Werner Mansfeld, the influential Labor Ministry official who drafted most of the 1942/43 Göring-Speer ordinances, for instance, believed in it because he, too, was convinced that helping the inventor—making it possible for genius to do its work—was a moral dictate, with the added benefit of working economic miracles. "It should never be forgotten," Mansfeld wrote in the summer of 1942, "that—as the Führer himself has pronounced—the most valuable thing about every invention is primarily the inventor as a personality." To be sure, this fact was temporarily eclipsed by the exigencies of war and pressures to increase performance in the armaments industry, which put the invention ahead of its creator. "But even today, one should not therefore forget the inventor himself." Protecting and encouraging the inventor was a worthy goal in its own right, the realization of which "will simultaneously bring about the greatest possible increase in the inventor's efforts, in the same way that the best social policy, precisely in today's circumstances, has time and again proved to be the best economic policy."[26]

Engineer Josef Dapper, principal author of the compensation formula for employee inventions, wrote in 1942 that "the community of the *Volk* has a great interest, not merely in the appropriate development of inventions, but above all in the making of inventions. From this it follows that it is a nation's vital interest to encourage the inventor in person, to recognize and protect him, because doing so is above all a stimulus to creative achievement."[27] Dapper merely paraphrased what had long been the core belief of fellow engineers such as Riemschneider and Riemschneider's colleague and co-author, Judge Heinrich Barth. As early as 1935, long before anyone even imagined the emergencies of war and armaments between 1941 and 1944, Barth had pointed out that "the Führer places the inventor next to the political leader as the prototype of the creative human being." Germany depended for its future survival on the inventor, so it was crucial to support him in all possible ways. "The inventor is an especially valuable people's comrade, who deserves special consideration and special support," in particular against all the legal tricks that employers and capitalists used to deprive him of his rights—which demoralized the inventor and killed the nation's principal hope for technological progress.[28]

Rationalization or Innovation?

Analyses of Nazi Germany's astonishing increase in armaments production under Speer typically emphasize the achievement's quantitative dimension. Rationalization, standardization, and various other measures promoting efficiency made possible vastly larger volumes of output, which were the essence of the "production miracle."[29] In the light of this finding, what was the relationship between quantitative improvement and the qualitative change—invention and innovation, "qualitative superiority"—that was the first concern of the regime's inventor crusaders in the Bureau of Technical Sciences?[30] These men recognized that rationalization and increased output themselves depended on a great deal of qualitative, creative technological input—in simplifying design, production, and assembly, for instance, but also in conceiving of labor-saving and material-saving operations. This was the invention of the technology of efficient production, which occurred all along the line, from the design laboratories to the factory floor, and potentially involved contributions by each and every employee. At the same time, the architects of inventor policy always remained very much alert to the danger of too much rationalization and efficiency. This was especially true if they thought the emphasis on quantity came at the expense of innovation—innovation understood as new and superior technology, whose introduction corporate management might block for reasons ranging from capitalist greed to bureaucratic inertia to fear of not meeting production quotas.

Ambivalence about production efficiency and distrust of capitalism, which went back to critiques of rationalization in Weimar and figured in Walter's discussion of his *Semperit-System* as well, manifested themselves with particular clarity in the wartime writings of Riemschneider. In a 1942 article on "Inventor Trusteeship in the Factory," Riemschneider emphasized the technological benefits of introducing the inventor trustee, who would make sure that the employer did what he was supposed to do. Locked in a momentous struggle for "survival and living space," Riemschneider wrote, Germany needed new technology now more than ever before. Increasing output by rationalization could not succeed without "the creative participation and the intellectual powers of every German, regardless of where he might have been posted." This insight had led to the "mobilization of all intellectual powers, because the true urge to create, combined with the intuitive talents of the inventor ..., are capable of realizing what seems impossible at first." The pressures on the armaments economy to develop new technology required "not just the creative assistance of engineers, designers, and chemists, but also the assistance of all inventive forces in the workforce (*Gefolgschaft*)."[31]

Creative participation by the workforce was "secure only if the individual worker (*Gefolgsmann*) experienced care that was designed especially for him," Riemschneider continued. This was the reason for the introduction of inventor trustees in the factories. Inventor trusteeship had to function in such a way that even before the "workforce member" experienced the desire to help invent, he would know in advance that whatever he discovered would be appreciated as a special achievement. His effort would always be recognized as "an exceptional achievement that receives its rightful symbolic acknowledgment and, when this exceptional achievement is applied in practice, also the appropriate material recognition." The usual promotional campaigns to increase invention and generate product improvements, Riemschneider pointed out, remained ineffective if the employer did not back up his words with deeds— i.e., with guarantees of authorship recognition and financial rewards for employees who made inventions. The employer was therefore obligated to "guarantee the inventor's honor and the award of appropriate financial compensation." To increase the likelihood of managerial compliance, the inventor trustee was given special Nazi party backing, holding the rank of "Political Leader" in the office of the Labor Front's official representative on the factory floor (*Betriebsobmann*). He received his appointment from the local Labor Front leader, who picked him from a slate of candidates proposed by the factory manager, upon approval by the party's local *Gau* Department of Technology.[32]

If Riemschneider here was concerned mostly about possible sabotage by the employer and downplayed the differences between rationalization and innovation, at other times he singled out those differences for special attention. In the 1944 edition of their legal commentary on the inventor ordinances, Riemschneider and his co-author Barth warned against too much emphasis on standardization and mass production because they threatened timely innovation. "One of technology's most difficult problems," they wrote, was the "balance between form-preserving, ordering forces, which result in standardization and uniformity, and the creative forces of progress …" Making a distinction akin to Hughes's contrast between "radical inventing" and "conservative inventing," the authors pointed out that "form-preserving standardization is a precondition for productivity increases in organized production. Progress, which presupposes change of form, is necessary to guard against obsolescence and against being overtaken by others." This was not just an academic point. It might determine the outcome of the war. "It is important," Riemschneider and Barth continued, "that the ordering forces do not smother the creative forces." The authors then wondered what might happen if, "owing to misguided standardization and uniformity," Germany's creative forces did not receive adequate breathing space. "… [O]ur future Siemens, Diesel, and Junkers, as yet undiscovered, will move to other firms or, out of utter desperation, to other countries. And if that, as in wartime, is impossible, then they waste away. Decisive technology remains unborn."[33]

It is difficult not to read these remarks as a critique of production rationalization as such, an echo of the old distrust of Taylorism and part of the long-standing primacy of "design culture" with which German engineering is often associated, and as an expression of faith in the miracles German inventive genius might still work even in 1944.[34] All this played a role, to be sure. But the main point was something else. The regime's inventor-policy experts emphasized the need for creativity and new ideas across the entire spectrum of technological activity, from design and development to production and assembly, from project management to blue-collar labor. They were quite flexible as to how and in what areas those creative ideas should be deployed. Their main goal was to overcome managerial inertia and bureaucratic attitudes, which resisted new ideas because they might upset established routines. Above all, the party's men were motivated by an undiminished suspicion that capitalism would always put profits ahead of technological rationality and innovation. Experience had shown, Riemschneider and Barth wrote, that "the forces employed to generate progress frequently lost out to the requirement of private enterprise that existing models, production methods, and markets, now threatened by progress, continue to exhaust their potential for profit."[35] The energy behind inventor policy, therefore, cut two ways. It might go in favor of increased manufacturing efficiency if this involved adoption of labor-saving, material-saving, or space-saving ideas in the production of a given, acceptable design. But it could also move against the dictates of production rationality if these took precedence over building a better technological product or new system. The choice, as Riemschneider and Barth indicated, was always a question of "balance" and judgment.

"The Number of Patent Applications … Grows Extremely High"

It is not entirely clear what were the practical results of all this theorizing and rule making. Inventor trusteeship, because it was anchored in notions of a special German talent for technology, had racist dimensions that were increasingly at odds with the reality of industrial labor during the war. Not only were Poles, Jews, Sinti, Roma, and countless numbers of *"Ostarbeiter"* explicitly excluded from the program's benefits, but Germany's wartime factories also depended on ever-larger numbers of foreign workers, most of them forcibly recruited by the plenipotentiary for labor, Fritz Sauckel, as well as on slave labor from the concentration camps. In theory, "biologically correct" foreign workers did qualify for support under the inventor program, but there is very little evidence to indicate how this functioned in practice—or whether it did so at all.[36]

While it is possible to measure the increased output and productivity generated by the sum total of all the efficiency measures, incentives, and reforms that had been instituted since the fall of 1941, it is difficult to isolate the contribution of specific programs such as the inventor ordinances.[37] It is equally difficult to determine how effective inventor policy was in keeping Germany competitive in the race for technological innovations, such as those in the submarine and air wars. The files of the Bureau for Technical Sciences that might have helped answer those questions no longer exist. Still, there are some highly suggestive clues.

The encouragement of innovation was supposed to function by financially rewarding employees for their inventions. The criterion of what constituted an invention was in practice an idea's patentability, which was decided by the Patent Office. This meant that the reward mechanism operated via the detour of a patent application and/or award. An employee who thought he had made an invention reported his idea to the employer. If the latter claimed the idea for the company, there were two consequences: the employee was entitled to special compensation and the employer was obligated to file a patent application.

What effect, if any, did this mechanism have on the rate of patenting? If the policy worked, one would expect to see an increase in patenting. It has been argued that the statistics prove the regime's inventor policy was a complete failure.[38] It is true that the total number of patent filings during the Third Reich was down from the peak reached during the last half of the Weimar Republic. The relevant figures, including for purposes of comparison those of the United States, are shown in Table 10.1. The same data in graphical form are shown in Figure 10.1.

Table 10.1   Patent Applications in Germany and the United States, 1919–1960

Year	Germany Total Applications	Germany Foreign Applications	Germany Domestic Applications	Germany % Foreign Applications	USA Total Applications
1919	43,279	4,736	38,543	11	76,710
1920	53,527	11,672	41,855	22	81,915
1921	56,721	10,720	46,001	19	87,476
1922	51,762	10,885	40,877	21	83,962
1923	45,209	9,127	36,082	20	76,783
1924	56,831	9,353	47,478	16	76,987
1925	65,910	10,508	54,402	16	18,208
1926	64,384	11,159	53,225	17	81,365
1927	68,457	12,827	55,630	19	87,219
1928	70,875	14,200	56,695	20	87,603
1929	72,748	15,226	57,522	21	89,752
1930	78,400	15,749	62,651	20	89,554
1931	72,686	14,225	58,461	20	79,740
1932	63,414	11,899	51,515	19	67,006
1933	55,992	11,011	44,981	20	56,558
1934	52,856	9,742	43,114	18	56,643
1935	53,592	8,912	44,680	17	58,117
1936	56,163	8,800	47,363	16	62,599
1937	57,139	8,629	48,510	15	65,324
1938	56,217	7,976	48,241	14	66,874
1939	47,555	n.a.	*47,555	n.a.	64,093
1940	43,479	n.a.	*43,479	n.a.	60,863
1941	49,885	n.a.	*49,885	n.a.	52,339
1942	53,386	n.a.	*53,386	n.a.	45,549
1943	49,060	n.a.	*49,060	n.a.	45,493
1944	n.a.	n.a.	n.a.	n.a.	54,190
1945	n.a.	n.a.	n.a.	n.a.	67,846
1946	n.a.	n.a.	n.a.	n.a.	81,056
1947	n.a.	n.a.	n.a.	n.a.	75,443
1948	n.a.	n.a.	n.a.	n.a.	68,740
1949	61,002	n.a.	n.a.	n.a.	67,592
1950	53,375	9,919	43,456	18	67,264
1951	55,457	9,749	45,710	18	60,438
1952	58,561	10,702	47,859	18	64,554
1953	60,202	12,643	47,559	21	72,284
1954	59,317	14,458	44,859	24	77,185
1955	54,865	15,862	39,003	29	77,188
1956	53,470	17,156	36,314	32	74,906
1957	53,002	18,216	34,786	34	74,197
1958	54,502	19,060	35,442	35	77,495
1959	56,611	21,375	35,236	38	78,594
1960	57,123	22,546	34,577	39	79,590

Source: J. Slama, *One Century of Technical Progress Based on an Analysis of German Patent Statistics* (Brussels, 1987), 122–23. Post-1945 figures are for FRG only. Figures marked with asterisks are estimates, based on the assumption that foreign patent applications disappeared with the outbreak of war. Data for 1944 are not available. The gap from 1945 to 1948 is due to Patent Office closure.

Figure 10.1 Patent Applications in Germany and the United
States, 1919–1960

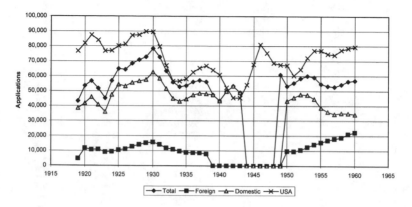

To measure inventiveness, patent applications are obviously of
greater interest than awards, the latter being primarily a function of
the efficiency of the Patent Office. Therefore, only applications are
shown. As indicated in Figure 10.1, the trends of total patent applications in Germany and the United States were remarkably similar
from 1919 to 1940 and, albeit to a lesser extent, again from 1950 to
1960. In both countries, the precipitous drop after 1930 and failure
to recover the heights scaled during the half decade before the Great
Depression stand out as the most striking development. From 1950
onward, both countries experienced comparable recoveries, followed by renewed slowdowns and leveling off during the mid- and
late 1950s. Patenting trends diverged sharply during the war—artificially prolonged by the German Patent Office's closure between
1945 and 1948—but their similarity both before and after suggests
that the overall health of the economy was the critical variable.[39]

The most interesting developments took place between 1940 and
1949. While the number of patent applications in the United States
kept falling until 1943, German figures sharply rebounded as of
1941, rising well above the American totals during the next two
years. There are no German data for 1944 and 1945, and until 1949
the Allies prohibited patenting in Germany. American patent applications began their own steep recovery in 1944, when they reached
Germany's 1943 level, and peaked in 1946 far beyond anything the
Nazi regime had ever attained. What is the explanation for the
rebounds beginning in 1941 in Germany and in 1944 in the United
States? Unlike patenting energy before and after the war, which

moved in tandem with the fate of the (mostly) civilian economy, the wartime spurts were in all likelihood a consequence of government policies, especially the creation of a vast military-industrial complex in the United States and inventor policy in Germany.

Whether Nazi inventor policy would have been able to sustain the upward trend in patent applications had Germany had not been overwhelmed by Allied military force after 1943 is anybody's guess. It should be noted, however, that while the United States had by 1965 surpassed its prewar peak of 1930, and kept on rising thereafter, patent applications in Germany to this day have not returned to their heights of the late Weimar Republic. This suggests that, at some point after 1930, a secular change in German patenting behavior took place. This phenomenon, as well as the true level of German patenting energy, becomes evident if, instead of looking at the aggregate numbers, one breaks down the German figures into applications by German nationals and foreigners. Before World War II, foreign filings made up an average of 18 percent of the total, falling from 20 percent in 1933 to 14 percent in 1938. At 28 percent, the average for foreign applications during the years 1949–1960 was far higher than before the war. Climbing to a remarkable 39 percent of the total in 1960, the high proportion of foreign patent applications indicates that the secular decline in domestic patent filings began as early as 1953—and might well have started earlier.

As for the relationship between applications by foreigners and by Germans during the Third Reich, Nazi inventor policy obviously did not aim to increase the former. On the contrary, autarky and substitution for foreign inputs in economic and technological matters became one of the regime's central themes. The encouragement of domestic inventing was a crucial part of this policy. Unfortunately, information on the number of applications by Germans and foreigners is not available for the years 1939–1945. The trend of foreign filings, however, was steadily downward from 1933 on, both in absolute numbers and on a percentage basis. While applications by foreigners undoubtedly continued during the first half of 1939, it is unlikely that the practice lasted to any significant degree beyond the outbreak of war in September. After 1939 the number of domestic applications increasingly accounted for the total of all applications.[40] If this is correct, the upswing in *German* inventive activity from 1939 through 1942 and 1943 shown above was considerably more powerful even than that suggested by the aggregate numbers. Moreover, the annual number of domestic patent applications from 1940 through 1943 also remained higher than at any time during the twelve years from 1949 through 1960. It is reasonable to conclude that patent law reform, inventor policy, and inventor propaganda did have much to do with the upswing.

The impression of increased inventiveness in connection with the war effort is reinforced if one looks beyond aggregate statistics to individual patent classifications. The German patent system has some eighty-nine major patent categories. Unfortunately, information about trends in the individual classes is available only through 1938, before the regime's most effective inventor incentives had been implemented. The trend during these first five years of the Third Reich is nonetheless revealing. Since applications plunged in all categories after 1930, the relevant question is not so much what happened to the absolute numbers as what were the patterns of recovery between 1930 and 1939. The results are not surprising. Applications in categories such as furniture, footwear, and sports, games, and amusements—in general, consumer goods for the civilian sector and a peacetime economy—never recovered and kept on falling. In contrast, applications in fields such as metallurgy, pumps, yarns, casting, engines, engineering elements, and presses made sharp recoveries, even if they did not quite return to the extraordinary levels of 1930. In other classifications there was a veritable flood of inventions, not merely recovering their prior heights but substantially exceeding them. Table 10.2 shows this last phenomenon for a few selected years and patent classifications.

Table 10.2 Patent Applications in Selected Categories, 1930–1938

Patent classification	1930	1934	(% of 1930)	1938	(% of 1930)
18 (metallurgy of iron)	1,142	886	(78)	1,178	(103)
39 (plastics and synthetics)	1,626	1,329	(82)	1,786	(110)
48 (metalworking)	591	460	(78)	815	(138)
61 (life saving and fire fighting)	466	480	(103)	730	(157)
62 (aviation)	1,518	1,409	(93)	2,309	(152)
72 (firearms)	798	1,548	(194)	2,166	(271)

Source: J. Slama, *One Century of Technical Progress Based on an Analysis of German Patent Statistics* (Brussels, 1987).

One can only guess at what happened from 1939 to 1944, but it seems safe to assume that before systematic air raids began to take their toll, the numbers in these and other patent classifications kept on rising steeply. All this happened, moreover, at a time when millions of German men—including thousands of talented engineers and inventors—had been called up to serve in the armed forces.

Data concerning the patenting behavior of individual firms during the war are not readily available. Information about the long-term patenting effort at one large company, however, is at hand and confirms the trend outlined above. The data in Table 10.3 and Figure 10.2 show patent applications by IG Farben's BASF unit.[41] They indicate that while four of the ten most productive years came during the Weimar Republic, which could also claim the number one and two spot, the remaining six years in the top ten all came between 1937 and 1943. In fact, the third highest number of applications during the entire period 1919–1955 came during the middle of the war, in 1943.

Table 10.3 BASF Patent Applications, 1919–1955

Year	Applications*	Rank	Year	Applications*	Rank
1919	41	–	1936	509	–
1920	68	–	1937	675	5
1921	95	–	1938	683	4
1922	134	–	1939	612	7
1923	187	–	1940	538	9
1924	232	–	1941	603	8
1925	349	–	1942	689	3
1926	426	–	1943	629	6
1927	516	10	1944	476	–
1928	720	2	1945	36	–
1929	679	5	1949	256	–
1930	759	1	1950	383	–
1931	411	–	1951	493	–
1932	353	–	1952	481	–
1933	342	–	1953	404	–
1934	436	–	1954	403	–
1935	448	–	1955	446	–

*German applications in Germany, recalculated

Source: Karl Holdermann, "Geschichte der Patentabteilung. Zusammenstellung von 1877 bis 1968," BASF B4/E 01/1.

Figure 10.2   BASF Patent Applications, 1919–1955

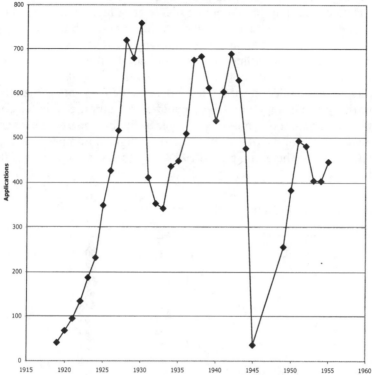

In addition to patent statistics, there is qualitative evidence to suggest that Nazi inventor policy in industry did a great deal to stimulate inventiveness. An internal Patent Office memorandum from 1948 commented that the inventor-trusteeship system had shown that "the number of patent applications ... grows extremely high." To avoid complaints of noncompliance, the employer was "necessitated to file patent applications for questionable matters, even when he is not convinced of their merit." As a result, "tens of thousands of applications were on hold at the companies and waiting to be filed with the Patent Office. Only the fast and furious events of the war at that time prevented the arrival of this flood of applications."[42]

In 1950 the West German federation of employer associations, hoping to undo the wartime inventor ordinances, less charitably observed: "every practitioner knows that under certain circumstances a certain overproduction of new, non-inventive ideas sets in—as confirmed again and again by the experience in the years since the ordinance of 1942 went into effect."[43] As early as 1943, Riemschneider and his collaborators in the Reich Working Group for Inventing became concerned about the pressures the Göring-Speer ordinances were putting on the Patent Office. There was no good solution to the problem. Imposing a filing stop, for instance, would have had "extraordinarily fateful consequences," according to IG Farben's Redies. "The National Socialist regime," Redies pointed out, "ha[d] done everything in its power, especially so during the war, to stress the importance of the inventor and strengthen his position." The only practical means of doing this was to use the incentive of a patent, which was, indeed, a "terrific spur to invention." For that reason, "the idea of instituting an temporary patenting moratorium, even as an emergency measure, is completely out of the question."[44]

Finally, there is the postwar assessment of Heinrich Kirchhoff, the former inventor trustee at Argus Motor Company in Berlin. Kirchhoff, who held doctorates in political science and engineering, had been a project engineer and patent expert at AEG until 1942. He served as inventor trustee and director of the patent department at Argus, a builder of aircraft engines, from 1943 to 1945. Kirchhoff never joined the Nazi party, even though the relevant ordinance required it. After the war, he briefly headed the department for intellectual property questions in the central administration for industry in the Soviet sector. In the summer of 1946 Kirchhoff accepted a teaching assignment in patent and inventor law at the Free University, and soon afterward established a practice as patent attorney in West Berlin.[45]

In 1947 Kirchhoff published *The German Patent System: Retrospective and Prospect*, in which he surveyed the German patent system from a comparative perspective.[46] While he found much to object to in capitalist patent systems, especially the countless inefficiencies and delays associated with the search for the first and true inventor in the United States, Kirchhoff generally liked the simplicity of invention management under socialism. He had profound respect for the Soviet system of inventor rewards and its procedure for examining the usability of new conceptions, which under conditions of state monopoly took the place of a western-style patent system. Kirchhoff did not think the Soviet system was a suitable model for capitalist economies but nevertheless thought it made sense to know something about it. Calling the Russian approach creative and original, he argued that the USSR's inventor and patent system had been instrumental in transforming the Soviet Union from a mostly agricultural society in 1917 to the world's second greatest industrial power in 1945.[47]

As far as Nazi Germany's system of inventor trusteeship was concerned, Kirchhoff castigated it for its propagandistic and totalitarian aspects but also thought that it had so many good features that it should be retained and adapted to the conditions of democracy. To claim that the ordinances of 1942 and 1943 had "completely fulfilled" the long-standing demands of salaried inventors, Kirchhoff wrote, would be an exaggeration. But to argue—as did its enemies after the war—that they "had worked only to the disadvantage of the salaried inventors [was] entirely groundless" as well. Only people with years of experience in the field (such as the author himself) were qualified to separate the good from the bad. On the plus side, the former regime had mandated that every employee invention claimed by the employer automatically qualified for special compensation. "In the year 1942 this fact alone represented a significant step forward with respect to the law on employed inventors." On the minus side, the regime's "'inventor trusteeship' ... and all the propaganda surrounding it had been introduced only to spur inventors to greater participation in the war economy." Run by the Nazi party, the regime's inventor support system had also served ulterior, political motives. It had sought to "strengthen the influence of the NSDAP also in the sphere of technological ideas and so to contribute to the inventors' 'uniform political orientation' along the lines of the Führer principle." In spite of "all these negative measures of National Socialist 'inventor trusteeship,' however," one should "not overlook the fact that the basic idea of the ordinance of 7 July 1942, freed of its party-political gloss, is entirely healthy and suited to encourage inventor activity. It has merely been abused."[48]

Kirchhoff's assessment of inventor trusteeship is a rare glimpse at the world of wartime inventor trusteeship by an insider who was not also a Nazi ideologue. To be sure, Kirchhoff exhibited socialist leanings, and to all appearances he shared with National Socialism the conviction that the bureaucracy of corporate capitalism, if left to its own devices, tended toward technological inertia. But this insight, just because it was a favorite of Marxist Socialists and National Socialists, was not necessarily wrong. In sum, it appears that the regime's inventor measures in industry were quite effective in encouraging invention and—if the emulation of Walter's *Semperit-System* in other companies yielded even remotely comparable results—innovation as well.

Notes

1. Karl Kasper (Gauamtsleiter für Technik der NSDAP Berlin), "Erfinderbetreuung," *Rundschau Deutscher Technik* 20 (11 Jan. 1940), BAK R131/27; Priemer's foreword to *Berichte der Arbeitsgemeinschaft "Fragen des gewerbl. Rechsschutzes" und "Gefolgschaftserfinderrecht" des Amtes für technische Wissenschaften*, no. 1 (Munich, 1941), BAK, R22/630, Bl. 27ff.
2. Members included (1) engineer Karl Ahlemann, member of the SS since 1931 and head of the raw materials section of the Julius Pintsch Company, a mechanical engineering firm in Berlin; (2) Dr. Karl Brodersen, inventor of a successful mercerizing process and a chemist employed by IG Farben's Wolfen plant who joined the party in 1937; (3) patent attorney Hermann Idel, SS member since 1933 and mining engineer, head of the party *Gauamt für Technik* Essen; (4) Robert Kahlert, mechanical engineer, active in the party, author of several books on inventor counseling, in 1940 became chief of DAF Berlin office for the iron and metals industry; (5) Dr. Joachim Schimmelpfennig, head of DAF Department of Social Self-Responsibility Berlin; (6) engineer Christian Semenitz, head of DAF Berlin AftW and party member since 1925; (7) attorney Talbot, Berlin engineer and patent expert in private practice; (8) Dortmund patent attorney Lorenz Weber, party member since 1933; (9) Munich patent attorney Franz Weickmann, in the party since 1933; (10) engineer Ernst Weisse, party member since 1937 and head of patent department of Askania Works, a Berlin maker of precision instruments and guidance systems; (11) attorney Hans Zemlin, a VDI patent expert who joined the party in 1937; (12) Director Franz Redies, IG Farben chief patent expert Leverkusen, who did not join the party until 1943 owing to onetime membership in the Freemasons; (13) patent economist Willy Salge, Riemschneider's old Berlin associate, who died in 1942 without a party file; (14) engineer and patent attorney Paul Wiegand, Ludwig Fischer's successor as chief of Siemens's patent division, party member since 1931; (15) Robert Hans Walter, engineer and patent attorney from Vienna who did not apply to join the party until Feb. 1942 (membership was backdated to January 1941), *Berichte der Arbeitsgemeinschaft "Fragen,"* no. 1, BAK, R22/630, Bl. 27; NSDAP affiliations from BDC files.
3. Robert Hans Walter, "Die Erfinderbetreuung bei der Semperit AG," in *Berichte der Arbeitsgemeinschaft, "Fragen,"* BAK, R22/630, Bl. 27, pp. 23–42; also Walter, *Der Erfinderbetreuer im Betrieb* (Berlin, 1943), Bayer 23/54.
4. On Arnhold and DINTA (originally established in 1925 by the Verein Deutscher Eisenhüttenleute): Nolan, *Visions of Modernity*, 179–205, esp. 185–92, 227–35; Frese, *Betriebspolitik*, 251–332; Smelser, *Robert Ley*, 188–94; Mason, *Social Policy*, 170; Anson Rabinbach, *The Human Motor: Energy, Fatigue, and the Origins of Modernity* (New York, 1990), 284–88; in 1933 Ley incorporated DINTA into the DAF and renamed it Amt für Berufserziehung und Betriebsführung.
5. Walter, "Erfinderbetreuung bei Semperit," 23–24. On the psychological astuteness of other experts in the DAF's Arbeitswissenschaftliches Institut: Smelser, "Sozialplanung," 83.
6. Walter, "Erfinderbetreuung by Semperit," 24–25.
7. Ibid., 25–26.
8. Ibid., 26–27.
9. Ibid., 27–28.
10. Ibid., 28–29, 22–37.

11. Ibid., 29–30.
12. Ibid., 31–33, 37–39.
13. Walter's BDC file indicates that Vienna's vice-mayor, patent attorney and engineer Hanns Blaschke, had appointed him to manage the affairs of Austria's "non-Aryan" patent attorneys.
14. Walter, "Erfinderbetreuung bei Semperit," 40–41.
15. Ibid., 42.
16. On social policy as the means to increase productivity: Smelser, "Sozialplanung,"; idem, *Robert Ley*, passim; Rabinbach, *Human Motor*, 238–88; Frese, *Betriebspolitik*; Freyberg and Siegel, *Industrielle Rationalisierung*; Heidrun Homburg, *Rationalisierung und Industriearbeit: Arbeitsmarkt, Management, Arbeiterschaft im Siemens-Konzern Berlin 1900–1939* (Berlin, 1991).
17. An article entitled, "Workforce and Productivity Increases," published in *Der Vierjahresplan Zeitschrift für Nationalsozialistische Wirtschaftspolitik* 3, no. 19 (5 Oct. 1939), stressed not only the importance of efficiency, quality work, rationalization, orderliness, cleanliness, and accident avoidance, but also the importance of employee suggestions. Suggestions were not only being solicited, but would be studied impartially and professionally, and the useful ones would receive awards. The entire suggestion system was being built up and transformed into a "permanent attitude of special achievement ... with economic incentives to spur collaboration by the workforce." In August 1940, Arnhold, now chief of the department for occupational training and productivity increases in the RWM, published an article, "Concerning the Future of Europe's Armory," in which he pleaded for continued rapid technological progress. This was not only a function of research at the country's institutes of higher technical education, Arnhold wrote, "but also in the factories." There it would "have to become a 'sport' to find inventive minds among the workforce and to draw them into participating in creative work." Arnhold called for arrangements to enable testing of ideas management had not taken up. One would have to "deploy geniuses unburdened by traditions and detached from all constraints to find fundamentally new roads that will let technological development progress." Many so-called developments were but a circular movement. "But we must escape from this magic circle. The breakthrough to real development must be forced. For that we need those revolutionary forces of genius that are free from all constraints and presuppositions." DM, BPV VIII 288 (6). On Arnhold's use of sport-like competitions to increase productivity: Smelser, *Robert Ley*, 192–97; Rabinbach, *Human Motor*, 284–87.
18. Origins and passage of the ordinances are described in ch. 11. Establishment of inventor trusteeship represented a major victory for the NSDAP/DAF complex in its quest for power and official standing at the governmental level.
19. Richard J. Overy, "Rationalization and the 'Production Miracle' in Germany during the Second World War," in his *War and Economy in the Third Reich* (Oxford, 1994), 343–75; Gregor, *Daimler-Benz*, 92–100, 118–26, 250–51; Hans-Joachim Braun, *The German Economy in the Twentieth Century* (London, 1990), 127–38; cf. also Alan Milward, *The German Economy at War* (London, 1965), 131–61.
20. Overy, "Rationalization," 356–66; Gregor, *Daimler-Benz*; 92–100, 118–26, 162–71, 250–51; Siegel, "Wage Policy."
21. Gregor, *Daimler-Benz*, 99–100, 118.
22. Siegel, "Wage Policy," esp. 21, 29, 36; Gregor, *Daimler-Benz*, 169; Sachse, *Siemens*, 59–77, esp. 72–76; Carola Sachse, Tilla Siegel, Hasso Spode, and Wolf-

gang Spohn, *Angst, Belohnung, Zucht und Ordnung: Herrschaftsmechanismen im Nationalsozialismus,* intr. Timothy W. Mason (Opladen, 1982); Peukert, *Inside Nazi Germany,* esp. 236–42.

23. Albert Speer, "Vorwort," in Karl August Riemschneider and Heinrich Barth, ed., *Die Gefolgschaftserfindung: Erläuterungen über die Behandlung von Erfindungen von Gefolgschaftsmitgliedern,* 2d ed. (Berlin, 1944), 9–10.

24. Ibid.; Hitler, *Mein Kampf,* 445–46.

25. Ansprache Professor Speers vor den Reichs- und Gauleitern am 24. Febr. 1942 über seine Aufgaben als Nachfolger Dr. Todts," BAP, RMRK (46.03). no. 4, 15–19. On Speer's lack of interest in encouraging systematic research and development: Milward, *German Economy,* 72–99.

26. Werner Mansfeld, quoted in *Berliner-Börsen-Zeitung* no. 374, (Aug. 10, 1942), BAK R43 II, 1559, Bl. 115.

27. Josef Dapper, "Zur Frage der Betriebserfindung," *Deutsche Technik* 10 (1942): 8, 14.

28. Heinrich Barth, "Persönlichkeit und Volksgemeinschaft im Rechte der Erfinder und Erfindungen," *Zeitschrift der Akademie für Deutsches Recht* 2 (1935): 823–26.

29. Overy, "Rationalization," 343–75; Gregor, *Daimler-Benz,* 92–100, 118–26, 250–51; Milward, *German Economy,* 131–61.

30. Discussion of the tension between the philosophy of "qualitative superiority" and the need for quantitative improvements: Milward, *German Economy,* 100–130; Braun, *German Economy,* 129, 132, 134–38.

31. On *Gefolgschaft* and *Betreuung* as Nazi terminology: Victor Klemperer: *LTI: Notizbuch eines Philologen* (Leipzig, 1975), 250–59.

32. Karl August Riemschneider, "Die Erfinderbetreuung im Betrieb," *Deutsche Technik* 10 (1942): 292.

33. Riemschneider and Barth, *Gefolgschaftserfindung,* 53.

34. On the primacy of "design culture" over "production culture" in German technology: Köning, *Künstler;* distrust of Taylorism: Nolan, *Visions of Modernity,* 42–49; Rabinbach, *Human Motor,* 238–75; see also ch. 13.

35. Riemschneider and Barth, *Gefolgschaftserfindung,* 53.

36. Exclusion of Poles, Jews, "Gypsies," and *Ostarbeiter,* but inclusion of other foreign workers: Riemschneider to Gotthold Baatz, RMBM, 20 Feb. 1943, BAK R22/630, Bl. 51a. Also: correspondence of Reich Ministry for the Occupied Eastern Territories and RMBM, 27 Jan. and 4 Feb. 1943, BAK R22/630, Bl. 47, 49. On foreign labor during the war: Edward L. Homze, *Foreign Labor in Nazi Germany* (Princeton, 1967); Ulrich Herbert, *Hitler's Foreign Workers: Enforced Foreign Labor in Germany under the Third Reich* (Cambridge, 1997); Mommsen, *Volkswagenwerk.*

37. Overy, "Rationalization," 366–74, gives figures for increased output in a variety of fields. On the difficulty of isolating the exact contribution of different measures that resulted in increased output: Gregor, *Daimler-Benz,* 92–100, 118–32, 248–49.

38. Ludwig, *Technik und Ingenieure,* 227–29.

39. This is also the principal conclusion of J. Slama, *One Century of Technical Progress Based on an Analysis of German Patent Statistics* (Brussels, 1987).

40. In a discussion of his company's patent developments during the war, BASF's Karl Holdermann in March 1942 mentioned that the outbreak of war had interrupted the filing of patent applications abroad and also made it impossible to keep foreign patents in effect, owing to the impossibility of sending fee payments abroad. At home, patent applications had hardly suffered at all and, because of war-related inventive activity, even increased "strongly" in 1941. Holdermann to Voigtländer-Tetz, 20 Mar. 1942, BASF, B4/E 01/1.

41. The data were collected and assembled by Holdermann for a never-completed history of the company's patent department. Holdermann left behind two over-lapping but slightly different series. The first covers 1919 to 1940 and counts patent applications filed in Germany from all BASF units, including foreign subsidiaries. The second covers 1938 through 1955 and also counts only applications filed in Germany. This particular series excludes applications for inventions made in the company's foreign laboratories. My table represents a recalculation of the data in both series, reducing those in the first series by 5 percent (the average difference between the two series during the three years of overlap, 1938–40), to build one long-term series of German applications filed in Germany.

42. RPA, unsigned notation, 15 Aug. 1948, BAK R131/22.

43. Vereinigung der Arbeitgeberverbände to BJM, 23 Mar. 1950, "Begründung zu dem Entwurf eines Gesetzes über die Erfindungen von Arbeitnehmern," BAZwi, B141/2798, Bl. 109, 10.

44. Redies to Klauer, Kühnemann, and RMRK's Baatz, 4 Aug. 1943, BAK, R131/159. The regime addressed the problem with an ordinance of 12 May 1943, which limited the scope of the patent examination and eliminated the right of third parties to file a claim of interference (*Einspruchsverfahren*). See also Heinrich Kirchhoff, *Das deutsche Patentwesen: Rückschau und Ausblick* (Berlin, 1947), 97.

45. On Kirchhoff (born 1903): BDC and *Wer ist Wer*, 1948 (p. 119) and 1962 (p. 744).

46. Kirchhoff, *Patentwesen*.

47. Kirchhoff, *Patentwesen*, 5–8, 63–72. Idem, "Der Erfindungsschutz in der Sowjet-union," *Die Technik* 1, no. 2 (August 1946): 99–101, Bayer 22/7.

48. Kirchhoff, *Patentwesen*, 123–24.

Chapter 11

GERMAN TECHNOLOGICAL CULTURE AND THE INVENTOR ORDINANCES OF 1942 AND 1943

*I*nventor trusteeship was introduced by an "Ordinance Concerning the Treatment of Inventions by Members of the Workforce" of 12 July 1942. Signed by Göring but drafted by Speer's lieutenants in the party, the decree included statutory compensation for employee inventions. The 1942 ordinance was a terse document that provided few details. Specifics were not announced until 20 March 1943, when Speer issued an "Implementation Ordinance." Composed of thirteen articles, eleven of which dealt with the rights and obligations of employee inventors, the implementation decree put in effect legislation that the Ministry of Labor had unsuccessfully tried to get approved since 1934 and that before those efforts went back to the early 1920s.[1]

The Labor Ministry's "Draft Law On Inventions By The Workforce"

When it became clear in the course of 1934 that a new Patent Code would soon be reality, the Labor Ministry dusted off its old plans for employee-inventor legislation. In consultation with a labor-law committee of the Academy for German Law chaired by Professor Hermann Dersch, the ministry studied the question until the spring of 1936. It announced a draft bill in late June of that year—in time, it hoped, for passage along with the new Patent Code.[2]

Notes for this section begin on page 280.

Created under the guidance of Werner Mansfeld, co-author of the Law for the Regulation of National Labor (Gesetz zur Ordnung der nationalen Arbeit of 20 January 1934, AOG), the June draft granted employees statutory compensation, but on balance much favored big business.[3] It was influenced by corporate lobbying and by the AOG, which gave the employer, as the "Führer" of his company, exceptionally broad powers over workers and employees, whose principal duty was loyalty. But while inventor advocates felt cheated, employers thought the bill went too far the other way. The two key issues were mandatory compensation and the right to employee inventions, which together became the yardstick of where the regime stood in relationship to corporate management and engineering labor between 1936 and 1945.

When the 1936 Patent Code did away with the company invention in 1936, this not only represented a moral victory over German capitalism but also destroyed management's most powerful instrument for claiming title to employee inventions. From the employer's perspective it was therefore imperative to broaden the service invention and limit the scope of the free invention sufficiently to make up for the loss. The Labor Ministry's June 1936 draft accommodated the employers by defining the service invention broadly, borrowing the approach of the Chemists' Contract of 1920 and requiring an employee to assign all inventions that came "within the sphere of activity of the business (*Betrieb*)."[4]

The key problem centered on the meaning of the word *Betrieb*. Whereas the Chemists' Contract had used the unambiguous and broad term "enterprise" (*Unternehmen*) to delimit the sphere of business activity determining which inventions went to the employer, the term *Betrieb* was vague. It could mean a sphere of activity as narrow as an employee's immediate place of employment, such as a certain plant, building, or factory. But it could also designate a larger unit, division, or company, or even an entire conglomerate as large as the worldwide operations of the Siemens or the IG Farben concerns. The Labor Ministry officials who had drafted the bill were aware of the problem, for they had added a separate article to deal with it. The article in question allowed the employer and the employee to make a separate patent agreement, which imposed on the latter the duty to assign also those inventions that "fall in the sphere of activity of a *Betrieb* that is economically and technically related to the *Betrieb* where the employee belongs." The addition of this language could only mean that the term *Betrieb* used in the bill's first article was intended to be rather narrow—though precisely how narrow was left up in the air, since it could apparently expand indefinitely.[5]

The problem had been anticipated well before the ministry's bill was announced. In 1935 Waldmann of the party's legal department recommended defining the scope of inventions to be assigned to the employer as narrowly as possible. Riemschneider shared Waldmann's opinion. He addressed the subject in a 1936 article in the *Journal of the Academy for German Law*. Riemschneider worried that the Labor Ministry, influenced by the AOG, might define service inventions too broadly and "go too far." In the end, he wrote, "based on the duty of loyalty, an invention that otherwise falls completely outside the field of activity, not only of the inventor, but also of the *Betrieb*, might have to be offered merely because it can somehow be connected to situations inside the company." The right solution instead would be to "restrict" the scope of service inventions to just the "employee's professional responsibilities."[6]

Corporate management, not surprisingly, disagreed. Anyone who knew anything about industry, wrote Wiegand in May 1935, knew it was "absolutely necessary … that the *Betrieb* must have an unrestricted right to those employee inventions that fall in the sphere of the business of the *Betrieb*." To prove his point, Wiegand quoted company founder and national hero Werner Siemens. Siemens had pointed out as long ago as 1888 that no company could survive if it were constantly exposed to the danger of its own employees taking out patents in their own name for improvements based on the experiences gained at work and with company means. The only feasible solution, wrote Wiegand, would be that an employee, "as soon as [he] has completed an invention, is obligated to report this invention to the employer." The latter, for his part, had "the duty to notify the employee within a reasonable period if he does not want to claim the invention, so the employee then has the opportunity to utilize the invention himself. If the inventor does not receive notice, the invention counts as having been assigned to the employer."[7]

Unable or unwilling to decide the issue, the Labor Ministry had merely papered over the contradictions with the elastic concept *Betrieb* and with lack of specificity about the employee's reward entitlement and reporting duty. But no one was fooled, as an avalanche of criticism and countercriticism, a string of revisions, reworked bills, renewed criticisms, and an ever-receding time horizon soon made clear. The objections came principally from other ministries and government or party agencies, which acted as proxies for the socioeconomic interest groups with which they tended to identify. In this different form the infighting over inventor rights in the Nazi regime was a direct continuation of the conflict as it had been fought in the Weimar Republic.

Julius Dorpmüller, head of the *Reichsbahn* and Minister of Transportation from 1937, fired the first shot. Dorpmüller, an engineer by training, thought the bill conceded too much to the employee. It also set a dangerous precedent for civil servants, who were fundamentally different from salaried employees because they owed all their loyalties and energies to the state and therefore deserved no special compensation for inventions.[8] Dorpmüller's comments provoked a sharp rebuttal from Postal Minister Wilhelm Ohnesorge.[9] A true believer in National Socialism, Ohnesorge attacked any attempt to curtail the incentives and rewards for technological creativity as a violation of the movement's inner spirit. Dorpmüller's only concern, he wrote, was to strengthen the employer's hand. But what was really needed was to stimulate creative research, to strengthen the urge to invent, and to reach ever-higher goals. Dorpmüller's proposals "contradict[ed] the duties of the National Socialist state toward the creatively working inventor." Ohnesorge's ministry could therefore "not accept that these most valuable powers of its workforce are to be paralyzed by legislation based on bureaucratic anxieties."[10]

Judge Barth of the party's legal office agreed with Ohnesorge. In an article for the *Völkischer Beobachter*, Barth urged those employers who thought the new legislation might let them get away with compensating just a few special inventions to reread the relevant pages in *Mein Kampf*. He attacked any attempt that would make employees turn over all their inventions to the employer. If the law were to allow this, the large corporations would immediately pounce on the opportunity and claim any and all employee inventions, thereby "choking off the inventor's creative freedom."[11]

Critical reactions such as these prompted Mansfeld and his team to go back to the drawing board. In November 1936 they came up with revisions to make the bill more inventor friendly. In fact, the Labor Ministry, having read the signs of the times, now became a staunch advocate of the employee. It drastically restricted the scope of inventions to be assigned to the employer. It clarified the employee's right to special compensation and eliminated language that allowed exemptions. It specified the basic principles governing calculation of a reward's size.[12]

The November revisions were circulated to gauge reactions and gather additional input, which came in the form of only minor criticisms by Riemschneider and Todt.[13] Bormann weighed in with suggestions for less onerous reporting duties of the employee; he also sought further restrictions of employer rights and called for arbitration by the Labor Front.[14] The most serious objection came from the Ministry of Economics, which lined up with big business and demanded a return to the June draft. The current thinking of the Labor Ministry was totally unacceptable, wrote the ministry's Dr. Pohl, paraphrasing Siemens's and IG Farben's arguments that trust and order in industry would disintegrate if employees received the rights now being contemplated.[15] But with Hess signaling the party's approval of the new draft, the Labor Ministry ignored Pohl and in April 1937 presented a draft bill that incorporated the November revisions and some of the party's additional suggestions.[16]

The April draft caused considerable apprehension in big business circles, in particular at IG Farben, Siemens, and the Ministry of Economics. At an interministerial meeting in late April, officials from Economics and Transportation clashed repeatedly with their counterparts from Labor, Justice, and Hess's office.[17] It proved impossible to work out the differences, as neither side was willing to budge. A twelve-month stalemate ensued, during which the leading high-technology firms undertook a concerted effort to produce a bill more to their liking.

Managers at IG Farben, who carefully monitored the regime's every patent move, concluded that the latest developments were "unbearable" and "not compatible with the interests of industry." But the company kept a low profile, preferring instead to work the many inside contacts and channels available through its central position in the Four-Year Plan.[18] Still, the company gave Redies permission to publish a scholarly critique of the Labor Ministry's latest plans. Redies's article, which appeared in June 1937, was a broadside against the inventor entitlements that the Nazi government proposed to introduce. Redies reasserted the reality of the company invention, regardless of what the new Patent Code said. He argued that the employee had an obligation to turn over almost all inventions, only some of which should be rewarded with special compensation. The method of calculating the inventor reward should be left up to the employer or allowed to vary with the different branches of industry. The word *Betrieb* should be defined as broadly as possible. In short, Redies called for a return to the conditions under which his industry had operated in the Weimar Republic.[19]

If Redies's paper left little room for doubt as to where IG Farben stood, it was nevertheless presented as merely one expert's opinion, a think piece that was not official company policy. This strategy contrasted with the policy of Siemens. At a meeting in Berlin's Kaiserhof Hotel in May 1937, Heinrich von Buol, general manager of Siemens & Halske from 1932 to 1945, told Mansfeld that the bill threatened the very survival of German industry.[20] When Mansfeld asked for substantiation of this extraordinary claim, Buol sat down to prepare a fundamental critique of the government's approach. It took Buol several months to write his paper, which was based on extensive research, study of Ludwig Fischer's work, and consultation with Köttgen and Wiegand. The final draft was ready by the second week of September 1937 and arrived at the ministries of Labor and Economics shortly afterward.[21]

In his cover letter, Buol, himself an accomplished engineer and physicist, chose to resuscitate the industrialists' old arguments from the Empire and the Weimar Republic as the best proof of the issue's fundamental importance, not just to his own company, but to the nation as a whole. His own experience, he began, had taught him that there was "no real, practical need to tighten the rules currently in effect concerning the employee invention." On the contrary, the only thing that legislation could accomplish was to "drive a wedge between company and inventor." Perhaps there were a few small firms where the position of inventors left something to be desired, but it would be a terrible mistake to "sacrifice the larger part of industry that is more important for the national economy" for their sake. Of the 12,245 electromechanical patents in effect in 1936, he pointed out, 8,513 belonged to Siemens, AEG, and their respective subsidiaries. Of the remainder, other large firms such as Bosch and Lorenz and foreign companies such as Western Electric, Brown Boveri, and Philips owned another 3,700. Compared to this, the inventive role of small firms and independent inventors amounted to almost nothing. The giants of the industry were decisive for progress in the field, and they invented and patented thanks to the work of their organized staffs of engineers and physicists, who had generated the 67,000 patent applications that Siemens alone had filed between 1900 and 1937. Did the government really think that all this represented the "products of creative personalities?" Of course not. It came from hard, systematic, well-remunerated work, which the company's employees had produced without "one serious conflict" and without ever even hinting that they had "significant wishes for improvement." Was not this fact "sufficient reason to approach the problem with the greatest caution?"

With respect to the company invention, Buol repeated Redies's argument. "One can abolish the *concept* of the company invention," he wrote, but no one familiar with industrial conditions could "seriously contest that doing so fails to get rid of the type of invention known until now as *company invention.*" Siemens could live with the new rule if the whole issue were reduced to the question of an inventor's appropriate reward. But the ownership of employee inventions was not negotiable; the company had to have it, come what may. Unfortunately industry's arguments—and they had been the same forever—had always been ignored and misunderstood. As evidence, Buol included a copy of Wilhelm von Siemens and Ernst Budde's 1908 article, "The Right of Employees to Inventions," which already covered everything now being raised again. As Fischer had done in 1923, so Buol in 1937 invited the Labor Ministry's top officials to visit the Siemens laboratories and spend "an hour or so with a couple of inventors from our company, to hear from them in their own words what their work is like in practice." He was sure that a great many of Siemens's best inventors—"i.e., those who have really achieved something creative—would be happy to make themselves available for such a discussion."

In the paper itself, Buol attacked the fundamental principle of German inventor compensation: using the patent as the criterion for special creative achievement and therefore entitlement to a reward. Though they were the stuff of everyday life in industry, the words "inventor" and "patent" cast a magic spell on lay people such as politicians, lawyers, and even many businessmen. But patents were nothing special, merely an affidavit by the Patent Office that, in its opinion, the underlying idea had never before been published nor used publicly and that one could not disprove that the idea represented economic progress. Whether it really was progress or could even be made to work usually remained an open question. This did not prevent patent holders from being regarded as a special breed of people. Meanwhile, the Patent Code had no way to protect the true geniuses of this world: those who made fundamental discoveries about the natural world. Wilhelm Röntgen never got a patent for the discovery of x-rays, nor did Robert Koch for discovering the tubercle bacillus, nor Fritz Schaudinn for finding the spirochete that causes syphilis. But there were at least 14,000 patents for medical cures about which the history of medicine knew nothing at all.

Conversely, the story of vacuum tubes showed the worthlessness of many patents and the inherent unfairness of the whole patent system. Vacuum tubes went back to Robert von Lieben's 1906 patent. Germany's large electrical engineering firms spent vast sums trying to make the Lieben invention work as promised, but failed until the American inventor Irving Langmuir determined in 1913 that the vacuum had to be greater by a factor of one hundred. Langmuir's discovery launched radio, television, and countless other technologies on their march of victory throughout the world, but in terms of German patent rights it was a disaster. The German Patent Office refused to grant Langmuir a patent on the grounds that the basic idea of his invention had been published before. Thus Lieben's patent went on to survive and make money, owing to the achievements of someone else who was left empty-handed. Buol presented a number of other examples, including Coolidge's 1914 inventions of ductile tungsten for lamp filaments and x-ray tube improvements, the German patent application for which was rejected. The unexpected and undeserving beneficiary was Siemens & Halske, owing to a hunch of its patent director, Ludwig Fischer, who years earlier had included tungsten in a patent application for an x-ray tube based on tantalum.

The point was to show "how poorly creative achievement and patent protection match[ed] up." The nominal inventor or the person or entity receiving the patent often got it on the basis of ordinary professional work or systematic development, after others had done the fundamental research that remained unpatented. Such was typically the case with employee inventions, Buol wrote, echoing Fischer's *Company Inventions*. "Precisely in companies that approach the work of progress systematically, inventions are very rarely the work of a brilliant flash of insight, but almost always the result of very solid and systematic work." He knew "hundreds of examples" where the inventor did not even realize he had made a patentable invention and had to be told and prodded by the company's patent engineers. These were the conditions, in addition to lots of money and a willingness to take risks, under which real technological progress was generated. It was the result of organized and systematic research and development—a task at which Siemens excelled. Besides Siemens, there were only a few such companies in the world. None of them could function without strong patent protection, for it was the only thing that defended their huge efforts and risk-taking from imitation and theft. The government should think twice before it decided to fiddle more with the patent system, hurting an industry that was vital for Germany's economy.

That the proposed inventor legislation would in fact hurt the industry was an absolute certainty, Buol continued. The law would never find all those for whom it was targeted and would include many whom it should not. This would "produce the most serious disturbances among the company's staff." And that, in turn, would have "the most fatal consequences for a company seeking to generate [technological] progress in a systematic fashion." Above-average creativity deserved its proper reward, but it was "fundamentally absurd to assume," as the government did, "that the same companies spending dozens of millions on research and development make it a point not to take care of the most valuable part of this development, [and] do not want to satisfy the creative employee." There was no need for "such a dangerous experiment as that represented by the bill." Besides destroying inventor morale and undermining a harmonious working environment, the law would cause inventors to look out for themselves instead of the company and make patentable inventions instead of solving the problems they had been assigned. It would cause a surfeit of patenting, which was hugely expensive and slowed down the work that needed doing, adding to the friction between company and employees. Employees would try to sell or license to third parties patents their employer did not really want, causing fundamental conflicts of interest. Mentioning several additional complications the inventor legislation would cause, Buol concluded with a long list of criticisms of the bill's individual articles.

How did the Labor Ministry react to the arguments of Buol, Redies, and the Ministry of Economics? Contrary to interpretations emphasizing the collaboration of industry and Nazi movement in a "power cartel," neither the Labor Ministry nor party functionaries such as Hess and Bormann allowed themselves to be swayed. On the contrary, Mansfeld proceeded to ignore all but one or two minor details raised by the industrialists and others such as Dorpmüller. By April 1938 the Labor Ministry issued a third and final draft of the bill. It differed from that of the year before only in that it made things easier for the employer with respect to foreign patents, eliminating some of the potential problems Buol had mentioned in his paper. In all other respects the bill stuck firmly to its pro-employee position.[22]

In the official commentary, which went out in early July 1938, Mansfeld explained the decision to ignore industrial objections. Like the new Patent Code, the current bill fulfilled a central promise of National Socialism, which was to help the nation's inventors with a variety of social measures. It was the "duty of the National Socialist state," he wrote, "to promote the unfolding of the creative personality and to protect its work against exploitation." Weimar's abusive free-for-all in this regard could no longer be tolerated. The "employee-inventor as an essential carrier of the creative development of the nation must be guaranteed the results of his know-how and his labors." While the employee's duties of loyalty to his employer were indeed substantial, it would nevertheless be "a perversion of the idea of the company community (*Betriebsgemeinschaft*), if the duty to surrender were indiscriminately extended to all inventions that fall within the scope of the company's activities." Whenever the invention was unrelated to the employee's professional responsibilities and did not rest on the company's experience, state of the art, or facilities, it was free.[23]

Mansfeld next turned to special compensation. Here, too, he left no doubt about the ministry's pro-employee stand. "As soon as an employee's achievement exceeds what can be required of him in terms of the substance of the employment relationship, he is entitled to special compensation, because his regular salary is based merely on this [required minimum] achievement." This meant that any invention that an employee was obligated to turn over to the employer and that proved patentable was, by definition, a special achievement, above and beyond the employee's normal responsibility as defined in the employment relationship. This special achievement, demonstrated by its patentability, deserved special, "appropriate compensation," regardless of whether the invention was precisely that for which the employee had been hired, or whether it was merely the result of applying ordinary professional skills.[24]

The question of what constituted "appropriate compensation" would soon emerge as the next battleground between government and industry. But for now, the Ministry of Labor, supported by most other ministries and the Nazi party, retained the initiative and momentum necessary to make the bill into law. In fact, it seemed nothing could stop speedy enactment of the legislation. When Dorpmüller, Admiral Canaris, and Education Minister Bernard Rust raised questions about treatment of inventions by government employees, in particular civil servants and professors in state research institutes and universities, Mansfeld simply deleted this issue from the bill, promising separate legislation for *Beamte* later.[25] Nor was the ministry derailed by the efforts of Albert Pietzsch, a Nazi official and influential businessman with personal ties to Hess and Hitler.[26]

In August 1938 Pietzsch had a conversation about the bill with Johannes Krohn, Secretary of State at the Ministry of Labor. According to Krohn's notes, Pietzsch stated that the legislation was not being drafted for "the good inventors, but for the driveling idiots." The law was "a great danger and [would] introduce strife in the factories." Instead of all the detailed regulations proposed by the government, a single and simple rule would suffice: an invention that was applied and brought results should give the inventor an appropriate share in the result. The courts should decide what was appropriate.[27] Pietzsch requested another interministerial meeting, which took place in November 1938.

Pietzsch repeated the arguments of Redies and Buol: most patented inventions were worthless anyway; the law would only encourage dreamers and crackpots to cause trouble; real inventors needed no law to protect them; there was no need for the law, which in any case was much too complicated. But if there was going to be a law in spite of all this, the employee should be required to assign to the company all inventions that came within the broadest possible scope of the company's activities. Compensation should be "appropriate," but it should be left to the employer to decide what that meant and apply it only when the invention was gainfully used.[28]

Pietzsch's efforts to change the bill were in vain. Although the November meeting triggered another furious round of memoranda and papers from various sides and Mansfeld wavered on some issues, Krohn sent the bill unchanged to Hitler in March 1939. The available documents are silent about the decision-making process that led to this point, but it appears Krohn, and perhaps Seldte, decided no greater consensus was possible beyond that already reached. The sources contain no evidence of any effort among the bill's opponents or supporters to coordinate their efforts. For every attempt to change the bill one way, there was a countereffort the other way, and vice versa. In the absence of direct orders from the regime's top, many of these last-minute moves seem to have taken place in a vacuum. This likely gave Mansfeld and Krohn an opportunity to ignore the stalemate and proceed on their own.

Whatever the precise circumstances, Krohn submitted the bill and official commentary to Heinrich Lammers, chief of the Reich Chancellery, on 17 March 1939. He requested enactment in time for the law to be promulgated on Hitler's birthday, 20 April.[29] The Economics Ministry, a question mark to the end, gave its approval in late March, explicitly disassociating itself from Pietzsch and disavowing the arguments that until recently it had itself advanced. On 15 April Lammers informed the Labor Minister that Hitler had approved the bill. Accordingly, it had been "decided that the draft of a Law Concerning the Inventions of Employees (*Gefolgsmänner*) presented by the Reich Minister of Labor has been enacted."[30]

It is unclear precisely what happened next. But three weeks later Lammers sent another letter to the Ministry of Labor, in which he wrote that Hitler had indeed approved the bill just as his earlier letter had indicated. However,

> prior to executing the law, the Führer once more repeatedly perused its contents. While doing so, he made remarks to the effect that he considered another thorough review of the law necessary, because in spite of everything its content in part caused him to have serious reservations. In particular, the Führer thought, it is doubtful whether the law is simple and comprehensible enough that the group of people for whom it is basically intended would be in a position to really familiarize themselves with the contents of the regulations that are supposed to protect their rights. Whether an invention falls in the sphere of activity of a company—this question alone will be difficult to decide for many workers and employees. The rules for reporting duties and their conditions, too, appear to be in need of simplification, if the ultimate goal—securing for the employees their rights to the invention—is really to be achieved.

Lammers made suggestions as to how to simplify the bill, so it would meet with Hitler's future approval. The basic idea was to make things easier for the ordinary workers and "people's comrades ... in the humbler positions" from whom Hitler expected inventive miracles. "[A] large part of the employees will not be in a position to obtain legal counsel but are dependent on having to protect their own rights." In short, there was a need for "simplification of the procedure and more easily understood language [to] improve the chances for employees to safeguard their rights." Until that was done, Lammers said, he would keep the original copy of the law in his safe.[31] That is where it remained, until Speer resurrected the legislation in 1942 in connection with his promulgation of inventor trusteeship.

There is no record of the reactions to Hitler's verdict at the Ministry of Labor or among the Labor Front's inventor crusaders in the spring of 1939. But there can be little doubt that they must have been profoundly frustrated and perplexed, considering that defeat was snatched from the jaws of victory by the very figure, that self-professed "fool for technology," who was the ultimate legitimization of the regime's pro-inventor efforts.[32] Conversely, the industrialists were surely relieved, even though this latest turn of events was not due to anything for which they could take credit. On the contrary, when Pietzsch, emboldened by the news of Hitler's rescission, sent a memorandum to Lammers in May 1939 restating why the law should remain locked up forever, the official in the Reich Chancellery whose job it was to analyze such papers, summarily dismissed it.

Pietzsch's arguments created "the impression of being a very one-sided representation of the entrepreneurial point of view" and should be ignored. Lammers agreed.[33]

In the meantime, Mansfeld and his chief assistant, *Ministerialrat* Georg Steinmann, rushed back to work, trying to satisfy Hitler's demands as soon as possible. The solution they came up with, besides adopting different, clearer language, was to separate the bill into two parts, an easily understood first part that set forth the legislation's principles, and, for the second part, an implementation ordinance that contained the more technical details. To meet Hitler's objection concerning the lack of clarity about reporting duties, they extended the latter to include any and all inventions made by the employee. Other than that, there were no substantive changes.

The new version was once again circulated to the different ministries for comment. Inevitably, this took up more time.[34] Additional delays resulted from a new round of discussions over a plan to include civil servants, soldiers and other government employees in the bill after all.[35] Most ministries liked the proposal, but some had doubts and needed further persuasion, while the change also necessitated more new draft language and commentary. The upshot was that the legislation was still in limbo when World War II broke out in September 1939.

Speer's Solution

Not surprisingly, the war brought new and different priorities and disrupted the peacetime rhythm of the bill's progress since 1936. Even so, by November 1939 Seldte thought the time had come to go back to trying to enact the bill. Seldte acted in part because the Labor Front had urged him to move forward and not to disappoint the thousands of inventors who had been energized by Göring's recent inventor appeals but also expected something tangible in return.[36]

The fall of 1939 was precisely the time when Riemschneider and Priemer, disappointed with the meager harvest of inventions by small independents, were beginning to focus their attention on employees. It was also during this same time that the Bureau for Technical Sciences established the predecessor of the Reich Working Group for Inventing and Riemschneider began to listen more carefully to the arguments of industry experts such as Redies and Wiegand, big business's principal representatives in the new consultative body. Seldte and Mansfeld, who had no way of noticing this gradually evolving shift in the bureau's orientation, stuck with their bill, telling the other ministries there were signs the employers had recently stiffened their attitude toward inventor compensation and needed the law to remind them of their obligations.[37]

While most ministries were willing to go along with Seldte and resumed negotiations in early 1940, the Supreme Command of the Armed Forces did not. Confident of Germany's technological superiority and counting on a short war, the military contended that the legislation was irrelevant to the conduct of the war and should be postponed until hostilities had been concluded and the country returned to peace. Despite repeated requests from the Ministry of Labor, the Supreme Command persisted in this attitude and blocked all movement on the bill for almost another year. Finally Todt was able to break the deadlock. Though the evidentiary record is slim, it appears that the new Minister of Armaments—whose ministry was not established until March 1940—had become convinced of the legislation's urgency. Todt in October 1940 managed to persuade Field Marshal Keitel to relinquish the army's stranglehold and took charge of the effort to get the legislation moving again.[38]

In November 1940 Todt convened a small meeting at his ministry with delegates from Hess's office, the Supreme Command, and the ministries of Labor, Justice, and Interior. Todt's own representatives at the meeting were *Ministerialdirektor* Günter Schultze-Fielitz and Riemschneider. The agenda was limited to working out the details concerning the rights of inventors in the public sector and the military. This proved difficult, because the military initially took the position that everything an inventor in the civil service or the military came up with belonged by definition to the state. The matter was turned over to a committee, which wrestled with the issue for many more months. No agreement proved possible, but eventually the reality of protracted war and the combined weight of other ministries caused the Supreme Command to give in.[39]

In September 1941 the bill was ready to go to the Reich Chancellery for Hitler's signature once again. Signed by State Secretary Syrup, the Labor Ministry's cover letter explained why it had taken two years for the legislation to make it back to Hitler and why rapid execution now was essential: "The legal protection of the inventing employee and worker indirectly serves the German people's struggle of defense," which was the condition Hitler had put on the passage of any legislation since the spring of 1941. The law would stimulate ingenuity, "which is especially desirable in war time," as both Todt and Bormann made it a point to emphasize.[40]

The Reich Chancellery did not think the new bill was substantially better than the old one. Nevertheless it reluctantly decided the matter was important enough to go forward. The bill, "considering the special interest the Führer takes in the rights of the creative human being," should be presented.[41] This was easier said than done. By now, Hitler was totally absorbed in the conduct of the war in Russia and apparently in no mood to deal with anything besides that. The relevant documents in the files of the Reich Chancellery, at any rate, indicate that Lammers never actually presented the bill to Hitler. The documents in question repeatedly bear the notation, "not suitable for presentation to the Führer," in October and November 1941, and again in January 1942. Finally, in late February 1942, about six weeks after Todt had died in an airplane crash and Speer had taken his place, someone from Bormann's office telephoned to find out what was the reason for the delay, and did the Führer know about this? Again, Lammers did nothing, until Bormann spoke to him about it in March and mentioned that Speer "place[d] great value on the further pursuit of the bill." Lammers countered that Bormann and Speer knew nothing about the particular circumstances pertaining to the delay, such as Hitler's critique of 1939, the bill's subsequent revision, and its lukewarm evaluation by his staff just a few months ago. In a follow-up letter in April 1942 Lammers further explained that owing to the military situation he had decided not to bring up the matter with Hitler. He remained convinced it was "doubtful whether the bill in its current version will meet with the Führer's approval."[42]

There is no record in the files of Bormann's response, but Speer wrote back on 14 May 1942. His assignment to "increase the productivity in all the works of the German economy, in order to implement the armaments program ordered by the Führer, ma[de] it absolutely essential to decide the employee-inventor right without further time-consuming discussions ..." To that end, Speer had put together a team of experts from the Party Chancellery, the party's Legal Office, and the Department of Technology. This group, which included Priemer and Riemschneider, had drafted a brief "Führer Order" that combined the basic idea of the Labor Ministry's bill with their own plans for inventor trusteeship. The document in question would give Speer, in consultation with Bormann, the power to promulgate the specific rules and detailed regulations necessary to implement the ideas that the general order merely outlined. Since Bormann approved of both the Führer Order and the implementation ordinance, copies of which accompanied his note, Speer requested Lammers to present the order to Hitler for his signature.[43]

The Reich Chancellery had no problems with the substance of Speer's proposed regulations, even though they departed in significant ways from the Labor Ministry's bill. Whereas the latter had sought to assist the employee by reducing the scope of the service invention as much as possible and maximizing the extent of an employee's free invention, Speer's text followed the opposite course. By deleting crucial language that stated the invention had to be within the scope of the company's (or the relevant government agency's) activities, the Armaments Minister's regulations increased the extent of the service invention well beyond most prior definitions and left only a small circle of free inventions. While this secured the employer's control over the employee's inventions, it also enlarged the number of inventions for which the latter received mandatory compensation. In addition, Speer's ordinances included detailed regulations and guidelines as to the calculation of the inventor reward. This, for all practical purposes, took away the employer's freedom to determine the size of the reward and made sure most employees received at least something in return for their inventions.[44]

Obviously, this shift had implications for the balance of power between employees and employers as well as for the kind of technological culture that inventor legislation, one way or another, sought to encourage. But this was not what concerned Lammers and his aides. Nor did they object to the most obvious difference from the Labor Ministry's bill: that Speer's ordinance amalgamated it with the party's totalitarian system of inventor trusteeship and broke down the division between state and party, which until then had been carefully observed. Instead, the chancellery objected to the form and the procedure Speer had chosen. At a time when millions of people were dying under the most horrific circumstances without any regard for the rule of law, chancellery officials worried that the Labor Minister might be offended. The latter would "surely not find it a pleasant experience if, without his participation, a subject matter about which he has already introduced a bill" were decided by a Führer Order drafted by another ministry. A different scenario would therefore have to be acted out, one that would include the Labor Ministry but bypass the need for Hitler's signature altogether. Instead of a Führer Order, Speer's draft would form the basis of a bill co-sponsored by the ministries of Labor and Armaments that would become law via a decree issued by Göring in his capacity as chief of the Four-Year Plan.[45]

Lammers presented this plan to Speer and Schultze-Fielitz at a meeting in early June, but the Armaments Minister rejected it. He insisted on the original Führer Order with one small amendment to acknowledge the temporary, wartime nature of the decree. They settled on the following plan: Göring would be recruited to make the recommendation for a Führer Order, and Lammers would then explain it to Hitler and have the papers ready to try to get his signature.[46] In the end, however, this did not happen. What did happen is not entirely clear. Göring did receive the material that would enable him to recommend a Führer Order, but the latter never materialized. Instead, the *Reichsgesetzblatt* of 22 July 1942 suddenly carried the text of an "Ordinance Concerning the Treatment of Inventions by Members of the Workforce," dated 12 July and signed only by Göring. Except for a brief explanatory preamble and substitution of the Minister of Labor for the Party Chancellery as the consulting agency, it was identical to the Führer Order drafted by Speer's assistants during the spring. The likely explanation is that Göring and Lammers were unable to get Hitler to listen when they tried to bring it up, or that they decided doing so was too risky. If Hitler turned them down, the possibility of enacting the legislation would be lost entirely. Considering the importance Speer attached to the matter, it was considerably safer to choose the path that was in fact taken: simple promulgation by Göring. This also avoided the need for further, undoubtedly contentious, discussions with the Labor Ministry.[47]

Speer's Solution Assessed: Compromise and Conservative Inventing

Thus, after almost half a century of conflict, did German workers finally acquire a right to special compensation for their inventions. The accompanying Implementation Ordinance was not promulgated until some eight months later.[48] But this delay did not result from any serious infighting over the all-important details by which the Implementation Ordinance filled in the framework order of July 1942. The differences between Speer's 1942 drafts of the former and the final version published in the *Reichsgesetzblatt* were all minor. The only remaining area of real controversy concerned the guidelines for calculating the size of an employee's inventor reward, which caused a series of rear-guard skirmishes that lasted until the very end of the regime.

Before examining this final issue, it is worthwhile to comment briefly on the ordinances of 1942 and 1943. The first point concerns the reasons for the differences between Speer's approach and that of the Ministry of Labor. Why did the Speer ordinances employ a sweeping definition of the service invention when the Labor Ministry, prodded by the very inventor crusaders who also worked for Speer, had striven since late 1936 to make it as narrow as possible? Much had changed since the mid-1930s, when the party's inventor crusaders still thought the best way to encourage technological progress was to give the employee as much independence as possible from the employer's grip on his inventions. This outlook in the early years had translated into reducing the scope of the service invention and maximizing that of the free invention; and that was precisely the policy they had successfully foisted upon the Ministry of Labor.

From late 1939, however, the Bureau for Technical Sciences had begun to cooperate more closely with industry and established the two committees that would eventually become the Reich Working Group for Inventing. Inevitably, given the lack of success of their own early policies, Riemschneider, Priemer, Barth, and others had come under the influence of industry's patent experts.[49] This was evident not just in the warm reception given to Walter's *Semperit-System*, but also in the weight carried by another Reich Working Group paper of August 1941, "The Delimitation of the Concept '*Betrieb*' and of Bound and Free Employee-Inventions in Employee-Inventor Law," by IG Farben's Redies. The underlying idea of this paper, which made the case for a broad definition of the service invention, reappeared in the Speer ordinances. At another Reich Working Group meeting of July 1942, Askania's engineer Weisse presented "The Valuation of Employee Inventions," a discussion of how to arrive at the financial worth of an employee's invention. In September 1942 Siemens's Wiegand published "The Inventions of Employees," an article outlining the details of Speer's Implementation Ordinance that would not be published until seven months later.[50]

The evolution of a working relationship between big business and the Bureau for Technical Sciences also showed up in internal company discussions about the regime's inventor legislation. In early August 1942, Redies informed the other members of IG Farben's Patent Committee that he and Wiegand had tried very hard to persuade Riemschneider not to include in the Implementation Ordinance a clause about mandatory compensation for nonpatentable ideas. Riemschneider, according to Redies, would try his best to persuade Bormann's office to drop the regulation—and in fact it duly disappeared from the final version. In December 1942, when he gave another report on the status of the Implementation Ordinance, the text of which had meanwhile been finalized, Redies singled out Wiegand for honorable mention as one of industry's most effective lobbyists. The Siemens patent chief, he said, had been particularly effective in suggesting language for the compensation guidelines "that eliminated many of the occasions for objections."[51]

Considering the extent of their influence, it is not surprising that the employers voiced few complaints when the Speer ordinances became law in 1942 and 1943. In fact, they welcomed certain parts of the regulations with open arms.[52] In contrast, officials at the Labor Ministry were decidedly unhappy. The fact that other agencies had hijacked their bill, of course, had something to do with this. But Labor officials were also upset that the Labor Front's and the party's earlier approach to protecting the inventor had been abandoned. That policy, which they had made their own from late 1936, was now being sacrificed to corporate interests and strengthening the employer's hand. In a published review of Speer's first ordinance, Mansfeld in July 1942 thought it "regrettable" that his ministry's bill had been shelved in favor of a mere ordinance. There was also the "danger that the protection of the inventing employee, which until now has stood in the forefront of all deliberations on this subject, must temporarily take a backseat to economic dictates."[53]

Officials in the Postal Ministry were also critical of the Speer ordinances. In August 1942 *Ministerialdirigent* Neugebauer met with Riemschneider to express Ohnesorge's objections to the new rules, and State Secretary Nagel sent a long, critical memorandum. Like most other government institutions and the military, postal officials objected to the plans for putting inventor trustees appointed by the party and the Labor Front in state agencies.[54] The chief complaint centered on the changes in the definitions of free and service inventions. Nagel denounced the rules entitling the employer to claim "*all* inventions of an employee 'that grow out of his professional responsibilities at work,' even when they 'do not come within the scope of the company's activities' in any way." This gave the employer "unjustified advantages, to the detriment of the employee inventor ..." It was "unclear why such a deterioration in the position of the creatively active inventor vis-à-vis the current legal situation is needed today." On the contrary, the regulations should have given the employee greater rights to his inventions, not fewer. "A reduction in the rights of the creative human being to his inventions ... is difficult to reconcile with the state's sociopolitical duty—especially important during the war—to stimulate inventive effort and promote the creative powers and talents of people who are actively engaged."[55]

Criticisms such as these accurately zeroed in on the fact that, compared to both the Labor Ministry's bill and prior jurisprudence, the Speer ordinances drastically reduced the scope of the employee's free invention and widened the service invention. This was, indeed, a crucial gain with respect to management's control over employee inventions. But the gain was bought at a high price. Industry paid by accepting mandatory compensation for *all* employee inventions, party-controlled trusteeship with its grasp for the company's technological know-how, and obligatory party mediation of disputes concerning the size of the inventor reward. Considering all this, it would be difficult to argue that, on balance, the industrialists came out ahead. This became particularly clear after the war.[56]

Another question is whether—as postal and labor officials contended—the new system harmed employees, and if so to what extent. It is true that employees had fewer opportunities to develop or market privately inventions that used to be free. But doing so had always been a very uncertain proposition to begin with, while under the new situation they had a guarantee of at least some financial compensation. As one anonymous commentator correctly pointed out in the *Reichsanzeiger*, the "expansion of the scope of bound inventions is of great significance for the entrepreneur. For the employee it is harmless, because on the basis of the new ordinance ... he is now entitled to appropriate compensation in every case the invention is claimed [by the employer]."[57] Much depended, of course, on what was considered "appropriate compensation," which was why that question became the next battleground between the Bureau for Technical Sciences and the industrialists.[58]

There is another consideration. Compared with nations such as the United States, German technological culture had never been particularly hospitable to independent inventors (including employees who owned free inventions). The absence of a tradition of venture capitalism and, in general, a different, more closed form of capitalism, as well as a patent system that had never been friendly to the small or independent inventor, had much to do with this. The Nazi regime's inventor crusaders had started out by trying to change this last aspect of Germany's technological culture, by expanding the rights of small independents and employee inventors. But their attempt had failed—not so much, perhaps, because attacking the historical biases of the German patent system was the wrong strategy, but because it would have taken much more to make Germany's technological culture more receptive to the type of "radical inventing" the inventor activists envisioned.

In addition, the war, with its requirements of increased productivity and efficiency and its increasing reliance on the largest and strongest technology companies, made the idea of independent development of inventions by employees even more illusory. So the inventor crusaders changed course. Instead of working against big business, they chose to work with it. The point of this shift, however, was not that it harmed the employee or gave the employer free reign, but that it encouraged employee productivity and guaranteed the employee's rights within the context of the firm rather than outside of it.[59] In other words, the criticisms of a Mansfeld or an Ohnesorge largely missed the point. The new strategy was not to encourage "radical inventing" in the sense of new start-up technologies emerging outside the confines of the large companies. Rather, it was to stimulate "conservative inventing" in the sense of making improvements, innovations, and breakthroughs within the context of the firm, by securing the inventor's right to special compensation for doing so.[60] Judging by Germany's surging productivity, its remarkable technological dynamism in the last years of the war, and the "production miracle" associated with Speer's reforms, it would appear that the new policy was indeed a success.

In short, rather than disadvantaging employees, the principal consequence of the Speer ordinances was to reinforce the traditional bias of German technological culture toward "conservative inventing," which was achieved largely by the combination of expanding the service invention and strengthening the legal rights of the employee-inventor inside the firm. Looked at only from the point of view of employee rights, there can be no doubt that this represented progress, even if it came in the form of draconian, wartime regulations, amalgamated with inventor trusteeship and other racist and totalitarian National Socialist overlays. But this totalitarian and racist apparatus was stripped away after 1945, while the underlying legal solution (and the consequences for the nation's technological culture) remained. One may agree with Heinrich Kirchhoff's postwar assessment that this aspect of Speer's ordinances "represented a significant step forward with respect to the law on employed inventors."[61]

Notes

1. Verordnung über die Behandlung von Erfindungen von Gefolgschaftsmitgliedern vom 12. Juli 1942 (*RGBl* 1942, I, 466-67), in *Nachrichten über die Erfinderbetreuung* 1 (15 November 1942): 1; Durchführungsverordnung zur Verordnung über die Behandlung von Erfindungen von Gefolgschaftsmitgliedern vom 20. 3. 1943 (*RGBl* 1943, I, 257), in *Nachrichten über die Erfinderbetreuung* 3 (1 July 1943): 5-7. See ch. 4 for attempts at legislation in early 1920s.

2. Kammergerichtsrat Herbert Kühnemann of the RJM to Professor Dersch and Privy Councillor Feig of the RAM, 14 May 1935, BAK, R22/629, Bl. 2; Entwurf eines Gesetzes über die Erfindungen von Gefolgsmännern, 25 Jun. 1936, BAK, R41/78, Bl. 18-21.

3. On origins of the AOG: Timothy W. Mason, "Gesetz zur Ordnung der nationalen Arbeit vom 20. Januar 1934," in *Industrielles System und politische Entwicklung in der Weimarer Republik*, ed. Hans Mommsen, Dietmar Petzina, and Bernd Weisbrod, vol. I, (Kronberg, 1977), 322-51; Frese, *Betriebspolitik*, 93-113.

4. See ch. 3 for the Chemists' Contract; Paul Wiegand, "Zur Frage der Angestelltenerfindung," *GRUR* 40 (May 1935): 261-70; Entwurf, 25 Jun. 1936, pp. 1-6.

5. Entwurf, 25 Jun. 1936, Article 9.

6. Text of Waldmann draft in Kühnemann to Dersch, 14 May 1935, BAK R22/629, Bl. 2; Riemschneider, "begriffliche Abgrenzung." RPA official Karl Müller agreed with Riemschneider, describing the broad definition contemplated by the RAM as "disturbing." Müller to Redies, 12 Jun. 1935, BAK, R131/158.

7. Wiegand, "Zur Frage," 267-68.

8. Dorpmüller to RAM and Rudolf Hess, 5 Aug. 1936, BAK R22/629, Bl. 27. Dorpmüller also argued, though, that inventors who were civil servants should be entitled to file in their own name, so long as they gave the employing state agency a shop right.

9. On Ohnesorge, see ch. 8, note 49.

10. Ohnesorge to all ministries, 14 Aug. 1936, BAK R22/629, Bl. 30.

11. Heinrich Barth, "Neuregelung der Angestelltenerfindung," *Völkischer Beobachter*, Berlin ed., no. 284 (10 Oct. 1936), unpaginated clippling in BAK R22/629, Bl. 34.

12. RAM to various ministries, party, DAF, and Hitler's deputy, interim draft of 6 Nov. 1936, BAK R22/629, Bl. 40-41.

13. Riemschnieder and Todt objected that an employee dissatisfied with his inventor reward had no recourse but to go to court. The party or Labor Front should be brought in to mediate first.

14. Riemschneider to DAF Amt für Rechtsberatungsstellen, 19 Nov. 1936, DM, BPV/DAF VIII 292 (2); Bormann to RAM, 15 Jan. 1937, BAK R22/629, Bl. 51a.

15. RWM to RAM, 29 Jan. 1937, BAK R22/629, Bl. 43. Riemschneider, in a 10 Feb. 1937 letter to Hess's assistant Bärmann, wrote that he possessed confidential documentation showing the RWM had received its marching orders in this matter from the industrialists, DM, BPV/DAF VIII 290 (1, pt.2).

16. RAM to all ministries and Hess, 6 Apr. 1937, BAK R22/629, Bl. 50-51.

17. Niederschrift über das Ergebnis der Ressortbesprechung am 28. April 1937 über den Entwurf des Gesetzes über die Erfindungen von Gefolgsmännern, BAK, R22/629, Bl. 53.

18. Niederschrift über die 35. Sitzung der Patentkommission der I.G. am 25. Januar 1937, Anlage; Niederschrift über die 36. Sitzung der Patentkommission am 10. Juni 1937, Bayer, 23/2.3, vol. 2. The strategy of keeping a low profile and cau-

tiously lobbying from the inside is well documented in surviving records of the Patent Committee meetings, ibid., vols. 1–3.

19. Franz Redies, "Zum Recht der Angestellten-Erfinder," *GRUR* 42 (Jun. 1937): 410–24, esp. 422.

20. Buol to Mansfeld, 18 Sep. 1937, SAA, 11/Lg671/v. Buol.

21. Ibid.; copies of Buol letter and paper in Mansfeld to Kühnemann, 22 Sep. 1937, BAK, R22/629, Bl. 60-61.

22. Objections by RWM and RAM's Steinmann to all ministries and Hess's office, 3 Jan. 1938; Frank to RAM, 24 Jan. 1938; Ministry of Transportation to RAM, 1 Feb. 1938; withdrawal of opposition: Frank to RAM, 19 Feb. 1938; RWM to RAM, 11 Apr. 1938; copy of revised bill in RAM to all ministries and Hess's office, 30 Apr. 1938, BAK, R22/629, Bl. 70, 74, 75, 77, 79, 80.

23. Vorläufige Begründung zu dem Entwurf eines Gesetzes ..., 4 Jul. 1938, BAK, R22/629, Bl. 101–2.

24. Ibid.

25. Objections from various ministries regarding the question of civil-service inventions in BAK, R22/629 passim, esp. Bl. 100a, 104, 106, 117, 123, 129; bracketing of the issue by RAM: Krohn to all ministries, 17 Mar. 1939, BAK, R22/629, Bl. 140b–c.

26. An electrochemical engineer and entrepreneur, Pietzsch had worked for Hitler on and off since 1925. He had first joined the party in 1927 and renewed his membership in 1930. In 1938 Pietzsch was head of the Reich Economic Chamber and president of the Munich Chamber of Commerce. He held a position as economic adviser on Hess's staff, and from 1936 to 1945 was the Deutsche Bank's representative on the supervisory board of Siemens-Schuckert. That Pietzsch was a determined opponent of labor is manifest in his unrelenting hostility to the DAF and the RAM inventor bill. Albert Pietzsch, "Wirtschaftslenkung durch den Staat," presented to Hitler June 1938, BAK R43 II/547; *Wer Ist's* 1935, 1213–14; BDC, Pietzsch file.

27. Notation by Krohn, 10 Aug. 1938, BAK, R41/652, Bl. 115.

28. Minutes of meeting, 10 Nov. 1938, BAK, R41/652, Bl. 130–35.

29. Krohn to all ministries and Krohn to Lammers, 17 Mar. 1939, BAK, R22/629, Bl. 140b–c.

30. RWM to all ministries, 28 Mar. 1939; Lammers to RAM, 15 Apr. 1939; BAK, R22/629, Bl. 143, 144.

31. Lammers to RAM, 8 May 1939, BAK R22/629, Bl. 147.

32. Hitler about himself, 2 Feb. 1942, quoted in Zitelmann, *Hitler*, 327; cf. Mommsen, *Volkswagenwerk*, 51–114, esp. 57.

33. Pietzsch to Lammers, 23 May 1939; analysis of Pietzsch memorandum, 7 Jun. 1939; Lammers to Pietzsch, 9 Jun. 1939; Pietzsch to Lammers, 25 Jul. 1939; Reich Chancellery to Pietzsch, 2 Aug. 1939, BAK R43 II/1559, Bl. 2–7, 8–9, 17–22, 25.

34. Copy of revised bill in RAM to all ministries, Jun. 9, 1939, BAK, R22/629; criticisms by Kerrl, Rust, and Frank in BAK R43 II/1559, Bl. 23, BAK R22/629, Bl. 152, 154, 157.

35. RMI to all ministries, 25 Jul. 1939, BAK, R22/629, Bl. 155.

36. Seldte to all ministries, 17 Nov. 1939, BAK R22/629, Bl. 162.

37. Ibid.

38. Correspondence of RAM and OKW, 28 Feb. and 1 Apr. 1940, BAK R22/629, 173; Ministerialdirektor Schultze-Fielitz of the RMBM to RJM, 31 Oct. 1940; minutes of RMBM meeting, 13 Nov. 1940, BAK R22/630, Bl. 1, 3.
39. RAM to all ministries, 22 Jan., 20 Feb., 26 Jul. 1941, BAK, R22/630, Bl. 14, 15, 22a.
40. Text of legislation and implementation ordinance (dated 26 Jul. 1941); Syrup to Lammers, 5 Sep. 1941, BAK, R43 II/1559, Bl. 74–81.
41. Evaluation of the bill, by either Kritzinger or Boley, for Lammers, BAK R43 II/1559, Bl. 90–91.
42. Reich Chancellery notations, 25 and 26 Feb. 1942, and various handwritten comments and instructions; Lammers to Bormann and Speer, 11 Apr. 1942, BAK, R43 II/1559, Bl. 93, 94–97.
43. Speer to Lammers, 14 May 1942, BAK, R43 II/1559, Bl. 99.
44. Ibid.
45. Boley or Kritzinger to Lammers, 23 May 1942, BAK R43 II/1559, Bl. 107.
46. Lammers notation, 2 Jun. 1942, BAK, R43 II/1559, Bl. 109.
47. Reich Chancellery's Dr. Killy to Lammers, 29 Jul. 1942, BAK, R43 II/1559, Bl. 110; Verordnung über die Behandlung von Erfindungen von Gefolgschaftsmitgliedern (see note 1; also: BAK, R22/630, Bl. 32).
48. Durchführungsverordnung zur Verordnung über die Behandlung von Erfindungen von Gefolgschaftsmitgliedern vom 20. 3. 1943 (see note 1).
49. See ch. 10.
50. Franz Redies, "Die Abgrenzung des Begriffs 'Betrieb' und der gebundenen und freien Gefolgschaftserfindungen im Sinne des Gefolgschaftserfindergesetzes," *Berichte der Arbeitsgemeinschaft "Fragen des gewerblichen Rechtsschutzes" und "Gefolgschaftserfinderrecht" des Amtes für technische Wissenschaften der Deutschen Arbeitsfront*, no. 1 (September 1941), in Priemer to Kühnemann, 11 Sep. 1941, BAK R22/630, Bl. 26; Ernst Weisse, "Die Bewertung von Gefolgschaftserfindungen," in DAF to RPA, 17 Feb. 1942, BAK R131/22; Paul Wiegand, "Die Erfindungen von Gefolgschaftsmitgliedern," *Zeitung des Vereins Mitteleuropäischer Eisenbahnverwaltungen* 82, no. 36 (Sep. 1942): 463–67; idem, "Die Erfindungspflege im Industriebetrieb," in NSDAP, AftW, ed., *Tagung der Reichsarbeitsgemeinschaft Erfindungswesen im Hauptamt für Technik der Reichsleitung der NSDAP. Amt für Technische Wissenschaften, am 11. Januar 1944 im Haus des Deutschen Rechts in München* (Munich, 1944), 31–48.
51. Redies to IG Farben Patent Committee, 3 Aug. 1942, Bayer 22/12; IG Farben, 60th Patent Committee meeting, 11 Dec. 1942, Bayer, 23/2.3.
52. In a commentary that Walter Beil, patent director at Hoechst from 1936 to 1946, prepared for distribution within IG Farben, the new, broader definition of the service invention was discussed in glowing terms: Walter Beil, "Die Erfindungen von Gefolgschaftsmitgliedern (Verordnung vom 12. Juli 1942 und Durchführungsverordnung vom ...," IG Farbenindustrie Aktiengesellschaft, Patentabteilung Frankfurt a. M.-Hoechst," e.g., 48–49, Hoechst, file "Angestelltenerfindungen."
53. Mansfeld remarks reported in *Berliner Börsen Zeitung*, no. 374, 10 Aug. 1942, BAK, R43 II/1559, Bl. 115.
54. This rule was, in fact, watered down in the DVO's final version, where state and military bureaucracies were allowed to appoint their own inventor trustees.
55. RJM note, Aug. 28, 1942; RPM to RMBM, Aug. 21, 1942, "Bemerkungen zum Entwurf einer Durchführungsverordnung zur Verordnung über Erfindungen von Gefolgschaftsmitgliedern vom 12. Juli 1942," pp. 2–3, BAK, R22/630, Bl. 35–37.

56. See ch. 13.
57. *Reichsanzeiger*, no. 107, 11 May 1943, p. 2, BAK, R131/22.
58. See ch. 12.
59. As Askania's Weisse put it, "It is hardly surprising that the large majority of employee inventions generated in a company should be considered 'inventions made in the course of professional duty' (*erfinderische Berufsleistungen*). Typically, these concern the solution of problems—either newly assigned or already being worked on—with regular professional means, by heads of the relevant departments and their designers and laboratory and development engineers, i.e., the inventive steps that are part of everyday technical work. The most important and pressing task of invention compensation is precisely this: to promote and encourage this routine work. The reason is that the success of a company depends far more on the high level of this everyday work—i.e., on the quantity and the quality of the inventions made all along the way—than on single inventions that represent top-class achievements. It is, therefore, not permissible to simply exclude from inventor compensation the inventions designers and development and laboratory engineers make in the course of their professional duty, on the grounds that these occupations were 'hired to invent' and their professional inventing work is compensated by their salary. To do so would be to fail to give encouragement and impetus to precisely those occupational groups to which one looks—and from which one may indeed expect—inventions in the first place." Ernst Weisse, *Kernfragen der Erfindungskunde für den Gefolgschaftserfinder* (Berlin, 1943), 59. On the parallels with new incentives and pay for blue-collar labor, see note 60.
60. This is not to deny that there were obvious parallels with Nazi social policy toward blue-collar labor. The new course was clearly also designed, in Siegel's words, to "bind the [employees] to 'their' company" and simultaneously encourage individual mobility through productivity, all of which was in keeping with broader Nazi and DAF social policy. Siegel, "Wage Policy," 30; idem, "It's Only Rational"; idem, *Leistung und Lohn*; there is a related, more nuanced criticism in Kirchhoff, *Patentwesen* (see note 48, ch. 10).
61. Kirchhoff, *Patentwesen*, 122–23.

Chapter 12

"APPROPRIATE COMPENSATION"

With the exception of the new definition of the service invention, the Speer ordinances were mostly bad news for the employers. Of course, they were not quite as bad as the bill the Labor Ministry had hoped to enact until Speer hijacked and changed it. But short of rolling back the legal clock to Weimar, which the industrialists tried but failed to do after the war, it is difficult to see how they could have gotten more. The distribution of employer and employee rights in the Speer ordinances was probably the closest to a real compromise the two sides could ever have achieved, even under the best of circumstances. It proved impossible, at any rate, to improve upon the quid pro quo at the heart of the Speer ordinances in the 1957 Act on Employee Inventions. Although both sides tried their best to make additional gains, the 1957 law retained exactly the same definitions of service and free inventions, and the very same principles for employee compensation, as those first laid down in the 1942 and 1943 ordinances.[1]

If employees were net gainers under the Speer ordinances, the fact remains that in practice almost everything hinged on the details of the as yet unspecified mechanism for determining the amount of the inventor reward. This was a problem that had first emerged in the Chemists' Contract of 1920. Both the latter and early draft legislation had merely mentioned the need for "appropriate compensation."[2] But "appropriate" was a term management and employee were unlikely to interpret the same way, and the remedy of taking the matter to court was not conducive to restoring a harmonious relationship between them.[3] When the Labor Ministry resumed work on inventor legislation in the fall of 1936, therefore, it decided to start spelling out the meaning of "appropriate."

The system the Labor Ministry opted for was fundamentally different from the profit sharing or royalty arrangements in the 1920 Chemists' Contract or the 1928 understanding between the BUDACI and the RDI. Those arrangements had rewarded only those inventors whose inventions actually made a profit—if the latter could be calculated, which was much more difficult in the electrical and mechanical engineering industry than in the chemical industry. As a consequence, the percentage of inventions that earned rewards was only a fraction of the total number patented. In line with the thinking of the Bureau for Technical Sciences and the Labor Front's philosophy of a "just wage" in general, the Labor Ministry chose criteria for calculating the reward that were largely independent of an invention's success in the marketplace and would reach a far larger group of inventors. It also sought to match the size of the reward to the level of intellectual achievement the invention represented—to accommodate the occasional scenario in which compensation could go all the way down to zero, but also to ratchet it up to whatever amount the particular situation merited.[4] Compensation would be based on the following four criteria: (1) the "usability of the invention" (i.e., its potential rather than actual value), (2) "the extent of the creative achievement," (3) "the employee's pay scale," and (4) "the employee's position in the company."[5] This approach survived all subsequent modifications and recurs, albeit with slightly different formulations, in the Speer ordinances, the wartime compensation guidelines, and their postwar successors. Current law specifies the reward criteria as "economic usability, the assignments and the position of the employee in the company, and the share of the company in the process of making the invention."[6]

The mere enumeration of the different factors to be taken into account was only the beginning. The criteria themselves, in turn, would have to be defined more precisely and converted into some kind of quantitative scheme, which could then be applied across the board, in all situations, companies, and industries. This was a low priority so long as enactment of the legislation as such was in doubt. But in the fall of 1941, as Todt and his aides in the Bureau for Technical Sciences were getting more and more involved in the rationalization and productivity effort, the compensation formula began to loom larger on the agenda. By the spring of 1942, as Speer was scheming to get his framework ordinance enacted, the issue became of paramount concern. Accordingly, engineer Ernst Weisse prepared a long paper on the subject, scheduled for presentation at a meeting of the Reich Working Group for Inventing in March 1942. The meeting did not take place until July, when Weisse's paper, "The Valuation of Employee Inventions," had circulated for six months.[7]

Trying to figure out the value of inventions before they had proved their worth in the marketplace was not an entirely new science. In 1936, Salge had come up with a spreadsheet-like system to arrive at the monetary value of inventions by independents. In 1928, when management and engineering labor were negotiating the terms of a voluntary patent agreement modeled on the Chemists' Contract, the BUTAB had devised its own method for calculating the value of employee inventions.[8] And companies that bought or licensed inventions from outsiders also had to have some way of arriving at a price. If Weisse's paper was therefore not without antecedents, it nevertheless surpassed in scope and rigor anything that had existed in the field until then. This is not to say that Weisse delivered a ready-made formula that inventor trustees or employers could simply plug in to calculate the size of the inventor reward. But he did lay a foundation for this, by carefully disaggregating two of the evaluation criteria, "usability" and "creativity," and creating a rank order for them.

The goal Weisse had set himself was to come up with a scrupulously fair and objective system of evaluating inventions across different companies and industries. The law could then be applied with justice and equality for all, and every inventor would get precisely the special compensation to which he was entitled, no more and no less. As he put it in his introduction, the "determination of the reward must be preceded by the kind of 'valuation' of the invention that makes it possible to compare with each other inventions of differing content and invention histories with differing facts." There was, according to Weisse, "an absolute necessity to assess according to the same criteria the value of the [different] employee inventions taking place within one company as well as those of different companies. Only then is it possible to arrive at a just solution of the compensation question—both within the individual company and in jurisprudence." Finding correct valuation criteria was, "therefore, the precondition for a uniform and objective treatment of the compensation question."[9] The author then proceeded to do precisely that—in a thirty-seven-page, double-spaced paper that was a model of lucidity and objectivity. Weisse sketched an eminently fair and judicious procedure, which went a long way toward giving the employee inventor his due but also took into account employer concerns. By carefully mediating between the interests of both groups, his work became the basis for all subsequent activity in the field.

There is, of course, considerable irony in this phenomenon. Here in the midst of World War II was engineer Weisse—a party member since 1937 and an eager adviser to perhaps the most criminal regime the world has ever known, a mid-level manager at a company that built gyroscopic and other guidance systems for rockets that would soon be produced by slave labor under the most inhumane conditions—puzzling over how to create a truly just and fair compensation mechanism for employee inventors, and doing so without any reference to ideology or politics one way or another. For Weisse, his task was strictly a technical one, the only difference from the usual engineering problem being that, this time it was a question of human engineering.[10]

Weisse began by analyzing the meaning of "usability." His point of departure was that usability could not be discussed abstractly but only in context, specifically that of the company where the invention had originated. If this put certain limits on the scope of usability, it still was considerably broader than actual application, because usability included the invention's potential as a blocking patent against competitors or as a storage patent to be used at some future time. A company's decision not to apply an employee invention in production, therefore, did not constitute grounds for classifying it as unusable and denying the inventor compensation.

A better way to approach usability was to consider an invention's properties in detail, in particular its various advantages and disadvantages with respect to improving a given function, streamlining a given manufacturing process, or expanding a given range of applicability. An invention might be an improvement in one or more areas but simultaneously a step back in one or more of the others. The resulting measures had to be balanced to arrive at an accurate assessment of usability. In addition, one had to take into account an invention's other technical properties. These included whether it filled a technical need; whether it could be made to work in practice; whether its introduction would complicate or interrupt the production line, and if so, by how much; whether it fit with the company's existing direction of technological development, or went against the grain. Then there were the various business factors to consider, including development and advertising costs and how the invention was likely to affect sales and profits. Weisse arranged the above factors in a table, so that for each criterion an invention earned a positive, negative, or neutral rating and thus, in the end, a type of usability score. He gave three examples from his own company to show how easy and revealing it was to work with his system. One particular invention—concerning a bearing mechanism for gyroscopes that used air pressure—was conceptually elegant, but when measured according to the table virtually unusable. But another invention—for improving the production of vacuum-based components of altimeters and other precision gauges—proved eminently usable, even though conceptually it seemed at first to be no particularly great achievement.[11]

Weisse next turned to the evaluation of creative achievement. In contrast to the usability table, which did not become quite as popular, his scheme for measuring creativity was immediately adopted. It appeared and reappeared in all the instructions and official regulations issued by the Ministry of Armaments between 1943 and 1945 and survived the war to become the basis for the current system of compensation calculation. Weisse began by debunking as a "dead end" the usual references to a company's internal state of the art, which employers typically used to deny employees special compensation for their inventions.[12] Instead, inventions should be classified according to the "degree of achievement" they represented. Some inventions were obviously nothing more than the result of "normal achievement." Others could be considered "overachievement" and yet others "special achievement." The point was to define these terms precisely enough so one could work with them objectively and quantitatively.

Weisse did so by breaking down the various elements involved in the inventive achievement. First there was the employee's assignment, which yielded five separate categories, ranging from "inventions of the first degree," i.e., those that resulted from systematic trials, to "inventions of the fifth degree," which resulted from a "conceptual solution of an entire problem that the inventor has posed for himself." Then there were the technological concepts and solutions used to make the invention, which came in three categories. The first and lowest of these represented concepts and means that were a normal part of the inventor's intellectual tool kit. The second included ideas and concepts from another of the company's divisions or departments, while the third comprised ideas that were entirely new, or new to the company. Finally, there was the position of the employee inventor in the company, ranging from senior project manager to blue-collar worker.

Weisse arranged the various categories in a series of tables, one each for the particular position in the company the inventor occupied. The basic scheme is presented in Figure 12.1.

Figure 12.1   Weisse's Schema for Measuring Creative Achievement

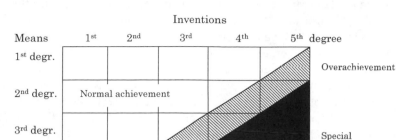

The diagonal lines moved either left or right, depending on the position of the employee in the company, in such a way that the black and gray areas were smaller for senior personnel, and larger for employees and workers lower down in the company's hierarchy of ranks. In Weisse's own illustrations, the "normal achievement" for a blue-collar worker included merely the box in the upper left-hand corner plus half of the boxes to the right and underneath. The normal achievement of a top-level technical executive, on the other hand, filled almost the entire frame, excepting the bottom corner of the lower right-hand box.[13]

Weisse's establishment of quantifiable measures of "usability" and "creativity" was a significant step forward. By itself, however, his work did not yet constitute a formula for calculating the inventor's reward. Devising such a formula was the achievement of engineer Josef Dapper, Riemschneider's right-hand man in the Bureau for Technical Sciences. A patent engineer who did not join the party until January 1941, Dapper worked up a first version of his formula in the spring and early summer of 1942. He took into account the level of pay if it was disproportionately high or low in relation to the employee's rank or position in the company, but ordinarily assigned it a neutral factor of one. He also refined and broadened Weisse's usability criterion by introducing an additional scale from one to five for the "patent-technical ranking" of an invention, to quantify the differences in licensing values, which were greater for some inventions than for others. Finally, he assigned values to each of the creative-achievement possibilities outlined in the Weisse scheme shown above, combining all of them in one very large table. Dapper then produced the following formula: $C = (L \times A \times P \times U \times PTR)/240$, expressed as a percentage of gross revenues. In this formula, C stood for Compensation; L for a Constant representing the customarily highest licensing or royalty fee, as a percentage of unit sales; A for the level of creative achievement; P for the level of pay; U for the usability score; and PTR for the patent-technical ranking. The denominator of 240 was the product of the highest possible scores of the various numerators. Assuming that L was 6, Dapper calculated the following inventor rewards. The first inventor was a blue-collar worker who received standard pay (P = 1). For this manual worker, the particular invention represented a special achievement (A = 8). The invention was used as a blocking patent of the first degree (U = 2) and had a patent-technical rank of four (PTR = 4). The reward was (6 x 8 x 1 x 2 x 4)/240 = 1.6% of the revenues generated by the invention. Another hypothetical inventor was an underpaid engineer with little responsibility (P = 2). For this worker the invention represented an over-achievement (A = 3), which the company applied in its production lineup (U = 3); it had a patent-technical ranking of 1 (PTR = 1). The reward was (6 x 3 x 2 x 3 x 1)/240 = 0.45% of the invention's gross revenues.[14]

Dapper presented his formula for the first time at the July 1942 meeting of the Reich Working Group for Inventing, where it met with a great deal of interest—as well as loud cries of rejection—on the part of some of the industrialists, especially Redies. According to one participant's report, Redies contended the figures were "far too high." Employees would henceforth base all their claims on the unrealistically high numbers of the formula and present "exaggerated demands."[15] The party's Department of Technology and the Ministry of Armaments ignored such complaints. The ministry published Dapper's formula, along with a detailed explanation and a summary of Weisse's work, in the September issue of *Deutsche Technik*, and Speer rapidly moved ahead with plans to implement the scheme. The Dapper formula, however, proved too controversial to be adopted as a mandatory part of his ordinances, at least initially. The military refused to adopt the formula at all. IG Farben, too, was highly critical. While it accepted the qualitative part of the guidelines for determining inventor compensation, the Dapper formula itself was said to be "unusable." Not only did it yield rewards that were too large, but it caused problems with regard to the wage freeze and necessitated a huge amount of extra bookkeeping and accounting, since instead of a standardized inventor bonus every single invention would have to be calculated separately.[16]

Not everyone rejected the formula out of hand. Redies, who reported on the matter to his IG Farben colleagues at a Patent Committee meeting in December 1942, mentioned that Krupp and

Siemens had a more positive view, while Askania was strongly supportive. He himself thought it might be possible to whittle down the amounts to the approximately 3% of an invention's profit that had long been standard in the large chemical companies. But Patent Committee chairman August von Knieriem ordered Holdermann and Mediger to draft a letter to Dapper, outlining the concern's objections and seeking modifications or some kind of voluntary compliance, short of having to revamp the concern's accounting procedures.[17] Dapper's reply, which was discussed at the next Patent Committee meeting in February 1943, failed to satisfy Knieriem, who ignored Redies's warning that continued opposition could only lead to harsher regulations. He told Mediger to come up with a different, more usable scheme, while Beil in his internal commentary on the ordinances devised all sorts of ways to circumvent or minimize the effects of the compensation rules.[18]

It is not known how other companies reacted to the Dapper formula, which was modified in minor ways before being put into effect as a semi-official guideline in the spring of 1943.[19] However, compliance was undoubtedly unsatisfactory from the point of view of its architects. For that reason, and to give inventors even greater incentives, Speer in September 1943 obtained the Finance Ministry's approval to make the employee-inventor reward partially tax-exempt—on the condition of exactly following the Dapper formula and the bureau's guidelines. For all practical purposes this made the formula mandatory and forced employers to comply.[20] The earlier mood of defiance in the Patent Committee meetings of IG Farben, at any rate, had disappeared by the fall of 1943. Instead, managers such as Knieriem, Redies, Mediger, and Beil now discussed with profound frustration but also resignation all the things they had to do to achieve compliance, "although it causes an almost unmanageable amount of extra work."[21] But comply—and pay—they did, even as the Dapper formula was revised and refined several more times before the war came to an end in the spring of 1945.[22]

Notes

1. Act on Employee Inventions of 25 July 1957, Articles 1–12, in Bernhard Volmer and Dieter Gaul, *Arbeitnehmererfindungsgesetz, Kommentar*, 2d ed. (Munich, 1983), 2–5.
2. See ch. 2.
3. See ch. 5.
4. On the functions of matching reward and achievement in Nazi social and wage policy, see Hachtmann, *Industriearbeit*; Siegel, *Leistung und Lohn*; idem, "Wage Policy"; also literature cited in note 10.
5. RAM to various ministries, 6 Nov. 1936, draft bill of 6 Nov. 1936, Article 4, BAK R22/629, Bl. 41.
6. Act on Employee Inventions, Article 9, quoted in Volmer and Gaul, *Arbeitnehmererfindungsgesetz*, 4.
7. Ernst Weisse, "Die Bewertung von Gefolgschaftserfindungen," in DAF to RPA, 17 Feb. 1942, BAK R131/22.
8. Salge's table in Riemschneider, *wirtschaftliche Ausnutzung*, 40, 86–87; BUTAB, "Entwurf eines Tarifvertrages über die Rechte und Pflichten der Angestellten aus ihren Erfindungen," SAA, 11/Lf237/238.
9. Weisse, "Bewertung," 1–2.
10. For similar approaches in other areas of DAF social policy planning: Smelser, *Robert Ley*, 149–217; idem, "Sozialplanung," 71–92, esp. 77–83; Sachse, *Siemens*, 59–77, esp. 76; Siegel, *Leistung und Lohn*, 13–21, idem, "Wage Policy."
11. Weisse, "Bewertung," 2–21.
12. Ibid., 24.
13. Ibid., 21–37.
14. Josef Dapper, "Die Gefolgschaftserfindung und ihre Bewertung, *Deutsche Technik* 10 (Sep. 1942): 372–76.
15. Report on meeting of RAG Erfindungswesen of 31 Jul. 1942 by RPA's *Dr.-Ing.* Schuster, 4 Aug. 1942, BAK R131/22.
16. IG Farben, Report on 60th Patent Committee meeting, 11 Dec. 1942, Bayer 23/2.3, vol. 3.
17. Ibid.
18. IG Farben, Report on 61st, 62nd, and 63rd Patent Committee meetings, 25 Feb., 9 Apr., 22 Jun. 1943, Bayer, 23/2.3, vol. 3.
19. Discussion and analysis of the 1943 version of compensation guidelines in Eduard Reimer, *Das Recht der Angestelltenerfindung: Gegenwärtiger Rechtszustand und Vorschläge zur künftigen Gesetzesregelung* (Berlin, 1948), 31–46.
20. Ruling by RFM, 10 Sep. 1943 (*Reichssteuerblatt* 1943, 701), in "Richtlinien für die Vergütung von Gefolgschaftserfindungen. Aufgestellt vom Reichsministerium für Rüstung und Kriegsproduktion, dem Hauptamt für Technik der Reichsleitung der NSDAP und der Reichsgruppe Industrie," RMRK's Gotthold Baatz to RJM, 17 May 1944, BAK, R22/630, Bl. 77–77a, 10; Volmer and Gaul, *Arbeitnehmererfindungsgesetz*, 46.
21. IG Farben, Report on 65th Patent Committee meeting, 26 Oct. 1943, Report on 66th Patent Committee meeting, 6 Dec. 1943, "Niederschrift über die Steuerbegünstigung der Erfindervergütungen in Frankfurt a. M., 15 Dec. 1943, Bayer, 23/2.3, vol. 3; Knieriem to Dr. Binder of the tax division of the Reichsgruppe Industrie, 29 Nov. 1943, Hoechst file "Steuervergünstigung bei Erfindungen."

22. On paying: there was a testy dispute between Dr. Karl Wilke, who synthesized the field-gray color of the German army's uniforms, and Hoechst management about the size of the inventor reward, 20 Dec. 1943 and 6 Jan. 1944, Hoechst file, "Verwaltung, Vertraulich 1936–1944, Erfindervergütungen." Whether the electrical and mechanical engineering industries were as conscientious about paying as the chemical industry is unclear. One postwar inventor advocate asserted the latter "hardly changed their methods." Stephan Deichsel, "Betrachtungen der Erfinder zur Reform des Erfinderrechts für Arbeitnehmer," p. 7, BA-Zwi, B141/2792, Bl. 28–39. On revisions: The final version of the wartime compensation guidelines and Dapper formula were "Richtlinien für die Vergütung von Gefolgschaftserfindungen. Fassung vom 10. Oktober 1944," published in *Deutscher Reichsanzeiger und Preußischer Staatsanzeiger* 271 (5 Dec. 1944), BAK, R131/22; reprinted in Reimer, Schade, and Schippel, *Recht der Arbeitnehmererfindung*, 670–80; analysis in Reimer, *Recht der Arbeitnehmererfindung*, 2d exp. ed. (1951), 29–40. The guidelines remained in effect until July 1959, when the BAM issued modified rules somewhat more advantageous to employees.

Part IV

THE POSTWAR LEGACY

Chapter 13

THE POLITICS OF INVENTING AFTER 1945

May 1945 was a crucial turning point in modern German history, but it was not the famous "hour zero," the fresh start of a new society unencumbered by its troubled past. Instead, the combined effect of Nazi social change and German military defeat was to create—at least in the western zones of occupation and Federal Republic— conditions for the gradual development of a very different political culture and society. The country's true break with its past, therefore, emerged only slowly, against a background of substantial continuity in social, economic, legal, and administrative institutions.[1]

Continuity was the hallmark also of inventor policy, the patent system, and their effects on West Germany's technological culture. This is not to say inventor policy and the patent system in the Federal Republic were not modified to make them compatible with parliamentary politics and the "social market economy." But the changes were relatively minor and nothing like the radical restructuring undertaken in the German Democratic Republic. The latter completely discarded the inherited Patent Code, replacing it with a Soviet-style "Inventor- and Innovator Law" that made Communist-led inventor trusteeship and concern with blue-collar "innovators" (*Neuerer*) its chief organizing principles.[2]

Needless to say, continuities in patent and inventor policy in the West were embedded in a radically new context and formed but one part of a whole series of initiatives aimed at resuscitating the economy and stimulating growth. At its most general level, the new climate was characterized by a decision in favor of the "social market economy," essentially economic liberalism in a free market system, modified by policies designed to provide a necessary minimum of social security and political consensus. Insofar as the "economic miracle" of the 1950s and early 1960s was the product of this choice, it was an unqualified success. Its preconditions included not merely a largely intact capital stock, inherited from the Nazi regime's investment in industrial capacity during the late 1930s and early 1940s, but also a highly qualified, plentiful, and inexpensive labor supply, owing in large measure to the millions of refugees fleeing the Soviet-occupied parts of Germany. Against this background, CDU/CSU-led governments under Konrad Adenauer and Economics Minister Ludwig Erhard implemented measures to keep inflation low, encourage savings and investment, reduce consumption, promote competition, lessen reliance on cartels and monopolies, provide export subsidies, and reintegrate German industry into the world economy.[3]

Policies toward labor closely meshed with the general strategy of encouraging high growth and productivity while maintaining social peace. The 1951 Act on Co-Determination in the mining, iron, and steel industries, and the 1952 Code on Factory Constitution (*Betriebsverfassungsgesetz*) were milestones in this respect. They helped to integrate the trade unions and to "create a climate for much more effective cooperation between the two sides of industry than existed elsewhere in Europe."[4] At the same time, the government continued policies designed to encourage individual productivity and performance—policies whose origins went back to Weimar and which had been a crucial instrument of National Socialist labor policy as well. These centered on measures such as the highly differentiated wage scales first introduced in the metals industry in 1942, which not only rewarded individual merit but also counteracted class solidarity. They included incentives to acquire more training, which would facilitate upward mobility, as well as the revival of the labor rationalization movement and the science of work, whose values and criteria rapidly became the "norm of action" after 1945.[5]

The decision to retain large parts of Nazi inventor policy was an integral part of the above strategy. The latter was characterized less by dogmatic liberalism, intentional "atomization," or dehumanizing rationalization than by eclecticism, pragmatism, and political calculation.[6] A further, crucial consideration arguing for retaining inventor policy was an acute awareness of the loss of all foreign patents, licenses, and trademarks, which the victors of World War II had confiscated, and the need to replenish this intellectual capital as quickly as possible.[7] West Germany had become a net importer of technology (patents, licenses, management strategies), mainly from the United States.[8] Recovering the country's high technological standards and reducing the outlays for acquiring foreign technologies were part of the context that explains why the Speer ordinances of 1942 and 1943 became the Act on Employee Inventions of 1957.

Salvaging Nazi Inventor Policy

In the immediate aftermath of the war, and before any kind of patent or inventor reform took place, there was a period of inactivity as occupiers and Germans alike tried to figure out the future shape of the country. It was not until these larger issues had been decided and the outlines of separate West German and East German states begun to crystallize that patent law and inventor rights reappeared on the agenda. This happened in the fall of 1947, more than two and a half years after the Patent Office in Berlin had closed its doors. In October 1947, the British and American military governments ordered the reestablishment of a separate patent office in the western zones, as part of the broader effort to revive economic growth and technological progress in their sectors.[9] It took two more years before a new West German Patent Office in Munich finally opened its doors in October 1949, and even longer before the customary patent examination made its comeback. To give inventors an opportunity to establish their priority in the interim, however, two temporary "receiving stations," one in Darmstadt and the other in Berlin, began taking patent applications again in October 1948.[10]

Reconstruction of the patent system in the western sectors was greatly facilitated by an early decision to purge the 1936 code of its references to the National Socialist regime but otherwise to make no substantive changes. Expediency and the desire to return to normalized conditions as quickly as possible played a large role, but so did the argument that the 1936 Code was, in fact, a fine piece of legislation, one that had nothing specifically National Socialist about it. As early as October 1945, Fritz Lindenmaier, former presiding judge for patent cases at the *Reichsgericht* in Leipzig, contended that the 1936 patent reform had "nothing whatsoever to do with any kind of National Socialist ideas." A superior court judge and professor of patent law in Hamburg after the war, Lindenmaier had entered the Nazi party in May 1937 and joined Riemschneider's Reich Working Group for Inventing in early 1942. From 1951 to 1953 he served on the Federal Republic's Supreme Court in Karlsruhe.[11]

The National Socialists, Lindenmaier argued, had spread all sorts of wild propaganda about their great patent reforms. "In truth," he wrote, the law "brought relatively little that was new and certainly nothing of fundamental significance." The few innovations of 1936 centered on the effort to "strengthen the legal position of the inventor with respect to personal and material rights." The other changes had to do with defining the proper balance between the inventor's rights and those of state and society and with minor procedural matters. Far from bearing the imprint of National Socialist ideology, said Lindenmaier, these reforms were merely "the codification of ideas integral to the conceptual world of patent law as such and already under discussion in professional circles before 1933." There could be "no objections, therefore, to look upon the relevant paragraphs ... as continuing to remain in effect." The same was true with respect to Speer's ordinances of 1942 and 1943: they were "more advantageous to the employee and guaranteed greater certainty in the protection of his rights than the collective bargaining agreements" of the Weimar era. Lindenmaier concluded that, "basically, the solution adopted in the ordinances is quite acceptable and deserves to be kept in force."[12]

Lindenmaier's argument had a certain basis in fact. It was true that the pro-inventor ideas implemented by the Nazi regime had been around since the turn of the century. It was also true they had been part of the discourse in intellectual property law long before 1933. But so were most ideas and currents the Nazis appropriated and implemented. National Socialism was eclectic but not random. The point was not that some proposal or notion must have originated after 1933 to qualify as typically National Socialist, but that Hitler's regime opted for choices and decided on directions that the Weimar Republic and the Empire had always rejected. It is unlikely that any other German government—either before or since—could ever have been as decisive in institutionalizing notions of technological romanticism in the Patent Code and inventor law as was the Nazi dictatorship. In that sense, the reforms of 1936 and 1942/43 had everything to do with National Socialism.

Lindenmaier and the many others who argued like him after the war knew this better than most. If they nevertheless argued differently, it was because they believed the Nazi inventor reforms represented social progress, which should be rescued for the postwar world. Conversely, those who hoped to dismantle the reforms tended to vilify them as the essence of National Socialist abuse. Not surprisingly the Speer ordinances, which directly intervened in the relationship between employer and employee, bore the brunt of such attacks. The 1936 Patent Code, in contrast, though it provided the theoretical underpinning of inventor entitlements, came away largely unscathed, with no change in the inventor's-right principle.[13]

The only noteworthy fact associated with transfer of the Patent Code from the Third Reich to the Federal Republic related not to its content but to the Patent Office's move from Berlin to Munich. The relocation caused much bitterness in Berlin and its large community of patent experts, as did the appointment of Eduard Reimer to Patent Office President in 1949. Reimer, a respected patent attorney in Berlin, had never joined the Nazi party. Unlike most of his professional colleagues, he had always kept the most lucrative corporate clients at arm's length and emerged after 1945 as an outspoken advocate of protecting the inventor rights of employees. Reimer initially was among the ringleaders opposing the departure of the Patent Office from Berlin. But the impossible conditions associated with the city's blockade in 1948–49 and the tempting offer to head up the new Patent Office won him over. Reimer's desertion, in addition to the physical removal of the relevant databases and the systematic recruitment of former Patent Office personnel—many of them former Nazi party members—broke the Berlin resistance. The city retained only an auxiliary patent office and responsibility for the so-called "old patent rights," i.e., all the questions pertaining to applications filed and patents issued before Germany's unconditional surrender on 8 May 1945.[14]

In contrast to the transition of the 1936 code, which was finalized before the Federal Republic came into existence, adapting Speer's 1942/43 reforms became a tortuous struggle. The wrangling lasted more than a decade and ended only in May 1957, when the Bundestag finally passed the employee-invention act. Until then the Speer ordinances—less the parts that involved National Socialist institutions such as the Labor Front, the Bureau for Technical Sciences, and compulsory inventor trustees—remained in effect.

On its face, the 1957 act was impeccable testimony to the scrupulous fairness and delicate equilibrium between the interests of employers and employees that came to characterize the political economy of the Federal Republic. The thirteen brief paragraphs of Speer's dictatorial 1943 ordinance had been replaced with forty-nine elaborate articles, while an impressive official commentary set forth the different theoretical underpinnings of the new law. Instead of justifying special compensation for employee inventions with reference to unique, personal ingenuity and artist-like creativity—the uncertain results of which were said to be beyond the grasp of any employment contract—the 1957 act was based on the "monopoly principle." According to the monopoly principle, the employee was entitled to special compensation for patentable inventions assigned to the employer, not because of exceptional achievement on his or her part, but because the patent gave the employer a monopoly right in the market place. Monopoly yielded profits beyond the company's ordinary rate of return, and so it was only right that the windfall be shared with the person(s) responsible for it.

Beyond its new theoretical justification, the 1957 act differed from its immediate predecessor in several other points. The Nazi party–controlled mediation system between employer and employee in the 1943 ordinance was replaced with an impartial board of mediation permanently attached to the Patent Office. The wartime rule that excluded nonpatentable suggestions for technical improvements from special compensation was modified to include some of them: technical suggestions that resulted in the employer gaining a monopoly-like economic advantage now entitled the employee to compensation, just as did patented inventions. The Speer ordinance's rigid mechanism compelling the employer to choose between claiming an invention (which triggered a duty to file a patent application) and declining it (which made the invention free) was replaced with a more flexible rule.[15] Finally, the new guidelines for calculating the size of the inventor's financial reward, though not, strictly speaking, part of the 1957 act, were somewhat more advantageous to the employee than the October 1944 version of the Dapper formula.[16]

The changes represented only marginal differences from the wartime regulations. As one of the law's co-authors put it in 1955, the government's bill had "adopt[ed] the basic design" of the 1943 implementation ordinance.[17] Considering how minor the differences really were between the Speer ordinances and the 1957 legislation, why did it take more than ten years to produce this unremarkable result? The answer is that management and engineering labor were still unable, after all that had happened since the early part of the century, to find common ground. To be sure, they were not as far apart now as in the Empire and the Weimar Republic, at least not on the surface. The tone, mood, and language after 1945 were generally conciliatory rather than combative. The talk was of "social partners," who were merely working out the arcane points of a very complicated matter. With a few exceptions, both sides professed adherence to the principle that employees were entitled to special compensation for inventions assigned to the employer. Management abandoned its attack on the Patent Code's inventor principle, on which this right was based. But the appearance of harmony was deceiving. In truth, labor and management continued to jockey for advantage, sometimes on the smallest of points. With great tenacity they tried to swing in their own favor the delicate balance that the Speer ordinances had so indelicately imposed. The only result of those efforts was marginal change, not a decisive victory for one side or the other. In the end, moreover, it was the ministerial bureaucracy, in conjunction with a unanimous Bundestag vote, that once again imposed a solution on the "social partners."

The Past Revisited

Preparations for transferring the Speer ordinances to a new West German state began in October 1947. The Economic Council in Frankfurt called for denazification only, without change in the substance of the laws. This was more easily said than done, as the inventor regulations were deeply enmeshed in Nazi institutions. The trusteeship system and the mechanism for resolving management-inventor disputes, for instance, were functions of the party and Labor Front. Adaptation was therefore not possible without making substantive changes, no matter how often it was asserted that the core of the wartime inventor measures had nothing to do with Nazism.[18]

Chaired by Lindenmaier, the committee in charge of revising the ordinances began by soliciting input from the various parties.[19] The result was a large catch of opinions, drafts, and recommendations. Some portrayed the inventor measures as the essence of National Socialism and called for discarding them without further ado. Compulsory, party-appointed inventor trustees, mediation by the party's Bureau of Technical Sciences, and the Dapper compensation guidelines, argued former IG Farben official Harald Mediger, made the law an instrument of anticapitalist totalitarianism, driven by socialistic visions of centrally directed technology steering. "Its most energetic architects being leading functionaries of the Labor Front, the outcome was a peculiar blend of hollow pathos and crass materialism that characterizes all totalitarian proclamations, both of that time and today. Its most astounding aspect was the blunt-edged concept of a formulaic calculation for the monetary value of all technical creations, from an improved potato cooker to synthetic rubber ... There should be no doubt ... that in the new Germany, which is to become better, one cannot allow this particular piece of National Socialist legislation to be dragged along. Absolutely nothing of value will be lost." Instead of adapting the law, Mediger concluded, policymakers should "make sure that the company and its management once again have the full and unlimited right to dispose of the employee invention in every respect. Under no circumstances are any parts of this right to be separated out, to create therewith compensation for the inventor."[20]

Frontal attacks such as Mediger's and explicit references to the inventor ordinances' paternity were the exception.[21] Most other commentators, having no stomach for a return to Weimar conditions, were in favor of keeping a reformed version of the legislation on the books. They recognized the ordinances' potential to spur technological progress and bind an important constituency to the new political constellation. The most objectionable aspects of the Speer ordinances were, in fact, already gone. Centrally directed technology policy by way of information sharing through the inventor trusteeship program had disappeared along with the Department of Technology and other Nazi institutions. Without the latter, inventor trustees at company level were no longer a threat to private control and might well be beneficial. Labor and its experts, at any rate, tended to favor retaining inventor trustees in the factories, though they conceded to make them voluntary, subject to collective bargaining.

Agreeing in principle to adapt the legislation was not the same as agreeing on the specifics of how to do so. The most serious problem in this regard had nothing to do with finding substitutes for Nazi institutions but went to the heart of the compromise imposed in 1942/43: which inventions must the employee assign to the employer and under what circumstances? One of the first proposals, by patent director Werner Cohausz of Vestag, brought back the company invention, employed an even broader definition of the service invention than in the 1943 ordinance, and limited employee compensation to a small number of inventions. Other proposals stuck more closely to the Speer regulations, while yet others changed them in favor of the employee.[22]

Despite such disagreements, initial signs pointed to the likelihood of compromise. Representatives of the trade unions and the employers in the British sector met in early 1948 in Braunschweig, where they hammered out a tentative agreement on the basis of recommendations by Lindenmaier's committee. The Braunschweig agreement and resulting draft bill were decidedly more employee-friendly than the Speer ordinances.[23] It is not clear why the management representatives involved in the Braunschweig talks agreed to give away so much, but it may have been due to British influence,

which went in the direction of supporting labor. Whatever the precise circumstances, Lindenmaier's optimism that it would be possible to transfer the wartime inventor regulations on the basis of this
tentative agreement proved illusory.

Almost before the ink was dry, complaints started coming in.
Patent attorney Hans-Gerhard Heine of Metallgesellschaft, the
Frankfurt metallurgical research and trading conglomerate, was in
the forefront of those who opposed the Braunschweig plan. Heine
pointed out that the Braunschweig draft was not a simple transfer,
as the official mandate called for, but a substantive reform of major
proportions.[24] Similar objections from other companies and business interests, mostly in the American sector, indicated the Braunschweig agreement was in trouble. But the Frankfurt administration,
relying on Lindenmaier's assurances, moved ahead and recruited
Reimer to prepare final editorial revisions.[25]

Meanwhile, the trade unions also jumped ship. Joseph Brisch,
former mayor of Cologne and the unions' responsible official in the
British sector, demanded that the Braunschweig plan first be
approved by the rank and file. In November 1948, Brisch's successor Emil Bührig announced the compromise was unacceptable. In
February 1949 the Trade Union Federation presented a bill of its
own.[26] At the same time, the newly formed German Trade Union of
White-Collar Employees (Deutsche Angestellten-Gewerkschaft,
DAG) demonstrated the reemergence of separate white-collar
consciousness with a draft of its own.[27] The employers now also
objected loudly to the Braunschweig plan, calling it "intolerable."[28]
Lindenmaier clung to the hope that the Economic Council would
pass the bill before the Federal Republic came into being and the
legislative process returned to its customary, cumbersome pace. But
by March 1949 it was clear the efforts to transfer the inventor ordinances during the interregnum had failed.[29]

The task of reviving the legislation fell chiefly to the new Federal
Ministry of Justice, in particular its Department for Business and Economic law.[30] The head of that department was Ministerialdirektor
Günther Joël.[31] The official directly responsible for employee-inventor legislation was Senior Councillor Kurt Haertel.[32] Haertel, whose
career in patent and inventor law started in 1949 and 1950 with
rather pathetic calls for help and confessions of ignorance, went on
a self-taught crash course that would eventually take him to the very
top of his profession. He became the principal author of the 1957 Act
on Employee Inventions and in 1963 was appointed president of the
Patent Office, a position he retained until his retirement in 1975.[33]

The first thing Haertel did after being assigned responsibility for employee-inventor legislation in the summer of 1949 was to recruit an assistant. He chose Bernhard Volmer, a lawyer with the Hesse Ministry of Labor and Economics in Wiesbaden who had strong pro-inventor leanings.[34] Together, the two men began familiarizing themselves with the issues. They sent Reimer a first, tentative design only in November 1950, some eight months after the CDU fraction in the Bundestag had made a motion urging the government to present its inventor bill in a timely fashion.[35]

In the meantime, the various interest groups and professional organizations drafted new bills of their own or reprinted old ones.[36] The differences in the bills were predictable: the employers proposed a broad definition of the service invention, demanded reporting of all employee inventions, refused compensation for nonpatentable suggestions, wanted a right of first refusal for free inventions, and sought the right of limited assignment.[37] For their part, the employees wanted a narrow definition of the service invention, demanded restrictions of the reporting duties, insisted on compensation for adopted improvement suggestions, objected to the employers' demand for a right of first refusal, and opposed the right of limited assignment. The proposals of the managerial employees and the Green Society fell somewhere in between, the former being closer to the trade union plan and the latter nearer the employer proposal.[38]

In addition to all the energy spent on preparing bills, the parties also found time to criticize each other. The attacks and counterattacks showed how little real progress had been made since the bitter fights of the 1920s. In the fall of 1950 Volmer published an article in which he explained the need for special compensation for employee inventions with reference to the public interest in creating material incentives for those who generate new technology. It should be remembered, Volmer argued—not very differently from the relevant paragraphs in *Mein Kampf*—that "only relatively few human beings have the educational background and are so endowed that they bring forth technically creative achievements. Every step forward in civilization is dependent on them. A nation's entire industry would come to a standstill in relatively short order if new, technically creative impulses did not spring time and again from a few human beings."[39]

Volmer's observations provoked an immediate reaction from Metallgesellschaft's Heine. "If one sees things in that light," he countered, echoing Fischer's *Company Inventions*, one should, indeed, "place much greater value on the inventor." But on the basis of "decades of experience, we in industry do not see things that way at all."

> Practice has taught us that it is only the very rare exception when a unique creative spark springs from a specially endowed human being. In the vast majority of cases inventions are indeed the products of a talented and creative intellect, but typically the impulse comes more or less by necessity from problems of the environment. There is a general consensus that there is hardly a single invention that, sooner or later, would not also be made by a whole array of other inventors if the first inventor had not made it. The breadth of scientific and technological research today enables a vast circle of technicians to solve the problems they are assigned, and it is mainly a question of precisely who receives the assignment. The invention is therefore in most cases not the work of an individual person, but the ripe fruit of development.

Heine conceded that the "so-called creative achievement of the real inventor" should "not go completely empty-handed." But it would be best if the government were to de-emphasize its focus on the individual. It should think more about the "work's community," the need for unrestricted collaboration of production engineers and laboratory researchers, and the danger that individual employees would be tempted to try to go it alone to get their reward, which in turn would "infect the works."[40]

If the argument was the same as in the 1920s and 1930s, the legal and political situation was not. Employees now had the right to special compensation, and there was little chance the government would eliminate it. The best the employers could hope for was some marginal changes in their favor. In the event, Haertel and Volmer chose to go the other direction. Their bill, completed in May 1951, steered a careful middle course between the different demands, though it tipped the balance slightly in favor of the employees. The reactions were predictable. While labor adopted a wait-and-see attitude, management—especially in the metals and electrical sector—declared the bill unacceptable. Director Paul Ohrt, Wiegand's successor as chief patent expert at Siemens, attacked the pro-employee changes as a misguided concession to labor, which could only breed disillusionment among employees and force employers into lowering wages and choosing different patent strategies. This, in turn, would harm not the companies, but society at large and thus in the first instance the mass of workers and employees themselves.[41] Heine was even more critical. The bill contained so many problems that it would "probably take very, very difficult and protracted discussions before a general consensus is brought about—if it is not simply abandoned." Haertel, he wrote, had strayed far from the realm of compromise, making the bill "once again an object of combat across the entire line."[42]

Haertel and Volmer ignored Heine's warnings. Justice Minister Thomas Dehler (FDP) approved the bill as drafted; the cabinet adopted it in January; the Bundesrat voted for it in February 1952.[43] Then progress came to a halt. Management bombarded the Bundestag with objections. Pro-management experts such as Beil sought to resuscitate the company invention. Mediger told Dehler that Speer's 1943 ordinance was, on balance, a great deal better than the current bill.[44] The Union of White-Collar Employees, which had never accepted the compromises discussed by the other parties, called for adoption of its own bill.[45] The professoriate came out against it, demanding to be exempted from the law on the grounds that academic freedom would be violated if not all inventions by academics were automatically free.[46] The Trade Union Federation then countered with new proposals of its own, as did the Union of Managerial Employees.[47]

Fearing the government might reconsider, chemist Stephan Deichsel, a former BUDACI member and now one of the principal negotiators for the Trade Union Federation, urged Haertel to stand firm. "Our social partner must get used to the idea that the ordinances of 1942/43 canceled decades-old exploitation of the inventors," Deichsel wrote. "There is no doubt that in the future more must be paid, and more often, than has been the case until now." In a similar letter to the chairman of the Bundestag's patent-law committee, the Social Democratic lawyer F. W. Wagner from Ludwigshafen, Deichsel described Speer's 1943 ordinance as a "revolutionary act, which regardless of its legal form broke radically with a decades-old legal theory." Based largely on ignorance, Deichsel continued, the old theory had resulted in a "kind of intellectual serfdom." The first ones to "cast off those bonds" had been the chemists with their 1920 Chemists' Contract, which formed the basis of all subsequent inventor law. Then, "with the Speer ordinance of 1943, the 'hour of liberation' struck for all employed inventors." Although he had never joined the Nazi party, Deichsel compared the wartime measure to the Stein-Hardenberg reforms. Like the great peasant emancipation of the early nineteenth century, he said, the Speer ordinance had been the product of emergency in war, and also like its illustrious forebear, it had unleashed a reactionary drive to "cut back the rights that the ordinance of 1943 has brought us." [48]

Considering that the issue in 1952–53 was not about dismantling but about expanding employee rights, Deichsel's rhetoric should probably be taken with a grain of salt, though his comparison of the Speer ordinances with the Stein-Hardenberg reforms is telling. But neither this effort nor any of the other lobbying managed to break the deadlock in the Bundestag. The Federal Republic's first parliamentary session expired in the summer of 1953 without the bill having reached the floor.[49]

As soon as the second Adenauer government had formed, the Ministry of Justice resubmitted the bill. Since this only resulted in renewed criticism across the same broad front that had blocked progress before, the ministry decided to withdraw the bill and make revisions reflecting the wider political constellation—principally the massive electoral gains by the CDU/CSU. Volmer returned to his old job in Wiesbaden, while Haertel decided to pay closer attention to the employers' arguments. He got rid of mandatory compensation for nonpatented proposals and proposed an "unlimited right of limited assignment."[50] To prevent abuse, the employers' new rights were balanced by other new rules that reduced potentially adverse consequences for employees.

This time it was labor's turn to strike the alarm. The Union of Managerial Employees criticized the "180-degree turn" and expressed its "utter dismay" over the new course. The Union of White-Collar Employees rejected the changes out of hand. The Trade Union Federation spoke of the "grossest discrimination" against employees and "purposely preferential treatment" of the employer. It portrayed the new policy as a reactionary threat to individual liberties guaranteed by the Basic Law and argued that the Federal Republic's inventor legislation was "even worse than the authoritarian and coercive measures of Messrs. Göring and Speer." The Labor Ministry objected to the new draft as well, calling it a "complete hollowing out of the employee's legal rights." Labor Minister Anton Storch, speaking for the CDU's trade-union wing, called the new plan a "considerable deterioration from the employee's point of view, both as compared to the earlier bill and relative to current law." Initially Storch even refused to sign off on the draft.[51]

Once again, officials at the Ministry of Justice remained unfazed. In May 1955 they sent the bill to the cabinet, which approved it in early June. The Bundesrat followed suit in July, and the bill went to the Bundestag in August.[52] The relevant committees once more began reviewing the legislation in the midst of posturing by labor and management, which lasted into the spring of

1956. The more conservative political constellation and persuasive testifying by Haertel and Joël, however, prevented labor from making any headway.

Before the bill became law in May 1957, it underwent one final change in labor's favor. In the spring of 1956, Patent Office president Reimer launched a ferocious attack on the legislation and announced a totally different design of his own. This had two effects. First, Reimer succeeded in bringing together management and labor, both of which for different reasons opposed his ideas. Second, his recommendations focused attention on the exclusion of nonpatentable inventions from special compensation.[53] When this caused Joël to wonder whether invention-like ideas that "gave [the employer] a monopoly-like position in practice" should perhaps qualify for special rewards after all, the relevant Bundestag committee immediately embraced the suggestion and amended the bill accordingly.[54] Following some last-minute delays, the Bundestag unanimously approved the legislation on 3 May 1957.[55] The law took effect on 1 October 1957. Two years later the Ministry of Labor issued new compensation guidelines, including a method for calculating the employee's inventor reward that was more generous than the wartime Dapper formula but otherwise not very different.[56]

The Politics of Consensus and Technological Choice

In the 1983 edition of Volmer's *Commentary on the Employee-Invention Act*, Dieter Gaul, the current author, mentions that the 1957 law serves two purposes. First, there is the goal of establishing a "sociopolitically equitable balance" between labor and management with regard to "dividing the economic gain from employee inventions." The law is a great success in this regard, according to Gaul. It laid to rest a management-labor conflict that had raged ever since the turn of the century and put in place a small but important piece of the Federal Republic's vaunted consensus model. There is no question but that Gaul's observation is correct. The passage of the 1957 Act on Employee Inventions marked the end of the politics of inventing. It did so by replacing the reactionary modernism at the core of the pre-1936 patent system with the modernized and disguised romanticism of the inventor's right to special compensation. The rationalistic-bureaucratic notion of the company invention—the corporation's employees forming an automatic inventing machine—was replaced with the principle that such full rationalization of inventing cannot exist and that the human being who invents on behalf of his employer does something special—above and beyond that which can be agreed upon in advance in an employment contract—and is therefore entitled to special compensation. Clearly, this is one small escape from Weber's iron cage.[57]

The law's second and economic purpose, according to Gaul, is "even more significant than the sociopolitical dimension." This is the "encouragement of invention." With its export-oriented economy, the Federal Republic is highly dependent on a steady stream of technological progress, which calls for policies to "mobilize all economically available forces for further creative development." Special tax benefits for those who invent and innovate and state funding for research and development are two key measures to encourage technological progress. Another such instrument is the 1957 Act on Employee Inventions. It plays a crucial role in promoting the inventive energies of employees, who account for the overwhelming majority of all valuable inventions. The principal reason for the ever-growing predominance of the employee inventor in the generation of progress, according to Gaul, is the enormous cost of developing new technology nowadays, which is also why free inventors have been forced to the margins. By recognizing the employed inventor, providing financial inducements, and consciously choosing not to treat him as a "'stepchild of the nation,'" inventor legislation is one of the principal means of keeping German industry competitive internationally.

Gaul concedes it is difficult to give precise, quantitative measures of the law's beneficial consequences but nonetheless points to West Germany's high number of patent applications as evidence of its effectiveness. With some 53,500 patent applications in 1956 and 55,000 in 1955, he writes, the Federal Republic had almost twice as many as France and three times as many as Italy. The only nation surpassing West Germany at the time was the United States. The pattern continued after 1957 and repeated itself yet again in the statistics of the European Patent Office, where the Federal Republic with 29.1 percent of all filings in 1980 occupied first place.[58]

Is Gaul's positive assessment of the 1957 act correct? It makes intuitive sense that the law's mechanism of incentives and rewards should stimulate inventing and supply companies with a steady stream of ideas for innovation and product improvement. But much depends on how the system functions in practice, how it is enforced, and how well it is known and used by those at whom it is aimed. Quantifying its effect is for all practical purposes impossible.[59]

Despite these problems, I will conclude with a brief discussion of the long-term effect that inventor policy may have had on German society and technological culture.

The 1957 act and its wartime predecessor reflected a clear historical choice: to develop a highly elaborate system of assistance and support for employees, while doing nothing comparable for independents. The relative decline of the free, entrepreneurial inventor and the growth of organized research and development in the past century are phenomena that characterize all industrialized nations. This condition, however, is more pronounced in some countries than others, and Germany is a more extreme case than the United States. The question therefore is: To what extent might the marginalization of the free inventor in Germany also be a function of policy measures and choices, rather than strictly a consequence of the iron laws of history that are usually held responsible for it?

Before 1945, the only attempt to change Germany's technological culture in an "American" direction—to make it more susceptible to "radical inventions" and more hospitable to independent inventors—came from the motley crew of inventor crusaders who imagined that National Socialism would help them realize their dreams. Their ideas may have been rooted in a nineteenth-century, romanticized understanding of the role of inventors and creative genius. They may have been motivated by an amalgam of resentment and utopianism, which was common in many other areas of life as well during the 1920s and early 1930s. They may have been naïve to think that simply changing the patent system and establishing inventor assistance programs of one kind or another would suffice to achieve their objectives. But their diagnosis was by no means completely wrong.

Beginning in the last quarter of the nineteenth century, Germany developed a version of industrial capitalism whose features eventually combined to hinder a flexible embrace of new technological systems and timely adjustment to new structures of demand. Examples of this "conservative bias" include the relatively slow movement toward mass markets for automobiles, radios, and household consumer durables after World War I, and the relatively slow adoption of computer hardware and software, consumer credit, and efficient financial markets after World War II. Economic and business historians have analyzed the problem at great length. They point to phenomena such as the power of the great investment banks, industrial and agricultural protectionism, the political power of the industrial magnates associated with the first industrial revolution, exceptional levels of industrial organization and ownership concentration, heavy reliance on market controls such as cartels, insufficient entrepreneurial dynamism, and the dominance of the producer (of capital goods) over the consumer. The net effect was to produce a constellation that helps explain, among other things, the stagnation and "sick economy" of the 1920s.[60]

The problem of the 1920s, of course, was not just economic (and political, and social, and cultural). It also had had a technological dimension. First, there was the patent system inherited from the Empire, which sacrificed the independent inventor on the altar of big business and reinforced a tendency toward conservative inventing, i.e., a strategy of incremental change and gradual perfection of existing technologies in the context of the large bureaucratic firms that dominated the economic landscape.[61] The other side of this was a reluctance to make great leaps, adopt entirely new technological systems, or embrace radically new manufacturing methods such as mass production.

A second, closely related aspect was the primacy of "design culture" over the logic of American-style production efficiency. The primacy of engineering design was inextricably intertwined, not just with a history of relatively small and fragmented markets in the German context, but also with the need of many German engineers to demonstrate their standing as professionals with, among other things, a special talent for design and invention.[62] On the production side, the dominance of design culture translated into the long-standing German predilection for skilled labor operating universal machine tools, which could accommodate design changes and customized engineering installations far more easily than the specialized machine tools and less-skilled labor of American mass manufacturing. During the rationalization boom of the 1920s there was some movement toward adopting more efficient production methods, as in the well-known case of coal mining, for instance, but on balance the old system had too many advantages simply to be abandoned.[63] Nor was it abandoned during the Third Reich and World War II, despite another push toward mass production associated with phenomena such as the establishment of Volkswagen and Speer's "production miracle." The relative ease of making design modifications in Germany's sophisticated engineering products had and retained crucial advantages. It meshed with notions such as "qualitative superiority" and had powerful other benefits, in terms of both courting the home market and winning exports contracts. It also allowed for rapid conversion to the needs of a peacetime economy, which remained a key consideration for German industrialists throughout much of the war.[64]

To suggest that Germany's particular "technological tradition" possessed certain advantages that explain its long-term survival is not to say that it served the country particularly well in the 1920s, when economic growth stagnated and management responded with additional doses of caution, rigidity, and defensiveness.[65] In many sectors there were signs of "technological momentum," stagnation, and calcification, strongly reinforced by the wider legal and political framework. It was this disease that patent reformers and inventor crusaders such as Jebens, Riemschneider, Waldmann, Barth, and others set out to cure in the 1930s, with their plans for helping small and independent inventors and making the Patent Code more inventor-friendly. They partially succeeded with the Patent Code, but failed miserably when it came to creating a vibrant culture of technological progress driven by small independents, as witnessed by their frustration over the disappointing yield of worthless inventions and perpetual-motion machines by self-styled and crackpot inventors. Thus, under the pressure of war, the party's inventor activists shifted gears. After 1939 they devoted most of their energy to assisting employed inventors, along the lines first developed by the Austrian patent attorney Walter and subsequently integrated into Speer's economic war machine.

The Nazi inventor crusaders downplayed the significance of their new orientation, but there is no doubt that their new approach made the chances of solving the original problem even smaller than they already were. Instead of trying to promote radical inventions by outsiders, they now threw their weight behind fortifying conservative inventing in the existing high-technology triad of the machinery, electrical, and chemical industries. One might object that the lack of success in revitalizing independent inventing was itself proof of the utopian and reactionary nature of the ideas that motivated the policy, and that the era of the independent inventor had irretrievably passed when the Nazi inventor crusaders appeared on history's stage. The change of direction in 1939–40 could then be interpreted as a belated acknowledgment of an irreversible historical trend. This objection is valid only if rigid and mutually exclusive stages of historical succession—from the era of independent inventing to the age of the organized research laboratory—are the only possibility. In fact, however, technological progress receives its most dynamic impulses from the interplay of both independent and organized research.[66] One might therefore also argue that the initial Nazi measures for independents did perhaps not go far enough. To make Germany's technological culture more entrepreneurial, more dynamic, and less dominated by the large corporations would have required even more radical surgery, and along a considerably broader front, than the inventor crusaders were capable of undertaking.

In contrast, it proved quite feasible to implement policies that spurred technological progress in a direction and of a type benefiting the interests of large-scale industry as well as those of its technologically creative employees. Defining the compensated service invention broadly and the free invention narrowly, the Speer measures channeled technological creativity into existing industrial structures, rejuvenating and reinforcing the kind of conservative inventing that was and has, to a large extent, remained a hallmark of Germany's great engineering prowess. Once established, the pattern continued. The 1936 Patent Code and the inventor-compensation aspects of the Speer ordinances remained in effect after 1945. The 1957 act did nothing to change the basic thrust, which was to trade the right to special compensation for broad managerial control of employee creativity. In fact, it was precisely this central feature of the 1957 act—the virtual elimination of the employee's chance to do anything with even the nominally free invention—that critics found most objectionable.[67]

Protests went nowhere precisely because the tradeoff was widely perceived to be the only practicable way inventor policy could encourage technological progress. This was the fundamental lesson most experts thought they had learned from the endless struggles between management and employees. The only way to harmonize their interests and effectively promote innovation was to give the employer full control over the invention and the employee special compensation. Any other solution, including those he had himself pursued in the 1930s, Riemschneider commented in 1949, would "produce obstacles to intra-company exchange of information and, in the end, a slowdown of technological progress."[68]

Faith in the technological and sociopolitical benefits of employee-inventor legislation ran both deep and wide after 1945. In fact, it was powerful enough to overwhelm the only other initiative for stimulating invention during the 1950s that is documented in the sources. When patent and inventor issues resurfaced in the late 1940s, there was once again talk about doing something for the independent. Discussion was prompted by the cries for help from independents, who complained about the patent system just as they had in the 1920s and 1930s: the high costs of patent fees (which went up again in 1954) and tax treatment that compared unfavorably with the rules for employees.[69]

The clamor was such that in April 1951 Haertel asked Volmer to look into the problem. In December Volmer had a long discussion with a Dr. Glewitzki of the Hessian Ministry of Economics and Labor. A physical chemist who had just been to the United States to study American business and technology, Glewitzki told Volmer that "true inventions, which generate technological progress in leaps, do not exist anymore today." Technology had developed too far in too many directions for an independent to make any meaningful contribution. Modern machine tools and measuring apparatus were so sophisticated, and operated with such minute tolerances, that only large corporations or government-funded laboratories could afford them. Today's individual inventors, therefore, said Glewitzki, had been reduced to the species of "cigar box inventors and do-it-yourselfers." Even if they had good ideas, realizing those ideas was all but impossible. Glewitzki concluded with the advice that if any tax money was going to be spent promoting technological progress, it should go to fund research in the universities, which would yield much greater and more useful results than helping independents.[70]

Volmer was not persuaded. Shortly afterward he completed a long paper advocating broad government measures for independent inventors. In many ways, his report reads like an updated version of the same inventor mystique one finds in Eyth, Hitler, Jebens, and the early Riemschneider.[71] He quoted Eyth with approval and made a case for precisely the kind of assistance programs that the Labor Front and Nazi party had established in the mid-1930s.[72] Helping inventors, in particular the independent, was necessary, Volmer wrote, because the state had a responsibility to "utilize the technologically creative forces that are unquestionably present in the nation (*die zweiffellos im Volke ruhen*) but have not been fully exploited yet, and to create the next generation of inventors." Pessimistic about regaining and keeping former export markets with current technology, Volmer emphasized the "energy and inventive potential residing in our *Volk*" as the best way to "create new industries." The state should exploit this potential by establishing an institution to assist independent inventors and evaluate their ideas.[73]

Volmer's ideas found a surprising degree of resonance, even if in the end they went nowhere.[74] His best hope was the interest shown by Dr. Hennenhöfer of the desk for productivity enhancement and technology in the Ministry of Economics. Hennenhöfer made Volmer's proposal the basis of his own recommendations for state-supported encouragement of independent inventors. The fundamental reason this type of inventor policy was necessary, according to Hennenhöfer, was the inadequate receptivity of German industry to new inventions from outsiders. This failure had many different causes, but allowed for only one conclusion. It could not be denied that inventors were right when they made the accusation that, "for want of government interest and state support, creative technological talent in Germany has no chance to benefit the economy, or it benefits foreign countries to the detriment of the German economy."[75]

Hennenhöfer's recommendations were not followed up, at least not in the 1950s. The sources are silent as to why, but the reason undoubtedly had to do with the free-market orientation of Ludwig Erhard, the export boom, the astonishingly successful performance of West German industry at this time, the high levels of investment, and the gradual adoption of new technologies in plastics, synthetic fibers, petrochemicals, and motor-vehicle production—all of which appeared to justify the low priority given to state spending on R&D for industry.[76] Lack of interest in stimulating technological innovation for the private sector is demonstrated also by the early history of the Fraunhofer Society, which from the 1970s became the Federal Republic's principal agency for supporting technological innovation in industry. Founded in 1949 by Bavarian industrialists and officials, the Fraunhofer Society pursued some of the same goals as those articulated by Volmer and Hennenhöfer, and the Bureau for Technical Sciences before them. In the long run, its establishment was probably the most significant new departure for the Federal Republic's technological culture. But in the 1950s and during most of the 1960s, it struggled along, not so much helping others as needing help itself, and surviving on handouts from private industry and the state governments of Bavaria and Baden-Württemberg.[77]

In the end, then, there was little that was new about West German inventor policy in the postwar period. Instead, there was a great deal of continuity with structures established in the first half of the 1940s. True, centralized inventor trusteeship as an instrument of technology steering was dismantled, representing a crucial degree of liberalization. But with the exception of the Fraunhofer Society, independent inventors were also left to their own unenviable fate, much as they had been in Weimar—and in the Third Reich once the Bureau for Technical Sciences changed its priorities. Likewise, the industrial-relations aspect of the Speer ordinances was retained in all its essentials, and with that the effect of channeling technological ingenuity back into the firms whence it came in the first place.

This reinforced the country's predisposition to perfecting existing technologies, or conservative inventing, rather than going for entirely new technological systems.

The tendency to rely on the results of in-house inventing was further strengthened by the exemption from the employee-invention act that academics won in the 1950s. The professors, eager to disavow their entanglement with National Socialism, used the new freedom not to patent the results of their research, but rather to publish them in scientific journals. Since publishing by definition precludes patenting, the options of companies to adopt inventions from the outside were narrowed even more by this practice.[78] The overall result was a constellation that encouraged two closely related tendencies that have been at the core of West Germany's political economy for a long time now. On the one hand, there was the establishment of entitlements and social peace between management and labor in the country's industrial laboratories, forming an integral part of the Federal Republic's propensity for consensus politics and bureaucratic regulation. On the other, inventor policy also reinforced the technological vitality and manifest success of those "fulcrum" sectors—chiefly, machinery, machine tools, chemicals, automobiles, and electrical engineering—that had already dominated the era of World War II, and which were no longer new after 1945.[79] In sum, inventor policy also tied West Germany's technological destiny to mature and aging industries, without doing much to encourage the adoption of completely new ones.

Notes

1. Lothar Kettenacker, *Germany since 1945* (New York, 1997), 1–3; Müller, *Hitler's Justice*, 201–300; Gregor, *Daimler-Benz*, 1–2, 247–52; Ritter, *Transformation*; Volker R. Berghahn, "West German Reconstruction and American Industrial Culture, 1945–1960," in *The American Impact on Postwar Germany*, ed. Reiner Pommerin (Providence, 1997), 65–81, esp. 66–68.

2. On the "social market economy," see A. J. Nicholls, *Freedom with Responsibility: The Scoial Market Economy in German 1918–1963* (Oxford, 1994); GDR inventor policy: Hans Nathan et al., *Erfinder- und Neuererrecht der Deutschen Demokratischen Republik*, 2 vols. (Berlin, 1968); Martin Hartmann, *Die Neuererbewegung: Das betriebliche Vorschlagswesen in der DDR* (Cologne, 1988).

3. Overy, "Economy of the Federal Republic," 3–34; Braun, *German Economy*, 165–254; Kramer, *West German Economy*, passim, esp. 177–219; Stokes, "Wirtschaftswunder," 1–9; Volker R. Berghahn, *The Americanization of West German Industry, 1945–1973* (Cambridge, 1986).

4. Overy, "Economy of the Federal Republic," 32.

5. Siegel, "It's Only Rational," 53; idem, "Wage Policy," 21–22; Overy, "Economy of the Federal Republic," 33.

6. Braun, *German Economy*, 181; Torsten Oppelland, "Domestic Political Developments I: 1949–69," in *The Federal Republic of Germany since 1949*, ed. Klaus Larres and Panikos Panayi (London, 1996), 74–99, esp. 79–86.

7. On confiscation of intellectual property, see John Gimbel, *Science, Technology, and Reparations: Exploitation and Plunder in Postwar Germany* (Stanford, 1990).

8. Braun, *German Economy*, 231; Stokes, "Wirtschaftswunder," 9–10; Berghahn, *Americanization*, 247–48; idem, "Reconstruction," 72–77.

9. "Draft of 6 October 1947 Ordinance ... of the Bizonal Economic Council. Subject: Bizonal Patent Office," BAK, Z22/153, Bl. 84–85; VfW internal memorandum, 15 Oct. 1947, BAK, Z22/502, Bl. 1.

10. Ulrich C. Hallmann and Paul Ströbele, "Das Patentamt von 1877 bis 1977," in *Hundert Jahre Patentamt*, 422–24. The 1936 Code was carried over into the Federal Republic via six separate "Laws for Changing and Transferring Regulations in the Field of Industrial Property Protection," passed between 8 Jul. 1949 and 23 Mar. 1961. With the exception of the sixth law, which established a Patent Court with jurisdiction over all decisions by the Patent Office, the transition laws concerned only adjustments to the postwar order; Benkard, *Patentgesetz*, 45–47. For the underlying political process: report of VfW main patent law committee meeting, 3–5 Feb. 1948 in Königstein, BAK, Z22/502, Bl. 72–78.

11. Lindenmaier: *Wer ist Wer* 1955, 716, party membership: BDC, membership in RAG Erfindungswesen: IG Farben, Report on 58th Patent Committee meeting, 25 Mar. 1942, Bayer, 23/2.3, vol. 3.

12. "Überleitung des Patent- und Gebrauchsmusterrechts," Oct. 1945, BAK, Z22/502, Bl. 33–40, p. 1; E. Pallas, "Der gewerbliche Rechtsschutz: Alte Bestimmungen in Kraft—Noch ungeklärte Lage," *Allgemeine Zeitung*, 29 Aug. 1947, BAK, Z22/153, Bl. III.

13. Reimer, *Recht der Angestelltenerfindung*, 1–2; idem, *Recht der Arbeitnehmererfindung*, 1–2. Former IG Farben patent experts Walter Beil and Harald Mediger contended the company invention was still alive, Beil to Haertel, 25 Jul. 1950, BA-Zwi, B141/2799; Harald Mediger, "Gedanken zur Gestaltung des Rechts des nicht-selbständigen Erfinders im Lichte der Betriebsverknüpftheit seiner Erfindung," 6 Jun. 1948, BA-Zwi, B141/2793; Heine to Volmer, 6 Feb. 1954, BA-Zwi, B141/2812, Bl. 93–96.

14. On Reimer (1896–1957, no BDC file): Munziger Archiv and *Wer ist Wer* 1955, 948. On reestablishment of Patent Office and the move to Munich: Reimer to Bernhard Wolf, 1 Dec. 1947, BAK, Z22/156, Bl. 116–121; Haertel to Strauß, 9 Apr. 1948, BAK, Z22/167, Bl. 154–58; minutes of patent law committee meeting of 27 Apr. 1948, BAK, Z22/40, Bl. 123; correspondence in BAK, Z21/633 and Z22/250. "Old patent rights": first transition law of 8 Jul. 1949, and commentary, BAK, Z21/625, Bl.f.PMZ 1948/49, 256, and Wirtschaftsrat, *Gesetzblatt* 1949, 175. Judge Kühnemann, Mansfeld's counterpart at RJM during the Third Reich, headed the Patent Office's Berlin unit until 1957, when he succeeded Reimer as President.

15. With the right of "limited assignment" (*beschränkte Inanspruchnahme*), the 1957 act gave the employer a third choice: to acquire a nonexclusive right (shop right) in the invention without the requirement to file a patent application. If the employer did use the invention, the employee was entitled to special compensation.

16. Discussions of the differences between the Speer ordinances and Dapper formula on one hand and their postwar successors in the Federal Republic on the other in Reimer, Schade, and Schippel, *Recht der Arbeitnehmererfindung*, 556–624; Volmer and Gaul, *Arbeitnehmererfindungsgesetz*, 21–108; Hans Schade, "Zu Fragen des Arbeitnehmererfinderrechts," *GRUR* 60 (Nov. 1958): 519–27; idem, "Die neuen Richtlinien für die Vergütung von Arbeitnehmererfindungen, *Der Betriebs-Berater*, no. 11 (20 Apr. 1960): 449–51.

17. Bernhard Volmer, "Der neue Regierungsentwurf eines Gesetzes über Arbeitnehmererfindungen," 2, BA-Zwi, B141/2824, Bl. 111; report of 26 Apr. 1950 meeting of Frankfurt chapter of DVGRU, comment by Walter Beil, BA-Zwi, B141/2798, Bl. 141–42. The employee continued having to report every invention. Reporting was unnecessary when the invention was manifestly unusable by the employer (e.g., a typewriter invented by a bakery apprentice). The scope of free inventions remained small, and the employer retained broad powers to claim the vast majority of employee inventions. The employee remained entitled to special compensation whenever the employer claimed the invention, and the amount continued to be calculated according to official guidelines, which were based on the creativity principle supposedly replaced by the monopoly principle. The inventor's fifty-percent income and withholding tax advantages remained in place, and private patent agreements continued to be illegal. Continuation of the tax advantage, deemed necessary to spur invention and technological progress, was decided in 1951: "Verordnung über die steuerliche Behandlung der Vergütungen für Arbeitnehmererfindungen vom 6. Juli 1951," *Bundesgesetzblatt* 1951, I, 388/*Bundessteuerblatt* 1951, I, 184, reprinted in Reimer, Schade, and Schippel, *Recht der Arbeitnehmererfindung*, 655–56. The tax advantage was eliminated in 1988.

18. Lindenmaier in inventor-rights committee meeting, 26 Jan. 1948, BAK Z22/502, Bl. 32–40 and 43–63; Reimer, *Angestelltenerfindung*, preface, 15; Kirchhoff, *Patentwesen*, 123–25; Reimer, Schade, and Schippel, *Recht der Arbeitnehmererfindung*, 80–81, 558–60.

19. The committee consisted of Lindenmaier (chair); patent attorney Hans-Gerhard Heine (Metallgesellschaft), patent attorney Werner Cohausz (Vestag); Eduard Reimer; Joseph Brisch; Dr. Koselke (Department of Labor in the British sector, succeeded by Ministerialdirektor Georg Steinmann, Werner Mansfeld's right-hand man at the RAM between 1936 and 1941); Senior Privy Councilor Hans Müller-Pohle (author and wartime commentator on the Speer ordinances, working for the British in Braunschweig); Dr. Seuring (MAN patent chief); and Dr. J. Willems (Redies's successor, Bayer).

20. Mediger, "Gedanken zur Gestaltung," 28.

21. Mediger was not alone. In June 1950 engineer and patent attorney Otto Stürner wrote the BJM that "the Göring ordinance is a true child of the Third Reich. Even if one gives this child another name or a haircut according to the latest fashion, it still remains the typical expression of the former regime." BA-Zwi, B141/2799, Bl. 52–65.

22. Drafts and proposals in BAK, Z22/502, e.g., Cohausz draft, 28 Nov. 1947, Bl. 22–26; Association of Berlin Patent Attorneys, 1 Dec. 1947, Bl. 27–29; trade

unions, 27 Jan. 1948, Bl. 63–68; Lindenmaier committee, 5 Feb. 1948, Bl. 32–40; VfW, 6 Jul. 1948, Bl. 196–99; DAG, 17 Jan. 1949, Bl. 261–64; DGB, 25 Feb. 1949, Bl. 328–36.

23. Report on meeting of 24 Mar. in Cologne and 13 Apr. 1948 in Braunschweig and draft of revised inventor ordinance, BAK, Z22/502, Bl. 95–97, 107–114.

24. Heine to VfW, 23 Apr., 5 and 10 May 1948, BAK, Z22/502, Bl. 122–23, 129, 131–36; also, Rüdiger Tremblau (for Cologne employer group) to VfW, 23 Apr. 1948, BAK Z22/502, Bl. 122–25, 129, 131–36.

25. Reimer had the same criticism. Joël to Müller-Pohle, Reimer, and Lindenmaier, 12 May 1948; Lindenmaier to Joël, 15 May 1948; Joël to Lindenmaier, 22 May 1948, Reimer to Joël, 22 May 1948; BAK, Z22/502, Bl. 137, 148, 155, 157.

26. Brisch to VfW, 18 May and 2 Jun. 1948; DGB to VfW, 21 Aug. 1948; Haertel to DGB's Werner Hansen, 13 Sep. 1948; Bührig to Lindenmaier committee, 11 Nov. 1948; Gewerkschaftsrat bill in Bührig to Pünder, 25 Feb. 1949, BAK, Z22/502, Bl. 154, 164, 222, 241, 273–74, 328–36.

27. DAG draft, 18 Nov. 1948, BAK, Z22/502, Bl. 261–64; on revival of a separate white-collar identity: Prinz, *Vom neuen Mittelstand*, 328–33.

28. Zentralsekretariat der Arbeitgeber des Vereinigten Wirtschaftsgebiet to VfW, 19 Aug. 1948; Willems to VfW, 27 Aug. 1948; Standard Elektrizitäts-Gesellschaft A.G. Stuttgart to VfW, 14 Jan. 1949; Mineralölwirtschaftsverband e.V. to VfW, 28 Mar. 1949, BAK Z22/502, Bl. 211, 224, 296–98, 343.

29. Lindenmaier to Joël, 15 May 1948; Lindenmaier to Haertel, 18 Nov. 1948 and 16 Mar. 1949; Haertel to Lindenmaier, 18 Mar. 1948; Rechtsamt's Dr. Schwabe to engineer Heinz Lux, 31 May 1949; Haertel to H. Wellhausen (FDP, Bundestag deputy, MAN director), 11 Jul. 1949, Z22/502, Bl. 148, 259, 338, 339, 368–71, 373.

30. Officially, the BAM became co-author. In practice its role was minor, although Steinmann remained involved and was frequently consulted by BJM officials.

31. Joël's father was Curt Joël, Secretary of State in the RJM during Weimar and briefly Minister of Justice in the second Brüning cabinet. Of Jewish ancestry, both the elder and younger Joël survived the Third Reich. Although their names were identical, the son is not to be confused with another Günther Joël, also associated with the Ministry of Justice. This Joël, no relation, was a collaborator of Franz Schlegelberger and Roland Freisler, an SS officer, and eventually Prosecutor General in Hamm; see Müller, *Hitler's Justice*, 208–12.

32. Haertel to Wellhausen, 11 Jul. 1949, BAK, Z22/502, Bl. 373.

33. Haertel, not yet thirty-nine when Adenauer became Chancellor in the fall 1949, had obtained his law doctorate in 1935 and joined the Nazi party in the spring of 1937. It is not known what he did during the Third Reich, but in 1947 he was working on commercial and business law in the Frankfurt VfW; BDC; *Wer Ist's* 1970, 420; Hallmann and Ströbele, "Patentamt," 432, BAK, Z22/153, Bl. 122, Z22/502, passim; BA-Zwi, B141/2784–2850, passim. Confessions of ignorance: Haertel to Lindenmaier, 19 May, 14 and 27 Jul. 1950, BA-Zwi, B141/2799, Bl. 9, 115–16, 143; requests for assistance: Haertel to Volmer, 18 Jan. 1950, Haertel to Walter Strauß, 28 Jan. 1950, BA-Zwi, B141/2798, Bl. 73, 93.

34. Volmer transferred to BJM in Bonn and became Haertel's closest collaborator and co-author of the 1952 bill. He returned to Wiesbaden in 1953. Volmer became a leading authority on the subject and wrote the first edition of a standard commentary on the 1957 act: Volmer and Gaul, *Arbeitnehmererfindungsgesetz*. I have no information on Volmer's background.

35. Haertel to Lindenmaier, 28 Jul. 1950, BA-Zwi, B141/2799, Bl. 142–43; Haertel to Reimer, 25 Nov. 1950, B141/2801, Bl. 54. Bundestag Antrag betr. Vorlage eines Gesetzentwurfs über Diensterfindungen, Drucksache 805, 1. Wahlperiode 1949 (Mar. 29, 1950), BA-Zwi, B141/2798, Bl. 113. (The Bundestag passed the motion on 14 Feb. 1951, BA-Zwi, B141/2802, Bl. 22).

36. Leading contenders were plans of the DGB, Green Society, BDA, and ULA. The ULA was the successor to the VELA plus about half of the old BUDACI, the other half having joined the DGB; DGB draft, identical to draft submitted 25 Feb. 1949, in DGB to Bundestag, 24 Nov. 1950, BA-Zwi, B141/2801, Bl. 90ff.; bill of Green Society in Rudolf Friedrich to Haertel, 1 Feb. 1951, BA-Zwi, B141/2802, Bl. 1–10; employer bill in Vereinigung der Arbeitgeberverbände (predecessor of BDA), to BJM, 23 Mar. 1950, BA-Zwi, B141/2798, Bl. 105–118; draft of Union der oberen Angestellten in Bergbau und Industrie (forerunner of ULA), 18 Sep. 1950, BA-Zwi, B141/2800, Bl. 76ff. The ULA, based in Essen, was founded in September 1950. Its constituent organizations included Verband oberer Bergbeamten e.V., Verband oberer Angestellter der Eisen- und Stahlindustrie e.V., Verband angestellter Akademiker der chemischen Industrie e.V., and VELA; see *Der Leitende Angestellte: Monatsschrift der Union der Leitenden Angestellten* 10 (Sep. 1960) special issue, Hoechst, 12/305; also Deichsel to Volmer, 27 Feb. 1951, BA-Zwi, B141/2802, Bl. 54.

37. See note 15.

38. A first breakthrough was the Green Society's proposal to solve the mediation problem by attaching a permanent board of mediation to the Patent Office. This idea was accepted and became one of the more successful aspects of the 1957 act. Schneider memorandum for Dehler, 13 Feb. 1951, BA-Zwi, B141/2802, Bl. 13–15.

39. Bernhard Volmer, "Zur Reform des Arbeitnehmererfindungsrechtes," *Recht der Arbeit* no. 10 (1950): 367–71, esp. 368, BA-Zwi, B141/2800, Bl.74.

40. Heine to Volmer, 28 Oct. 1950, BA-Zwi, B141/2801, Bl. 3–4.

41. Ohrt to Haertel, 9 Jul. 1951, BA-Zwi, B141/2804, Bl. 89–95.

42. Heine to Haertel, 30 May 1951, BA-Zwi, B141/2803, Bl. 38–40.

43. Justice Minister Neumayer to Secretary of State Hans Globke, 15 Dec. 1953, BA-Zwi, B141/2812, Bl. 10; Bundesrat, *Verhandlungen*, 79th session, 29 Feb. 1952, pp. 66–68, BA-Zwi, B141/2807, Bl. 133.

44. Wellhausen to Dehler, 2 Feb. 1952, Ba-Zwi, B141/2807, Bl. 127–28; Walter Beil, "Die Betriebserfindung," unpublished paper, 25 Feb. 1952; report of DVGRU meetings, 29 Feb.–1 Mar., 19 Mar., BA-Zwi, B141/ 2808, Bl. 5–11, 77–78; DVGRU meeting, 4–5 Apr. 1952, BA-Zwi, B141/2809, Bl. 67–76; Harald Mediger, "Gedanken eines Praktikers zur Regelung des Angestellten-Erfinderrecht," *GRUR* 54 (Feb. 1952): 57–79, BA-Zwi, B141/2809, Bl. 27–35; BDA to Bundestag, committees for patent law and labor, 10 Nov. 1952, BA-Zwi, B141/2811, Bl. 75a–q; IHK Berlin to BJM and BAM, 16 Mar. 1954, BA-Zwi, B141/2816, Bl. 55–58.

45. DAG to BJM, 29 Jun. 1951, BA-Zwi, B141/2804, Bl. 65–70.

46. Prof. Helferich of chemical institute, Bonn University, to Dehler, 23 Feb. 1952; Prof. Felgenträger, chair of Hochschulverband, to BJM, 29 Mar. 1952, BA-Zwi, B141/2808, Bl. 26–30, 79–81; Hochschulverband to BJM, 26 May 1952, BA-Zwi, B141/2809, Bl. 39–44. The professors got their wish in the 1957 act.

47. DGB's Bührig and Fette to Bundestag committees for patent law and labor, 13 Jun. 1952, BA-Zwi, B141/2809, Bl. 51–64; DGB's Hauser (chair, patent-law

committee) to BJM, BA-Zwi, B141/2812, Bl. 147–65; ULA to Bundestag committees for patent law and labor, 4 Aug. 1952, BA-Zwi, B141/2810, Bl. 50–56.

48. Deichsel to Haertel, Deichsel to Wagner, 7 Jan. 1953, BA-Zwi, B141/2811, Bl. 9a–l.

49. BDA to Volmer, 30 Apr. 1953, BA-Zwi, B141/2811, Bl. 75.

50. See also note 15. *"Unbeschränkte beschränkte Inanspruchnahme"* meant the absence of any exclusions, which had been proposed earlier. Management and the government advanced two reasons for the change. One was concern about the Patent Office's inability to handle the volume of applications if the employer was forced to always file, even if interest in the invention was remote. The second was that forcing the employer to choose between applying for a patent and declaring the invention free was too inflexible. Deutscher Bundestag, 2. Wahlperiode 1953, Drucksache 1648, 15 Aug. 1955, *Entwurf eines Gesetzes über Erfindungen von Arbeitnehmern und Beamten*, 22–25, BA-Zwi, B141/2824, Bl. 1ff. Standard Elektrizitäts-Gesellschaft AG to BJM, 24 Sep. 1954, BA-Zwi, B141/2817, Bl. 74–75; "Dritter Vorläufiger Entwurf des Bundesjustizministerium und des Bundesarbeitsministerium," 1 Sep. 1954, article 9, BA-Zwi, B141/2817, Bl. 48.

51. Heydt to Volmer's successor, Albert Krieger, 9 Dec. 1954, BA-Zwi, B141/2818, Bl. 20–24. Grüll to BJM, 10 Nov. 1955, BA-Zwi, B141/2825, Bl. 126. Queißer to BJM, 8 Sep. 1955, BA-Zwi, B141/2824, Bl. 65–69. Krieger memorandum, 16 Dec. 1954, BA-Zwi, B141/2818, Bl. 25–27; Hauser to BJM, 27 Jun. 1955, BA-Zwi, B141/2822, Bl. 63, 67; Hauser, "Das neue Arbeitnehmer-Erfindergesetz," *Deutsche Angestellten-Zeitschrift* 9, no. 6 (Sep. 1955), section "Wirtschaft und Wissen," BA-Zwi, B141/2824, Bl. 108; Krieger memorandum, 2 May 1955, Storch to Neumayer, 13 Jan. 1955, BA-Zwi, B141/2818, Bl. 49–50, 105–7.

52. Joël to Neumayer, 3 May 1955, BJM and BAM to Bundeskanzleramt, 13 May 1955, Joël to "social partners," 8 Jun. 1955, BA-Zwi, B141/2820, Bl. 1–2, 8–15, 227–30; Auszug aus dem Kurzprotokoll der 144. Sitzung des Bundesrates, 8 Jul. 1955, pp. 210–12 (Bundesrat-Drucksache 202/55), BA-Zwi, B141/2822, Bl. 111–12; Bundeskanzleramt to Bundestag, 19 Aug. 1955, BA-Zwi, B141/2824, Bl. 1–31.

53. It appears that Reimer's attack was a tactical maneuver designed to restore statutory compensation for certain nonpatentable inventions. Krieger to Haertel, 3 and 5 Apr. 1956, BA-Zwi, B141/2827, Bl. 150–54; minutes of Bundestag patent law committee meeting, 12 Apr. 1956, pp. 3–7, BA-Zwi, B141/2828, Bl. 151–53; "Stellungnahme des Bundesministers der Justiz zu den beiden Gegenentwürfen von Professor Dr. Reimer," presented to Bundestag patent-law committee, 22 Jun. 1956, BA-Zwi, B141/2829, Bl. 133ff.; ULA to Bundestag patent-law and labor committees, 18 Jun. 1956, BA-Zwi, B141/2829, Bl. 66–74; BDA to Haertel, 15 Oct. 1956, BA-Zwi, B141/2830, Bl. 49–52.

54. Joël in testimony before Bundestag patent-law committee, 5 Dec. 1956, BA-Zwi, B141/2831, Bl. 15. Analyses in BAM to Haertel, 19 Jun. 1956, BA-Zwi, B141/2829, Bl. 77–79; Storch to Wagner, 12 Sep. 1956, BA-Zwi, B141/2830, Bl. 22–24; Hueck report, 18 Oct. 1956, BA-Zwi, B141/2830, Bl. 93–106; Haertel to Strauß, 7 Feb. 1957, BA-Zwi, B141/2832, Bl. 9–10.

55. Karl Hauser, one of the DGB's two inventor-rights negotiators (Deichsel was the other one), made a last-minute effort to block passage of the bill, claiming it represented a totalitarian subversion of the Basic Law and deprived employees of their fundamental rights. This created momentary panic but nothing else. Karl

Hauser, "Erfindungen—nur Routine?" *Frankfurter Allgmeine Zeitung*, no. 227 (28 Sep. 1956), p. 7, BA-Zwi, B141/2830, Bl. 28; DBG to Bundestag and BJM, "Abänderungsvorschläge zum Gesetzentwurf über Erfindungen von Arbeitnehmern und Beamten," 3 Dec. 1956, BA-Zwi, B141/2831, Bl. 105–45; Deichsel to Haertel, 6 Dec. 1957, BA-Zwi, B141/2847; Krieger to Willems, 23 Apr. 1957, BA-Zwi, B141/2833, Bl. 128–29.

56. "Auszug aus dem Bericht über die 206. Sitzung des 2. Deutschen Bundestages vom 3.5.1957," report by CDU deputy Hedwig Jochmus (Heidelberg) in *Verhandlungen des 2. Deutschen Bundestag, 206. Sitzung,* 3 May 1957, pp. 1815–23, BA-Zwi, B141/2833, Bl. 132–38; "Richtlinien für die Vergütung von Arbeitnehmererfindungen im privaten Dienst vom 20. Juli 1957," reprinted in Reimer, Schade, and Schippel, *Recht der Arbeitnehmererfindung,* 57–79.

57. Ironically, the escape is made possible only by bureaucracy itself, i.e., the complicated system of the 1957 law and 1959 compensation guidelines.

58. Volmer and Gaul, *Arbeitnehmererfindungsgesetz,* 21–25; similar figures in Braun, *German Economy,* 246–48; on the problem of using patents as a measure of technological progress, ibid., 232–33, 236 (note 16); see also Ulrich Schmoch, Hariolf Grupp, and Uwe Kuntze, *Patents as Indicators of the Utility of European Community R&D Programmes* (Luxemburg, 1991).

59. See literature cited in note 58.

60. James, *German Slump,* 146–47; Peukert, *Weimar Republic,* 107–28; Borchardt, "Constraints"; idem, "Economic Causes"; von Kruedener, *Economic Crisis*; Kershaw, *Why Did German Democracy Fail?*; Gregor, *Daimler-Benz,* 1–35; Mommsen, *Volkswagenwerk,* 27–92; for the post-1945 period: Overy, "Economy of the Federal Republic," 27–28; Berghahn, "Reconstruction," 71–72; Stokes, "Wirtschaftswunder"; Braun, *German Economy,* 231–32.

61. Herrigel, *Industrial Constructions,* rightly points out that Germany's economic landscape was characterized not just by large corporations, which have received the bulk of scholarly attention, but also by vibrant communities of small and medium-sized business. Technologically, however, large firms such as Siemens, AEG, IG Farben, and MAN did dominate, at least in terms of invention as measured by patent statistics (see, e.g., ch. 11).

62. On design culture see König, *Künstler,* passim, esp. 218–32. Consolidation of design culture was also related to German industry's response to the economic depression of the 1870s, when an early industrial strategy of "cheap but bad" was replaced with a philosophy of "expensive but good"; cf. Gispen, *New Profession,* 115–29; Radkau, *Technik,* 148–55.

63. On coal mining, see Shearer, "Talking about Efficiency"; idem, "The Politics of Industrial Productivity in the Weimar Republic: Technological Innovation, Economic Efficiency, and their Social Consequences in the Ruhr Coal Mining Industry, 1918–1929" (Ph.D. diss., Univ. of Pennsylvania, 1989); universal machine tools: Gregor, *Daimler-Benz,* passim, esp. 16–35, 247–52; Herrigel, "Industry as a Form of Order."

64. Gregor, *Daimler-Benz,* 98–100, 114–18, 247–52; cf. also Riemschneider and Barth's hesitation about mass production discussed in ch. 10.

65. For "technological tradition," see Stokes, "Wirtschaftswunder," 3, 19–22.

66. Jewkes, Sawers, and Stillerman, *Sources of Invention*; Wiener, *Invention*; Hughes, *American Genesis.*

67. See note 55.

68. Riemschneider's remarks were provoked by the DAG 1948 draft bill; Riemschneider to Strauß, 31 Jan. 1949, Haertel to Kommission für gewerblichen Rechtsschutz, 28 Feb. 1949, BAK, Z22/502, Bl. 309–20, 323; Verband Angestellter Akademiker to Rechtsamt of VfW, 21 Apr. 1949, BAK, Z22/502, Bl. 361–65; also Volmer's assessment of updated version of DAG draft, 30 Sep. 1950, BA-Zwi, B141/2800, Bl. 110–15.

69. Numerous examples in BAK, R131/159, for immediate postwar period, and BA-Zwi, B141/2759–62 (entitled *Erfinderförderung*), for early 1950s. Other resentful brochures and letters in BA-Zwi, B141/2759, B141/2793. Tax advantages: employees received their fifty-percent tax rebate without having to do anything (the employer did it when calculating deductions). Independents not only got significantly less favorable treatment but also had to follow complicated bureaucratic procedures; a year after the relevant ordinance had passed in May 1951 not a single independent had qualified for the tax cut; DVGRU meeting on encouragement of inventors, 4 and 5 Apr. 1952, BA-Zwi, B141/2761; text of tax regulations in Reimer, Schade, and Schippel, *Recht der Arbeitnehmererfindung*, 655–58.

70. Volmer memorandum, 15 Dec. 1951, BA-Zwi, B141/2760; cf. Radkau, *Technik*, 171–76.

71. Bernhard Volmer, "Entwurf eines Gutachtens über die Möglichkeiten einer wirksamen Förderung der Erfinder im Bundesgebiet" (31 Dec. 1951), BA-Zwi, B141/2761, pp. 1–113, esp. 4–9, 42, 56, 62, 70, 72–75, 87–90. 92.

72. Volmer's model was the British National Research Development Corporation (founded in 1948), which was itself a copy of the BPV and its Nazi successor, the AftW, ibid., pp. 10–42.

73. Ibid., 6–9.

74. Metallgesellschaft's Heine thought Volmer's recommendations were interesting and agreed that German industry was not as receptive to innovation as it should be. He concluded it was "fundamentally right to do something special for the encouragement of inventions that hail from so-called free inventors." The Green Society devoted several meetings to the issue and invited speakers from Britain to learn more about the NRDC; Heine to Volmer, 7 Apr. 1952; DVGRU report, 25 Apr. 1952; Heine to Haertel, 8 May 1952; BA-Zwi, B141/2761; DVGRU report, 11–12 Oct. 1952, BA-Zwi, B141/2840, Bl. 18–87.

75. Dr. Hennenhöfer, "Vermerk betr. Gründung einer Organisation zur Förderung von Erfindungen," 8 May, 1953, updated 20 Aug. 1953, BA-Zwi, B141/2762.

76. Figures for government spending on R&D in Braun, *German Economy*, 232; new technological departures: ibid., 232–34; Stokes, "Wirtschaftswunder," 17–19; high levels of investment: Kramer, *West German Economy*, 195–214; higher priority for R&D after 1957: Radkau, *Technik*, 314–26.

77. On the Fraunhofer Society, see Helmuth Trischler and Rüdiger vom Bruch, *Forschung für den Markt: Geschichte der Fraunhofer-Gesellschaft* (Munich, 1999). The Fraunhofer Society started out trying to support development of new technologies, assist independent research, and help with pooled technology efforts of small and medium-sized business organizations. One of its best-known early projects was the institute where Felix Wankel worked on the rotary engine. It also performed contract research and provided assistance with patenting and invention evaluation, much like the BPV and AftW had done earlier.

78. Cf. Gerhard Becher, Thomas Gering, Oliver Lang, and Ulrich Schmoch, *Patentwesen an Hochschulen: Eine Studie zum Stellenwert gewerblicher Schutzrechte im Technologietransfer Hochschule-Wirtschaft* (Bonn, 1996).
79. For the combination of technological conservatism and economic success, see Stokes, "Wirtschaftswunder," 15–22; Fischer, "Role of Science and Technology," 101–4; Radkau, *Technik*, 313–26; "fulcrum" sectors: Allen, "Competing Communitarianisms," 94.

WORKS CITED

A Primary Sources

I Government Archives

Bundesarchiv Koblenz
Hauptamt für Technik:	NS 022/000851
Neue Reichskanzlei:	R43 II/547, /1558a-b, /1559
Rechtsamt:	Z003/000081
	Z21/625-28, /633-36, /640-45
	Z21 Anh./58-62
	Z22/40, /151-72, /182-83,
	/186-90, /224, /349-50, /502
Reichsarbeitsministerium:	R41/78, /652
Reichsjustizministerium:	R22/625, /629-30
Reichspatentamt:	R131/19-22, /27, /76-77, /96,
	/111, /121-22,
	/153-63, /165
	R133/1571

Bundesarchiv Potsdam
Reichsarbeitsministerium:	3448-49, 3451-52, 10373-6,
	10375, 10378-84
Reichsinnenministerium:	25333-34
Reichsjustizministerium:	10129-33, 20608-13, 2336-45

Reichsministerium für Rüstung und Kriegsproduktion (before 1942: Reichsministerium für Bewaffnung und Munition): 46.03

Bundesarchiv Zwischenarchiv St. Augustin
Bundesjustizministerium:	B141/16688-89, /16757,
	/2757-84, /2788, /2792-95,
	/2798-2851, /49799, /49814

II Institutional Archives

Deutsches Museum
    General: 6.1
    Bayerischer Polytechnischer Verein:
                    II 36/1–2, III 97/1–2, V 128/6,
                    VIII 250, 260, 282/1–4, 283/1–2,
                    287/1–2, 288/6, 290/1–2,
                    291/1–3, 292/1–7, 293/1–2
Rheinisch-Westfälisches Wirtschaftsarchiv
    Handelskammer Duisburg: 20-119-12, 20-2015/1–2, 20-
                    287-3, 20-287-13, 20-477-9,
                    20-874/3-6, 20-875-1

III Industrial Archives

BASF:	B4/E 04/1; B4/1979
Bayer:	AS Duisberg
	22/1/4, 22/7, 22/8, 22/12,
	22/23, 23/2.3/1–3, 23/37,
	23/43, 24/1, 27/1.0, 28/1.1,
	28/10.1, 28/6.1, 28/6.2,
	213/2.12, 215/5
Hoechst:	4/63/1, 12/11/2, 12/35/4,
	12/305, 12/255/1–2
MAN (Augsburg):	150 (Nachlass Neumann),
	Nachlass Guggenheimer,
	Nachlass Lauster
Mannesmann:	M60.200, P2.25.45
Siemens:	4/Lk31–32, 4/Lk78, 4/Lk153,
	11/Lf36, 11/Lf215, 11/Lf237/238,
	11/Lf352/11, 11/Lf364, 11/Lf391,
	11/Lg671, 11/Lg709

B Secondary Sources

Abraham, David. *The Collapse of the Weimar Republic: Political Economy and Crisis*, 2d ed. New York, 1986.
Aitken, Hugh G. J. *The Continuous Wave: Technology and American Radio, 1900-1932*. Princeton, 1985.
Allen, Christopher S. "Germany: Competing Communitarianisms." In *Ideology and National Competitiveness: An Analysis of Nine*

Countries, ed. George C. Lodge and Ezra F. Vogel, 79–102. Boston, 1987.

Allen, Kenneth R. "Invention Pacts: Between the Lines." *IEEE Spectrum*, Mar. 1978: 54–59.

Ardenne, Manfred von. *Ein glückliches Leben für Technik und Forschung: Autobiographie.* Zurich, 1972.

Banse, Gerhard, and Siegfried Wollgast, eds. *Biographien bedeutender Techniker, Ingenieure und Technikwissenschaftler: Eine Sammlung von Biographien.* Berlin, 1987.

Barkai, Avraham. *Nazi Economics: Ideology, Theory, and Policy.* Translated by Ruth Haddass-Vashitz. New Haven, 1990.

Barth, Heinrich. "Persönlichkeit und Volksgemeinschaft im Rechte der Erfinder und Erfindungen." *Zeitschrift der Akademie für Deutsches Recht* (1935): 823–26.

Bauman, Zygmunt. *Modernity and the Holocaust.* Ithaca, 1989.

Becher, Gerhard, Thomas Gering, Oliver Lang, and Ulrich Schmoch. *Patentwesen an Hochschulen: Eine Studie zum Stellenwert gewerblicher Schutzrechte im Technologietransfer Hochschule-Wirtschaft.* Bonn, 1996.

Beckmann, Lothar. *Erfinderbeteiligung: Versuch einer Systematik der Methoden der Erfinderbezahlung unter besonderer Berücksichtigung der chemischen Industrie.* Berlin, 1927.

Beier, Friedrich-Karl. "Wettbewerbsfreiheit und Patentschutz: Zur geschichtlichen Entwicklung des deutschen Patentrechts." *GRUR* 80 (Mar. 1978): 123–32.

Belt, Henk van den, and Arie Rip. "The Nelson-Winter-Dos Model and Synthetic Dye Chemistry." In *The Social Construction of Technological Systems: New Directions in the Sociology and History of Technology*, ed. Wiebe E. Bijker, Thomas P. Hughes, and Trevor J. Pinch, 149–55. Cambridge, Mass., 1989.

Belz, Wolfgang. *Die Arbeitnehmererfindung im Wandel der patentrechtlichen Auffassungen.* Munich, 1958.

Benkard, Georg. *Patentgesetz, Gebrauchsmustergesetz; Kurzkommentar begr. von Dr. Georg Benkard; fortgeführt von Werner Ballhaus et al.,* 7th ed. Munich, 1981.

Benz, Wolfgang, and Hermann Graml, eds. *Biographisches Lexikon zur Weimarer Republik.* Munich, 1988.

Berghahn, Volker R. *The Americanization of West German Industry, 1945–1973.* Cambridge, 1986.

———. *Imperial Germany 1871–1914: Economy, Society, Culture and Politics.* Providence, 1996

_____. "West German Reconstruction and American Industrial Culture, 1945–1960." In *The American Impact on Postwar Germany*, ed. Reiner Pommerin, 65–81. Providence, 1997.

Berliner, Robert. "10 Jahre Verband angestellter Akademiker der chemischen Industrie e.V." *Der Leitende Angestellte* 9 (Apr. 1959): 58–60.

Bessel, Richard. *Political Violence and the Rise of Nazism: The Storm Troopers in Eastern Germany 1925–1934*. New Haven, 1984.

Beyerchen, Alan D. "On the Stimulation of Excellence in Wilhelmian Science." In *Another Germany: A Reconsideration of the Imperial Era*, ed. Jack R. Dukes and Joachim Remak, 139–68. Boulder, 1988.

_____. "Rational Means and Irrational Ends: Thoughts on the Technology of Racism in the Third Reich." *CEH* 30, no. 2 (1997): 386–402.

_____. *Scientists under Hitler: Politics and the Physics Community in the Third Reich*. New Haven, 1977.

Bijker, Wiebe E. "The Social Construction of Bakelite: Toward a Theory of Invention." In *The Social Construction of Technological Systems: New Directions in the Sociology and History of Technology*, ed. Wiebe E. Bijker, Thomas P. Hughes, and Trevor J. Pinch, 159–90. Cambridge, Mass., 1989.

Blackbourn, David, and Geoff Eley. *The Peculiarities of German History: Bourgeois Society and Politics in Nineteenth-Century Germany*. New York, 1984.

Blunk, Richard. *Hugo Junkers: Der Mensch und das Werk*. Berlin, 1943.

Borchardt, Knut. "Constraints and Room for Maneuver in the Great Depression of the Early Thirties: Towards a Revision of the Received Historical Picture." In his *Perspectives on Modern German Economic History and Policy*. Translated by Peter Lambert, 143–60. Cambridge, 1991.

_____. "Economic Causes of the Collapse of the Weimar Republic." In his *Perspectives on Modern German Economic History and Policy*. Cambridge, 1991.

Bracher, Karl Dietrich. *The German Dictatorship*. New York, 1970.

Braun, Hans-Joachim. *The German Economy in the Twentieth Century*. London, 1990.

Broszat, Martin. *Hitler and the Collapse of Weimar Germany*. Translated and foreword by V. R. Berghahn. Hamburg, 1987.

_____. *The Hitler State: The Foundation and Development of the Internal Structure of the Third Reich*. London, 1981.

Browning, Christopher R. *The Path to Genocide: Essays on Launching the Final Solution.* Cambridge, 1992.

Bruch, Rüdiger vom. "Wilhelminismus: Zum Wandel von Milieu und politischer Kultur." In *Handbuch zur "Völkischen Bewegung" 1871–1918*, ed. Uwe Puschner, Walter Schmitz and Justus H. Ulbricht, 2–21. Munich, 1996.

Bruch, Walter. "Ein Erfinder über das Erfinden." In *Hundert Jahre Patentamt*, ed. Deutsches Patentamt, 317–24. Munich, 1977.

BUDACI, ed. *Denkschrift zum Erfinderschutz, Sozialpolitische Schriften des Bundes Angestellter Chemiker u. Ingenieure e.V.*, 1st series, no. 6 (Sep. 1922).

_____, ed. *Kommentar zum Reichstarifvertrag für die akademisch gebildeten Angestellten der chemischen Industrie.* Berlin, 1920.

Bugbee, Bruce W. *Genesis of American Patent and Copyright Law.* Washington, D.C., 1967.

Burleigh, Michael. *Death and Deliverance: 'Euthanasia' in Germany 1900–1945* Cambridge, 1994.

Burleigh, Michael, and Wolfgang Wippermann. *The Racial State: Germany 1933–1945.* Cambridge, 1991.

BUTAB, ed. *25 Jahre Technikergewerkschaft, 10 Jahre BUTAB: Festschrift zum 25jährigen Jubiläum des Bundes der technisch-industriellen Beamten (Butib) und zum 10-jährigen Jubläum des Bundes der technischen Angestellten und Beamten (BUTAB) im Mai 1929.* Berlin, 1929.

Cahan, David. *An Institute For An Empire: The Physikalisch-Technische Reichsanstalt, 1871–1918.* Cambridge, 1989.

Caplan, Jane. "The Rise of National Socialism." In *Modern Germany Reconsidered 1870–1945*, ed. Gordon Martel, 117–39. London, 1992.

Caron, François, Paul Erker, and Wolfram Fischer, eds. *Innovations in the European Economy Between the Wars.* Berlin, 1995.

Cesarani, David, and Mary Fulbrook, eds. *Citizenship, Nationality and Migration in Europe.* London, 1996.

Childers, Thomas. "Inflation, Stabilization, and Political Realignment in Germany 1924 to 1928." In *The German Inflation Reconsidered: A Preliminary Balance*, ed. Gerald D. Feldman, Carl-Ludwig Holtfrerich, Gerhard A. Ritter, and Peter-Christian Witt, 409–31. Berlin, 1982.

_____. "The Middle Classes and National Socialism." In *The German Bourgeoisie*, ed. David Blackbourn and Richard J. Evans, 318–37. London, 1991.

_____. *The Nazi Voter: The Social Foundations of Fascism in Germany, 1919–1933*. Chapel Hill, 1983.

_____. "The Social Language of Politics in Germany: The Sociology of Political Discourse in the Weimar Republic." *AHR* 95, no. 2 (Apr. 1990): 331–58.

Childers, Thomas, and Jane Caplan, eds. *Reevaluating the Third Reich*. New York, 1993.

Costa, Jasper Silva. *The Law of Inventing in Employment*. New York, 1953.

Crew, David, ed. *Nazism and German Society 1933–1945*. London, 1994.

Croner, Fritz. *Die Angestellten in der modernen Gesellschaft*. Vienna, 1954.

_____. *Soziologie der Angestellten*. Cologne, 1962.

Dahrendorf, Ralf. *Society and Democracy in Germany*. Garden City, 1969.

Dapper, Josef. "Die Gefolgschaftserfindung und ihre Bewertung. *Deutsche Technik* 10 (1942): 372–76.

_____. "Zur Frage der Betriebserfindung." *Deutsche Technik* 10 (1942): 8, 14.

Deutsche Arbeitsfront, ed. *Berichte der Arbeitsgemeinschaft "Fragen des gewerbl. Rechsschutzes" und "Gefolgschaftserfinderrecht" des Amtes für technische Wissenschaften*, no. 1. Munich, 1941.

Deutsches Patentamt, ed. *Hundert Jahre Patentamt*. Munich, 1977.

Diner, Dan. *America in the Eyes of the Germans: An Essay on Anti-Americanism*. Princeton, 1996.

Domarus, Max. *Hitler: Reden und Proklamationen 1932–1945*. Munich, 1965.

DVSGE, ed. *Beschlüsse des Augsburger Kongress*. Berlin, 1914.

_____, ed. *Vorschläge zu der Reform des Patentrechts: Denkschrift der Patentkommission und der Warenzeichenkommission*. Berlin, 1914.

Eley, Geoff. "Conservatives and Radical Nationalists in Germany: The Production of Fascist Potentials, 1912–28." In *Fascists and Conservatives: The Radical Right and the Establishment in Twentieth-Century Europe*, ed. Martin Blinkhorn, 50–70. London, 1990.

_____. "Is There a History of the *Kaiserreich?* In *Society, Culture and the State in Germany, 1870–1930*, ed. Geoff Eley, 1–42. Ann Arbor, 1996.

_____. "What Produces Fascism: Preindustrial Traditions or a Crisis of a Capitalist State?" *Politics and Society* 12, no. 1 (1983): 53–82.

Elias, Norbert. *The Germans: Power Struggles and the Development of Habitus in the Nineteenth and Twentieth Centuries*. New York, 1996.

Engländer, Konrad. *Die Angestelltenerfindung nach geltendem Recht: Vortrag vom 24. Februar 1925*. Leipzig, 1925.

Eyth, Max. "Zur Philosophie des Erfindens." In his *Lebendige Kräfte, Sieben Vorträge aus dem Gebiete der Technik*, 4th ed. Berlin, 1924.

Feldenkirchen, Wilfried. *Siemens 1918–1945*. Columbus, 1999.

_____. *Werner von Siemens: Inventor and International Entrepreneur*. Columbus, 1994.

Feldman, Gerald D. "German Business Between War and Revolution: The Origins of the Stinnes-Legies Agreement." In *Entstehung und Wandel der modernen Gesellschaft: Festschrift für Hans Rosenberg zum 65. Geburtstag*, ed. Gerhard A. Ritter, 312–41. Berlin, 1970.

_____. *The Great Disorder: Politics, Economics and Society in the German Inflation, 1914–1924*. New York, 1993.

_____. *Iron and Steel in the German Inflation 1916–1923*. Princeton, 1977.

_____. "The Historian and the German Inflation." In *Inflation through the Ages: Economic, Social Psychological and Historical Aspects*, ed. Nathan Schmukler and Edward Marcus, 386–99. New York, 1983.

_____. "The Social and Economic Policies of German Big Business, 1918–1929" *AHR* 75, no. 1 (1969): 47–55.

Feldman, Gerald D., and Irmgard Steinisch. "Die Weimarer Republik zwischen Sozial- und Wirtschaftsstaat: Die Entscheidung gegen den Achtstundentag." *Archiv für Sozialgeschichte* 18 (1979): 353–439.

Fischer, Conan. *Stormtroopers: A Social, Economic, and Ideological Analysis 1929–1935*. London, 1983.

Fischer, Ludwig. *Die Arbeit des Patentingenieurs in ihren psychologischen Zusammenhängen*. Berlin, 1923.

_____. "Bericht über meine dienstliche Tätigkeit 1899–1934." Berlin, unpublished ms., 1934.

_____. *Betriebserfindungen*. Berlin, 1921.

_____. *Patentamt und Reichsgericht*. Berlin, 1934.

_____. *Werner Siemens und der Schutz der Erfindungen*. Berlin, 1922.

Fischer, Wolfram. "The Role of Science and Technology in the Economic Development of Modern Germany." In *Science, Technology, and Economic Development: A Historical and Comparative Study*, ed. William Beranek, Jr., and Gustav Ranis, 71–113. New York, 1978.

Fisk, Catherine L. "Removing the 'Fuel of Interest' from the 'Fire of Genius': Law and the Employee-Inventor, 1830–1930." *The University of Chicago Law Review* 65, no. 4 (fall 1998): 1127–98.

Förster, Michael. *Jurist im Dienst des Unrechts: Leben und Werk des ehemaligen Statssekretärs im Reichsjustizministerium, Franz Schlegelberger (1876–1970)*. Baden-Baden, 1995.

Frank, Hans. "Die Grundlagen des nationalsozialistischen Patentrechts." In *Das Recht des schöpferischen Menschen: Festschrift der Akademie für deutsches Recht anlässlich des Kongresses der Internationalen Vereinigung für gewerblichen Rechtsschutz in Berlin vom 1. bis 6. Juni 1936*, ed. Akademie für Deutsches Recht, 7–8. Berlin, 1936.

Frank, Hans, ed. *Nationalsozialistisches Handbuch für Recht und Gesetzgebung*, 2d ed. Munich, 1935.

Frank, Niklas. *In the Shadow of the Third Reich*. New York, 1991.

Frei, Norbert. "Wie modern war der Nationalsozialismus?" *GUG* 19, no. 3 (1993): 367–87.

Frese, Matthias. *Betriebspolitik im "Dritten Reich": Deutsche Arbeitsfront, Unternehmer und Staatsbürokratie im Dritten Reich*. Paderborn, 1991.

Freyberg, Thomas von, and Tilla Siegel. *Industrielle Rationalisierung unter dem Nationalsozialismus*. Frankfurt a. M., 1991.

Fritzsche, Peter. *Germans into Nazis*. Cambridge, Mass., 1998.

Fulbrook, Mary. *A Concise History of Germany*. Cambridge, 1990.

Galbraith, John Kenneth. *American Capitalism: The Concept of Countervailing Power*. Boston, 1952.

Gareis, Karl. *Über das Erfinderrecht von Beamten, Angestellten und Arbeitern: Eine patentrechtliche Abhandlung*. Berlin, 1879.

Geary, Dick. "The Industrial Bourgeoisie and Labor Relations in Germany 1871–1933." In *The German Bourgeoisie*, ed. David Blackbourn and Richard J. Evans, 140–61. London, 1991.

Geren, Gerald S. "New Legislation Affecting Employee Patent Rights." *Research & Development* (Jan. 1984): 33.

Gimbel, John. *Science, Technology, and Reparations: Exploitation and Plunder in Postwar Germany*. Stanford, 1990.

Gispen, Kees. *New Profession, Old Order: Engineers and German Society, 1815–1914*. Cambridge, 1989.

Grefermann, Klaus. "Patentwesen und technischer Fortschritt." In *Hundert Jahre Patentamt*, ed. Deutsches Patentamt, 37–64. Munich, 1977.

Gregor, Neil. *Daimler-Benz in the Third Reich*. New Haven, 1998.

Groehler, Olaf. "Hugo Junkers: Luftfahrtpionier, Industrieller und bürgerlicher Liberaler." In *Alternativen: Schicksale deutscher Bürger*, ed. Olaf Groehler, 57–90. Berlin, 1987.

Groehler, Olaf, and Helmut Erfurth. *Hugo Junkers: Ein politisches Essay*. Berlin, 1989.

Habermas, Jürgen. *Legimitation Crisis*. Boston, 1973.

Hachtmann, Rüdiger. *Industriearbeit im "Dritten Reich": Untersuchungen zu den Lohn- und Arbeitsbedingungen in Deutschland, 1933–1945*. Göttingen, 1989.

Hallmann, Ulrich C., and Paul Ströbele. "Das Patentamt von 1877 bis 1977." in *Hundert Jahre Patentamt*, ed. Deutsches Patentamt, 403–44. Cologne, 1977.

Hartmann, Martin. *Die Neuererbewegung: Das betriebliche Vorschlagswesen in der DDR*. Cologne, 1988.

Hartung, Günter. "Völkischer Ideologie." In *Handbuch zur "Völkischen Bewegung" 1871–1918*, ed. Uwe Puschner, Walter Schmitz, and Justus H. Ulbricht, 22–41. Munich, 1996.

Hauser, Karl. "Das deutsche Sonderrecht für Erfinder in privaten und öffentlichen Diensten." *Die Betriebsverfassung* 5 (Sep. 1958): 168–75.

Hayes, Peter. *Industry and Ideology: IG Farben in the Nazi Era*. Cambridge, 1987.

Heggen, Alfred. *Erfindungsschutz und Industrialisierung in Preussen 1793–1877*. Göttingen, 1975.

Herbert, Ulrich. "Good Times, Bad Times: Memories of the Third Reich." In *Life in the Third Reich*, ed. and intro. Richard Bessel, 97–111. New York, 1987.

_____. *Hitler's Foreign Workers: Enforced Foreign Labor in Germany under the Third Reich*. Cambridge, 1997.

_____. "Labor as Spoils of Conquest, 1933–1945." In *Nazism and German Society 1933–1945*, ed. David Crew, 219–74. London, 1994.

Herf, Jeffrey. *Reactionary Modernism: Technology, Culture and Politics in Weimar and the Third Reich*. Cambridge, 1984.

Herrigel, Gary. *Industrial Constructions: The Sources of German Industrial Power*. Cambridge, 1996.

_____. "Industry as a Form of Order: A Comparison of the Historical Development of the Machine Tool Industries in Germany and the United States." In *Governing Capitalist Economies: Performance and Control in Economic Sectors*, ed. J. Rogers Hollingsworth, Philippe Schmitter, and Wolfgang Streeck, 97–128. New York, 1992.

Heymann, Ernst. "Der Erfinder im neuen deutschen Patentrecht." In *Das Recht des schöpferischen Menschen: Festschrift der Akademie für deutsches Recht anlässlich des Kongresses der Internationalen Vereinigung für gewerblichen Rechtsschutz in Berlin vom 1. bis 6. Juni 1936*, ed. Akademie für Deutsches Recht, 99–126. Berlin, 1936.

Hilberg, Raul. *Victims, Perpetrators, Bystanders: The Jewish Catastrophe 1933–1945*. New York, 1994.

Hintze, Otto. "Der Beamtenstand." In *Soziologie und Geschichte: Gesammelte Abhandlungen zur Soziologie, Politik und Theorie der Geschichte*, ed. Gerhard Oestreich, 66–125. Göttingen, 1964.

Hitler, Adolf. *Mein Kampf*. Translated by Ralph Manheim. Boston, 1943.

Holdermann, Karl. *Im Banne der Chemie: Carl Bosch Leben und Werk*. Düsseldorf, 1953).

Holtfrerich, Carl-Ludwig. *The German Inflation 1914–1923*. Berlin, 1986.

Homburg, Ernst. "The Emergence of Research Laboratories in the Dyestuffs Industry 1870–1900." *British Journal for the History of Science* 25 (1992): 91–111.

Homburg, Heidrun. *Rationalisierung und Industriearbeit: Arbeitsmarkt, Management, Arbeiterschaft im Siemens-Konzern Berlin 1900–1939*. Berlin, 1991.

Homze, Edward L. *Foreign Labor in Nazi Germany*. Princeton, 1967.

Horkheimer, Max, and Theodor W. Adorno. *Dialectic of Enlightenment*. New York, 1975.

Hounshell, David A. "Invention in the Industrial Research Laboratory: Individual Act or Collective Process?" In *Inventive Minds: Creativity in Technology*, ed. Robert J. Weber and David N. Perkins, 273–90. New York, 1992.

Hughes, Thomas P. *American Genesis: A Century of Invention and Technological Enthusiasm 1870–1970*. New York, 1989.

———. "The Evolution of Large Technological Systems." In *The Social Construction of Technological Systems: New Directions in the Sociology and History of Technology*, ed. Wiebe E. Bijker, Thomas P. Hughes, and Trevor J. Pinch, 51–83. Cambridge, Mass., 1989.

James, Harold. *The German Slump: Politics and Economics 1924–1936*. Oxford, 1986.

Jamin, Mathilde. *Zwischen den Klassen: Zur Sozialstruktur der SA-Führerschaft*. Wuppertal, 1984.

Jarausch, Konrad H. "Illiberalism and Beyond: German History in Search of a Paradigm," *JMH* 55, no. 2 (June 1983): 268–84.

_____. *The Unfree Professions: German Lawyers, Teachers, and Engineers, 1900–1950*. New York, 1990.

Jebens, Heinrich. *Die Bezwingung des Herrschers Technik und die dadurch ermöglichte Erschliessung eines neuen Volks- und Welt-Wohlstandes.* n.p., privately printed, 1933.

_____. *Der Kleinsthofplan: Das Fundament zum Volksneubau.* Hamburg, 1948.

_____. *Die Philosophie des Fortschritts: Contra Spengler, Contra Hitler.* Hamburg, 1931.

_____. *Der Weg zur Allmacht: Durch Fortschritt und Nationalsozialismus.* Hamburg, 1933.

Jenkins, Brian, and Spyros A. Sofos. "Nation and Nationalism in Contemporary Europe: A Theoretical Perspective." In *Nation and Identity in Contemporary Europe*, ed. Brian Jenkins and Spyros A. Sofos, 9–32. London, 1996.

_____, eds. *Nation and Identity in Contemporary Europe*. London, 1996.

Jewkes, John, David Sawers, and Richard Stillerman. *The Sources of Invention*, 2d ed. New York, 1969.

Johnson, Jeffrey. *The Kaiser's Chemists: Science and Modernization in Imperial Germany*. Chapel Hill, 1990.

Jones, Larry Eugene. "The Dying Middle: Weimar Germany and the Fragmentation of Bourgeois Politics." *CEH* 5, no. 1 (March 1972): 23–71.

_____. *German Liberalism and the Dissolution of the Weimar Party System, 1918–1933*. Chapel Hill, 1988.

Kaelble, Hartmut. *Industrielle Interessenpolitik in der Wilhelminischen Gesellschaft*. Berlin, 1967.

Kahlert, Robert. *Erfindertaschenbuch*. Berlin, 1939.

Kershaw, Ian. *Hitler 1889–1936: Hubris* (New York, 1999).

_____. *Hitler 1936–1945: Nemesis* (New York, 2000).

_____. *The Nazi Dictatorship: Problems and Perspectives of Interpretation*, 3d ed. London, 1993.

_____, ed. *Why Did German Democracy Fail?* New York, 1990.

Kettenacker, Lothar. *Germany since 1945*. New York, 1997.

Kevles, Daniel J. *The Physicists: The History of a Scientific Community in Modern America*. Cambridge, Mass., 1995.

Kirchhoff, Heinrich. *Das deutsche Patentwesen: Rückschau und Ausblick*. Berlin, 1947.

Klauer, Georg and Philipp Möhring, eds., *Patentgesetz vom 5. Mai 1936, erläutert von Georg Klauer und Philipp Möhring*. Berlin, 1937.

Klemperer, Victor. *LTI: Notizbuch eines Philologen*. Leipzig, 1975.

Kocka, Jürgen. "1945: Neubeginn oder Restauration?" In
Wendepunkte Deutscher Geschichte 1848–1945, ed. Carola Stern
and Heinrich A. Winkler, 141–68. Frankfurt a. M., 1979.
_____. "Angestellter." In *Geschichtliche Grundbegriffe: Historisches
Lexikon zur politisch-sozialen Sprache in Deutschland*, vol. 1, ed.
Otto Brunner, Werner Conze, and Reinhart Koselleck, 110–28.
Stuttgart, 1972.
_____. "The Rise of Modern Industrial Enterprise in Germany." In
*Managerial Hierarchies: Comparative Perspectives on the Rise of the
Modern Industrial Enterprise*, ed. Alfred D. Chandler and Herman
Daems, 77–116. Cambridge, Mass., 1980.
_____. "White-Collar Employees and Industrial Society in Imperial
Germany." In *The Social History of Politics: Critical Perspectives in
West German Historical Writing since 1945*, ed. Georg Iggers,
113–36. Leamington Spa, 1985.
_____. *White Collar Workers in America 1890–1940: A Social-Political
History in International Perspective*. London, 1980.
Kolb, Eberhard. *The Weimar Republic*. London, 1988.
König, Wolfgang. *Ingenieurausbildung, Ingenieurberuf und
Konstruktionstechnik in Großbrittanien, den USA, Frankreich und
Deutschland seit der Industrialisierung: Ein vergleichender Essay*.
Berlin, 1990.
_____. *Künstler und Strichezieher: Konstruktions- und Technikkulturen
im deutschen, britischen, amerikanischen und französischen
Maschinenbau zwischen 1850 und 1930*. Frankfurt a.M., 1999.
König, Wolfgang, and Wolfhard Weber. *Netzwerke Stahl und Strom
1840 bis 1914*. Vol. 4 of *Propyläen Technikgeschichte*, ed.
Wolfgang König. Berlin, 1990.
Köttgen, Carl. *Das wirtschaftliche Amerika*. Berlin, 1925.
Kracauer, Siegfried. *The Salaried Masses: Duty and Distraction in
Weimar Germany*. Introduction by Inka Mülder Bach. Translated
by Quintin Horare. London, 1998.
Kramer, Alan. *The West German Economy, 1945–1955*. New York, 1991.
Kretzschmar, Hugo. *Die Technischen Akademiker und die
Führerauslese*. Berlin, 1930.
Kruedener, Juergen Baron von, ed. *Economic Crisis and Political
Collapse: The Weimar Republic 1924–1933*. New York, 1990.
Lambi, Ivo N. *Free Trade and Protection in Germany, 1868–1879*.
Wiesbaden, 1963.
Langewiesche, Dieter. "Die Angestellten in industriet-
kapitalistischen Systemen." *GUG* 6, no. 1 (1980): 283–96.

Layton, Edwin T., Jr. *The Revolt of the Engineers: Social Responsibility and the American Engineering Profession*. Cleveland, 1971.

Lederer, Emil. *The New Middle Class: (Der neue Mittelstand) in Grundriss der Sozialökonomik, IX. Abteilung, I. Teil. Tübingen, 1926*. New York, 1937.

Lindenfeld, David F. *The Practical Imagination: The German Sciences of State in the Nineteenth Century*. Chicago, 1997.

Lüdtke, Alf. "'Ehre der Arbeit': Industriearbiet und Macht der Symbole. Zum Reichweite symbolischer Orientierungen im Nationalsozialismus." In *Industrielle Welt: Schriftenreihe des Arbeitskreises für moderne Sozialgeschichte*, ed. Reinhart Koselleck and M. Rainer Lepsius, vol. 51 of *Arbeiter im 20. Jahrhundert*, ed. Klaus Tenfelde, 343–92. Stuttgart, 1991.

Ludwig, Karl-Heinz. *Technik und Ingenieure im Dritten Reich*. Düsseldorf, 1974.

Machlup, Fritz. *An Economic Review of the Patent System*. Washington, D.C., 1958.

Machlup, Fritz, and Edith T. Penrose. "The Patent Controversy in the Nineteenth Century." *JEH* 10 (1950): 1–29.

Maier, Charles S. "Between Taylorism and Technocracy: European Ideologies and the Vision of Industrial Productivity in the 1920s." *JCH* 5, no. 2 (1970): 27–61.

_____. *Recasting Bourgeois Europe: Stabilization in France, Germany, and Italy in the Decade After World War I*. Princeton, 1975.

Maier, Charles S. et al., eds. *The Rise of the Nazi Regime: Historical Assessments*. Boulder, 1986.

Manegold, Karl-Heinz. *Universität, Technische Hochschule und Industrie: Ein Beitrag zur Emanzipation de Technik im 19. Jahrhundert unter besonderer Berücksichtigung der Bestrebungen Felix Kleins*. Berlin, 1970.

Manegold, Karl-Heinz, and Wolfgang König, eds. *Technik, Ingenieure und Gesellschaft Geschichte des Vereing Deutscher Ingenieure 1856–1981*. Düsseldorf, 1981.

Mason, Timothy W. "Gesetz zur Ordnung der nationalen Arbeit vom 20. Januar 1934." In *Industrielles System und politische Entwicklung in der Weimarer Republik*, ed. Hans Mommsen, Dietmar Petzina, and Bernd Weisbrod, 322–51. Kronberg, 1977.

_____. *Social Policy in the Third Reich: The Working Class and the "National Community."* Translated by Joan Broadwin. Edited by Jane Caplan. Introduction by Ursula Vogel. Providence, 1993.

_____. "Whatever Happened to 'Fascism'?" In *Reevaluating the Third Reich*, ed. Thomas Childers and Jane Caplan, 253–62. New York, 1993.

Matschoss, Conrad. "Max Eyth zum hundertsten Geburtstag." *Abhandlungen und Berichte des Deutschen Museums* 8, no. 2 (1936): 29–41.

McClelland, Charles. *The German Experience of Professionalization: Modern Learned Professions and Their Organizations from the Nineteenth Century to the Hitler Era*. Cambridge, 1991.

McLeod, Christine. *Inventing the Industrial Revolution: The English Patent System, 1660–1800*. Cambridge, 1988.

Meiksins, Peter. "Professionalism and Conflict: The Case of the American Association of Engineers." *JSH* 19 (spring 1986): 403–21.

Meyer, Georg J. "Praktische Beispiele aus der Elektrotechnik zum Problem der Angestelltenerfindung." *Zeitschrift für Industrierecht* 15, 1/2 (15 Apr. 1921): 1–8.

Meyer-Thurow, Georg. "The Industrialization of Invention: A Case Study from the German Chemical Industry." *Isis* 73, no. 256 (Sep. 1982): 363–81.

Milde, Kurt. *Abbau, ein Schlagwort und seine tiefere Bedeutung*. Berlin, 1930.

Milward, Alan. "Fascism and the Economy." In *Fascism: A Reader's Guide: Analyses, Interpretations, Bibliography*, ed. Walter Laqueur, 379–412. Berkeley, 1976.

_____. *The German Economy at War*. London, 1965.

Moeller, Robert G. "The Kaisserreich Recast? Continuity and Change in Modern German Historiography." *JSH* 17, no. 4 (1983): 655–83.

Mommsen, Hans. "The Realization of the Unthinkable: The 'Final Solution of the Jewish Question' in the Third Reich." In Hans Mommsen, *From Weimar to Auschwitz*, trans. Philip O'Connor, 224–53. Princeton, 1991.

_____. *The Rise and Fall of Weimar Democracy*. Translated by Elborg Forster and Larry Eugene Jones. Chapel Hill, 1996.

Mommsen, Hans, and Manfred Grieger. *Das Volkswagenwerk und seine Arbeiter im Dritten Reich*. Düsseldorf, 1996.

Mommsen, Wolfgang J. *Der autoritäre Nationalstaat: Verfassung, Gesellschaft und Kultur des deutschen Kaiserreiches*. Frankfurt a. M, 1990.

_____. *The Political and Social Theory of Max Weber: Collected Essays*. Chicago, 1989.

_____, ed. *The Emergence of the Welfare State in Britain and Germany, 1850–1950*. London, 1981.

Müller, Ingo. *Hitler's Justice: The Courts of the Third Reich.* Translated by Deborah Lucas Schneider. Cambridge, Mass., 1991.

Müller-Pohle. *Erfindungen von Gefolgschaftsmitgliedern.* Berlin, 1943.

Nathan, Hans, et al. *Erfinder- un Neuererrecht der Deutschen Demokratischen Republik,* 2 vols. Berlin, 1968.

Neufeld, Michael. *The Rocket and the Reich: Peenemünde and the Coming of the Ballistic Missile Era.* New York, 1995.

Neumeyer, Fredrik. *The Employed Inventor in the United States: R&D Policies, Law, and Practice.* Cambridge, Mass., 1971.

_____. *The Law of Employed Inventors in Europe.* Washington, D.C., 1963.

Nicholls, A. J. *Freedom with Responsibility: The Social Market Economy in Germany 1918–1963.* Oxford, 1994.

Nirk, Rudolf. "100 Jahre Patentschutz in Deutschland." In *Hundert Jahre Patentamt,* ed. Deutsches Patentamt, 345–402. Munich, 1977.

Noble, David F. *America by Design: Science, Technology, and the Rise of Corporate Capitalism.* New York, 1977.

Nolan, Mary. *Visions of Modernity: American Business and the Modernization of Germany.* New York, 1994.

Nolte, Ernst. *Three Faces of Fascism: Action Française, Italian Fascism, National Socialism.* Translated by Leila Vennewitz. New York, 1965.

Oppelland, Torsten. "Domestic Political Developments I: 1949–69." In *The Federal Republic of Germany since 1949,* ed. Klaus Larres and Panikos Panayi, 74–99. London, 1996.

Orkin, Neal. "Innovation; Motivation; and Orkinomics." *Patent World* (May 1987): 32-35.

_____. "The Legal Rights of the Employed Inventor: New Approaches to Old Problems," *JPOS* 56, no. 10 (Oct. 1974): 648–62, and no. 11 (Nov. 1974): 719–45.

_____. "The Legal Rights of the Employed Inventor in the United States: a Labor-Management Perspective." In *Employees' Inventions: A Comparative Study,* ed. Jeremy Phillips, 152-79. Sunderland, 1981.

Orkin, Neal, and Mathias Strohfeldt. "Arbn Erf G—the Answer or the Anathema?" *Managing Intellectual Property* (Oct. 1992): 28–32.

Overy, Richard J. "The Economy of the Federal Republic since 1949." In *The Federal Republic of Germany since 1949: Politics, Society and Economy before and after Unification,* ed. Klaus Larres and Panikos Panayi, 3–34. London, 1996.

_____. "Rationalization and the 'Production Miracle' in Germany during the Second World War." In his *War and Economy in the Third Reich*, 343–75. Oxford, 1994.

Parsons, Gerald P. "U.S. Lags in Patent Law Reform," *IEEE Spectrum* 15 (Mar. 1978): 60–64.

Perrucci, Robert, and Joel E. Gerstl. *Profession without Community: Engineers in American Society*. New York, 1969.

Petroski, Henry. *Invention by Design: How Engineers Get from Thought to Thing*. Cambridge, Mass., 1996.

Petzina, D., W. Abelshauser, and A. Faust. *Sozialgeschichtliches Arbeitsbuch III: Materialien zur Statistik des Deutschen Reiches 1914–1945*. Munich, 1978.

Peukert, Detlev J. K. *Inside Nazi Germany: Conformity, Opposition and Racism in Everyday Life*. New Haven, 1987.

_____. *The Weimar Republic: The Crisis of Classical Modernity*. Translated by Richard Deveson. New York, 1992.

Pfirrmann, Fritz. "Ein neuer Erfinderrevers." *Deutsche Techniker-Zeitung* 11, no. 1 (4 Jan. 4 1929): 1–5.

Phillips, Jeremy, ed. *Employees' Inventions: A Comparative Study*. Sunderland, 1981.

Plaisant, Robert. "Employees' Inventions in Comparative Law." *IPQ* 5, no. 1 (Jan. 1960): 31–55.

Plumpe, Gottfried. *Die I.G. Farbenindustrie AG: Wirtschaft, Technik und Politik 1904–1945*. Berlin, 1990.

Potthoff, Heinz et al. *Rechtsprechung des Arbeitsrechtes 1914–1927: 9000 Entscheidungen in 5000 Nummern in einem Band systematisch geordnet*. Stuttgart, 1927.

Prahl, Klaus. *Patentschutz und Wettbewerb*. Göttingen, 1969.

Prinz, Michael. *Vom neuen Mittelstand zum Volksgenossen: die Entwicklung des sozialen Status der Angestellten von der Weimarer Republik bis zum Ende der NS-Zeit*. Munich, 1986.

Puschner, Uwe, Walter Schmitz and Justus H. Ulbricht, eds. *Handbuch zur "Völkischen Bewegung" 1871–1918*. Munich, 1996.

Rabinbach, Anson. *The Human Motor: Energy, Fatigue, and the Origins of Modernity*. New York, 1990.

Radkau, Joachim. *Technik in Deutschland: Vom 18. Jahrhundert bis zur Gegenwart*. Frankfurt a. M., 1989.

Redies, Franz. "Zum Recht der Angestellten-Erfinder." *GRUR* 42 (Jun. 1937): 410-24.

Reich, Simon. *The Fruits of Fascism: Postwar Prosperity in Historical Perspective*. Ithaca, 1990.

Reichshandbuch der Deutschen Gesellschaft, 2 vols. Berlin, 1930.

Reimer, Eduard. *Das Recht der Angestelltenerfindung: Gegenwärtiger Rechtszustand und Vorschläge zur künftigen Gesetzesregelung.* Berlin, 1948.

_____. *Das Recht der Arbeitnehmererfindung: Gegenwärtiger Rechtszustand und Vorschläge zur künfiten Gesetzesregelung,* 2d, exp. ed. Berlin, 1951

Reimer, Eduard, Hans Schade, and Helmut Schippel. *Das Recht der Arbeitnehmererfindung: Kommentar zu dem Gesetz über Arbeitnehmererfindungen vom 25. Juli 1957 und deren Vergütungsrichtlinien,* 5th ed. Berlin, 1975.

Retallack, James. *Germany in the Age of Kaiser Wilhelm II.* New York, 1996.

Reynolds, Terry S., ed. *The Engineer in America: A Historical Anthology from Technology and Culture.* Chicago and London, 1991.

Riedler, Alois. "Das deutsche Patentgesetz und die wissenschaftliche Hilfsmittel des Ingenieurs." *VDIZ* 48 (1898): 1319.

Riemschneider, Karl August. "Die begriffliche Abgrenzung der Angestelltenerfindungen," *Zeitschrift der Akademie für Deutsches Recht* 3 (1936): 1004–05.

_____. "Die Erfinderbetreuung im Betrieb," *Deutsche Technik* 10 (Jul. 1942): 292.

_____. *Die wirtschaftliche Ausnutzung von Erfindungen.* Berlin, 1936.

_____. "Zum Recht der Angestelltenerfinder," *GRUR* 42 (1937): 393–95.

Riemschneider, Karl August, and Heinrich Barth, eds. *Die Gefolgschaftserfindung: Erläuterungen über die Behandlung von Erfindungen von Gefolgschaftsmitgliedern,* 2d ed. Berlin, 1944.

Ritter, Gerhard A. *Die deutschen Parteien 1830–1914.* Göttingen, 1985.

_____. *The Transformation of German Society: Continuity and Change After 1945 and 1989/90.* Berkeley, 1996.

Rosenberg, Hans. *Große Depression und Bismarckzeit: Wirtschaftsablauf, Gesellschaft und Politik in Mitteleuropa.* Berlin, 1967.

Sachse, Carola. *Siemens, der Nationalsozialismus und die moderne Familie: Eine Untersuchung zur sozialen Rationalisierung in Deutschland im 20. Jahrhundert.* Hamburg, 1990.

Sachse, Carola, Tilla Siegel, Hasso Spode, and Wolfgang Spohn, *Angst, Belohnung, Zucht und Ordnung: Herrschaftsmechanismen im*

Nationalsozialismus. Introduction by Timothy W. Mason. Opladen, 1982.

Sauer, Wolfgang. "Das Problem des deutschen Nationalstaates." In *Moderne deutsche Sozialgeschichte,* ed. Hans-Ulrich Wehler, 407–36. Cologne, 1968.

Saunders, Thomas. "Nazism and Social Revolution." In *Modern Germany Reconsidered,* ed. Gordon Martel, 159–77. London, 1992.

Schade, Hans. "Zu Fragen des Arbeitnehmererfinderrechts," *GRUR* 56 (Nov. 1958): 519–27.

Schäfer, Hans Dieter. "Amerikanismus im Dritten Reich." In *Nationalsozialismus und Modernisierung,* ed. Rainer Zitelmann and Michael Prinz, 199–215. Darmstadt, 1994.

Schiff, Eric. *Industrialization Without National Patents: The Netherlands, 1869–1912; Switzerland, 1850–1907.* Princeton, 1971.

Schmelzer, Hermann. "Erfinder oder Naturkraftbinder?" *Der leitende Angestellte, Zeitschrift der Vereinigung der leitenden Angestellten in Handel und Industrie e.V.* 3 (1921), nos. 22–24 (15 Nov.–30 Dec. 1921): 170–72, 177–181, 184–85 and 4 (1922) no. 1 (2 Jan. 1922): 5–6.

Schmoch, Ulrich, Hariolf Grupp, and Uwe Kuntze. *Patents as Indicators of the Utility of European Community R&D Programmes.* Luxemburg, 1991.

Schmookler, Jacob. *Invention and Economic Growth.* Cambridge, Mass., 1966.

Schoenbaum, David. *Hitler's Social Revolution: Class and Status in Nazi Germany, 1933–1939.* New York, 1967.

Schubert, Werner, ed. *Akademie für Deutsches Recht 1933–1945, Protokolle der Ausschüsse.* Berlin, 1986–96 (vols. 1–4); Frankfurt a.M, 1997–99 (vols. 5–9).

———. ed. *Ausschüsse für den gewerblichen Rechtsschutz (Patent-, Warenzeichen-, Geschmacksmusterrecht, Wettbewerbsrecht), für Urheber- und Verlagsrecht sowie für Kartellrecht (1934–1943),* vol. 9 of *Akademie für Deutsches Recht 1933–1945, Protokolle der Ausschüsse,* ed. Werner Schubert. Frankfurt a. M., 1999.

Schumpeter, Joseph A. *Capitalism, Socialism and Democracy,* 3d ed. New York, 1975.

———. *The Theory of Economic Development: An Inquiry into Profits, Capital, Credit, Interest, and the Business Cycle.* Translated by Devers Opie. Cambridge, Mass., 1936.

Schwerber, Peter. *Nationalsozialismus und Technik.* Munich, 1930.

Scott, James C. *Seeing Like a State: How Certain Schemes to Improve the Human Condition Have Failed.* New Haven, 1998.

Shearer, J. Ronald. "The Politics of Industrial Productivity in the Weimar Republic: Technological Innovation, Economic Efficiency, and their Social Consequences in the Ruhr Coal Mining Industry, 1918–1929." Ph.D. dissertation, Univ. of Pennsylvania, 1989.

———. "Talking About Efficiency: Politics and the Industrial Rationalization Movement in the Weimar Republic." *CEH* 28, no 4 (1995): 483–506.

Siegel, Tilla. "It's Only Rational: An Essay on the Logic of Social Rationalization," *International Journal of Political Economy* 24, no. 4 (winter 1994/95): 35–70.

———. *Leistung und Lohn in der nationalsozialistischen "Ordnung der Arbeit."* Opladen, 1989.

———. "Rationalizing Industrial Relations: A Debate on the Control of Labor in German Shipyards in 1941." In *Reevaluating the Third Reich*, ed. Thomas Childers and Jane Caplan, 139–60. New York, 1993.

———. "Wage Policy in Nazi Germany," *Politics & Society* 14, no. 1 (1985): 1–51.

Simon, Leslie E. *German Research in World War II: An Analysis of the Conduct of Research.* New York, 1947.

Slama, J. *One Century of Technical Progress Based on an Analysis of German Patent Statistics.* Brussels, 1987.

Smelser, Ronald. *Robert Ley: Hitler's Labor Front Leader.* Oxford, 1988.

———. "Die Sozialplanung der Deutschen Arbeitsfront." In *Nationalsozialismus und Modernisierung*, 2d ed., ed. Michael Prinz and Rainer Zitelmann, 71–92. Darmstadt, 1994.

Speier, Hans. *German White-Collar Workers and the Rise of Hitler.* New Haven, 1986.

Stachura, Peter. *The Nazi Machtergreifung.* London, 1983.

Staudt, Erich et al. *Der Arbeitnehmererfinder im betrieblichen Innovationsprozess.* Bochum, 1990.

Steele, J. Rodman, Jr. *Is This My Reward? An Employee's Struggle for Fairness in the Corporate Exploitation of His Inventions.* West Palm Beach, 1968.

Stern, Fritz. *The Failure of Illiberalism.* Chicago, 1975.

Stokes, Raymond G. *Divide and Prosper: The Heirs of I.G. Farben under Allied Authority 1945–1951.* Berkeley, 1988.

———. *Opting for Oil: the Political Economy of Technological Change in the West German Chemical Industry, 1945–1961.* Cambridge, 1994.

_____. "Technology and the West German Wirstschaftswunder." *Technology and Culture* 32, no. 1 (1991): 1–22.

Thomsen, Andreas. *Denkschrift an den Deutschen Reichstag betr. Nutzbarmachung der im deutschen Volke vorhandenen Erfinderkräfte zum Wiederaufbau und zur Mehrung des Volksvermögens.* Münster, 1931.

Treue, Wilhelm. "Ingenieur und Erfinder: Zwei sozial- und technikgeschichtliche Probleme," *Vierteljahrsschrift für Sozial- und Wirtschaftsgeschichte* 54, no. 4 (Dec. 1967): 456–76.

Trischler, Helmuth, and Rüdiger vom Bruch. *Forschung für den Markt: Geschichte der Fraunhofer-Gesellschaft.* Munich, 1999.

Turner, Henry A., Jr. *German Big Business and the Rise of Hitler.* New York, 1985.

Van Laak, Dirk. *Weiße Elephanten: Anspruch und Scheitern technischer Großprojekte im 20. Jahrhundert.* Stuttgart, 1999.

Veblen, Thorstein. *Imperial Germany and the Industrial Revolution.* Ann Arbor, 1968.

Volkmar, Erich. "Rechtspflege und Rechtspolitik: Welche Umgestaltungen erfuhr das Privatrecht seit dem Siege der nationalsozialistischen Revolution?" *Deutsche Justiz* 97, no. 10 (1935), ed. A: 355–86.

Volmer, Bernhard, and Dieter Gaul. *Arbeitnehmererfindungsgestz: Kommentar,* 2d ed. Munich, 1983.

Waldmann, Kurt. "Auf dem Wege zu einem nationalsozialistischen Patentgesetz." In *Nationalsozialistisches Handbuch für Recht und Gesetzgebung,* 2d ed., ed. Hans Frank, 1032–49. Munich, 1935.

Walter, Johann. *Erfahrungen eines Betriebsleiters (Aus der Praxis der Anilinfarbenfabrikation).* Leipzig, 1925.

Walter, Robert Hans. *Der Erfinderbetreuer im Betrieb.* Berlin, 1943.

Walton, Richard E. *The Impact of the Professional Engineering Union.* Cambridge, Mass., 1961.

Wankel, Felix. "Patentamt, Erfinder, und Volkswirtschaft." In *Hundert Jahre Patentamt,* ed. Deutsches Patentamt, 324–32. Munich, 1977.

Weber, Max. "Bureaucracy." In *From Max Weber: Essays in Sociology,* trans. and ed. H. H. Gerth and C. Wright Mills, 196–244. New York, 1946.

Wehler, Hans-Ulrich. *The German Empire 1871–1918.* Translated by Kim Traynor. Leamington Spa/Dover, 1985.

Weigand, Wolf. "Emil Guggenheimer (1860–1925), Geheimer Justizrat." In *Geschichte und Kultur der Juden in Bayern. Lebensläufe.* Munich, 1988.

Weihe, Carl. *Max Eyth: Ein kurzgefasstes Lebensbild mit Auszügen aus seinen Schriften.* Berlin, 1922.

Weindling, Paul. *Health, Race and German Politics Between National Unification and Nazism, 1870–1945.* Cambridge, 1989.

Weisbrod, Bernd. "The Crisis of Bourgeois Society in Interwar Germany." In *Fascist Italy and Nazi Germany: Comparisons and Contrasts,* ed. Richard Bessel, 23–39. Cambridge, 1996.

——. *Schwerindustrie in der Weimarer Republik.* Wuppertal, 1978.

Weiss, Sheila Faith. "Pedagogy, Professionalism and Politics: Biology Instruction During the Third Reich." In *Science, Technology and National Socialism,* ed. Monika Renneberg and Mark Walker, 184–96. Cambridge, 1994.

Weisse, Ernst. *Kernfragen der Erfindungskunde für den Gefolgschaftserfinder.* Berlin, 1943.

Wiegand, Paul. "Die Erfindung von Gefolgsmännern unter besonderer Berücksichtigung ihrer wirtschaftlichen Auswirkungen auf Unternehmer und Gefolgsmänner." *Dr.-Ing.* diss., Technische Hochschule Hannover, 1941.

——. "Die Erfindungen von Gefolgschaftsmitgliedern." *Zeitung des Vereins Mitteleuropäischer Eisenbahnverwaltungen* 82, no. 36 (Sep. 1942): 463–67.

——. "Zur Frage der Angestelltenerfindung," *GRUR* 40 (May 1935), 261–70.

Wiener, Norbert. *Invention: The Care and Feeding of Ideas.* Introduction by Steve Joshua Heims. Cambridge, Mass., 1993.

Wilcken, Hugo, and Karl August Riemschneider. *Das Patentgesetz von 5. Mai 1936.* Berlin, 1937.

Winkler, Heinrich August. "Der entbehrliche Stand: Zur Mittelstandspolitik im 'Dritten Reich,'" *Archiv für Sozialgeschichte* 17 (1977): 1–40.

Winnacker, Karl. *Challenging Years: My Life in Chemistry.* London, 1972.

Zitelmann, Rainer. *Hitler: Selbstverständnis eines Revolutionärs.* Hamburg, 1987.

Zussman, Robert. *Mechanics of the Middle Class: Work and Politics Among American Engineers.* Berkeley, 1985.

INDEX

New Volumes in

Monographs in German History

"ARYANISATION" IN HAMBURG
The Economic Exclusion of Jews and the Confiscation
of their Property in Nazi Germany

Frank Bajohr, Forschungsstelle für Zeitgeschichte in Hamburg and lecturer at the Department of History at the University of Hamburg. *Translated from the German by* George Wilkes

"This searing book about "Aryanisation", the process by which the Nazis robbed Jews of their economic livelihood, presents a lucid and riveting analysis of a little investigated subject. Compassionate towards the victims of the Third Reich's "Aryanisation" program and enraged by the perpetrators, Dr. Bajohr has set a new standard for Holocaust scholarship. Integrating several narrative threads – the Nazi's economic policy, popular reactions, Jewish responses – this book is about people: people how harmed; who profited; who were robbed and exploited; who watched. Creatively conceived and meticulously documented, [this book] will become a classic work on this subject."
— **Debórah Dwork,** Center for Holocaust and Genocide Studies, Clark University

Volume 7. 2001. *ca.* 300 pages, 14 tables, bibliog., index
ISBN 1-57181-484-1 hardback **$69.95/£45.00**
ISBN 1-57181-485-X paperback **$25.00/£17.00**

THE POLITICS OF EDUCATION
Teachers and School Reform in Weimar Germany

Marjorie Lamberti, Professor of History, Middlebury College

Although the early history of progressive education is often associated with John Dewey in America, the author argues convincingly that the pedagogues in the elementary schools in the big cities of Imperial Germany were in the avant garde of this movement on the European Continent.

Far more than a history of ideas, this study provides the first comprehensive analysis of the culture wars over the schools in Germany in the 1920s. Going up to the Nazi seizure of power, the author's narrative sheds new light on the courageous defense of the republican state by the progressive educators in the 1930s and the relationship between the traditionalists' opposition to school reform and the attraction of certain sections of the teaching profession to the Nazi movement.

Volume 8. Spring 2002. *ca.* 240 pages, bibliog., index
ISBN 1-57181-298-9 hardback *ca.* **$69.95/£47.00**
ISBN 1-57181-299-7 paperback *ca.* **$25.00/£17.00**

www.berghahnbooks.com

Previous Volumes in
Monographs in German History

OSTHANDEL AND OSTPOLITIK
German Foreign Trade Policies in Eastern Europe from Bismarck to Adenauer
Robert Mark Spaulding, Department of History, University of North Carolina at Wilmington

"A long-term perspective on a topic of vital current and continuing importance, and a major contribution to the study of German-Russian relations." —Gerald D. Feldman

Volume 1. 1997. 544 pages 15 tables, 37 figs, bibliog., index / ISBN 1-57181-039-0 hardback **$89.00/£60.00**

A QUESTION OF PRIORITIES
Democratic Reform and Economic Recovery in Postwar Germany
Rebecca Boehling, Department of History, University of Maryland, Baltimore County

"... most welcome as the first detailed analysis of political reconstruction in major postwar German cities available in English." —Choice

Volume 2. 1996. 320 pages 12 photos, 1 map, bibliog., index
ISBN 1-57181-035-8 hardback **$59.95/£40.00** / ISBN 1-57181-159-1 paperback **$24.00/¢16.00**

FROM RECOVERY TO CATASTROPHE
Municipal Stabilization and Political Crisis
Ben Lieberman, Department of Social Science at Fitchburg State College, MA

"Ben Lieberman has contributed a valuable book on the Weimar welfare state... an attractive package with a full scholarly apparatus."—German Studies Review

Volume 3. 1998. 192 pages 13 tables, bibliog., index / ISBN 1-57181-104-4 hardback **$42.00/£30.00**

NAZISM IN CENTRAL GERMANY
The Brownshirts in 'Red' Saxony
Claus-Christian W. Szejnmann, European History Dept., Middlesex University

"...an impressive work of detailed scholarship and a valued contribution to 20th-century German history." —Midwest Book Review

Volume 4. 1998. 304 pages 12 tables, 6 figs, bibliog., index / ISBN 1-57181-942-8 hardback **$59.95/£40.00**

CITIZENS AND ALIENS
Foreigners and the Law in Britain and German States 1789-1870
Andreas Fahrmeir, Senior Research Fellow at the German Historical Institute, London

"...concisely written ... excellently researched." —Frankfurter Allgemeine Zeitung

Volume 5. 2000. 304 pages, 13 tables, 3 figs., bibliog., index / ISBN 1-57181-717-4 hardback **$69.95/£47.00**

www.berghahnbooks.com